Biogeochemical Element Cycling in Plant–Soil Systems: Implications for Ecosystem Dynamics and Services

Biogeochemical Element Cycling in Plant–Soil Systems: Implications for Ecosystem Dynamics and Services

Guest Editors

Daniel Puppe
Wajid Zaman

MDPI

Basel • Beijing • Wuhan • Barcelona • Belgrade • Novi Sad • Cluj • Manchester

Guest Editors
Daniel Puppe
Leibniz Centre for
Agricultural Landscape
Research (ZALF)
Müncheberg
Germany

Wajid Zaman
Department of Life Sciences
Yeungnam University
Gyeongsan
Republic of Korea

Editorial Office
MDPI AG
Grosspeteranlage 5
4052 Basel, Switzerland

This is a reprint of the Special Issue, published open access by the journal *Biology* (ISSN 2079-7737), freely accessible at: https://www.mdpi.com/journal/biology/special_issues/J0G0SQX25E.

For citation purposes, cite each article independently as indicated on the article page online and as indicated below:

Lastname, A.A.; Lastname, B.B. Article Title. *Journal Name* **Year**, *Volume Number*, Page Range.

ISBN 978-3-7258-6794-3 (Hbk)
ISBN 978-3-7258-6795-0 (PDF)
https://doi.org/10.3390/books978-3-7258-6795-0

Contents

About the Editors

Daniel Puppe

Daniel Puppe is a researcher at the Leibniz Centre for Agricultural Landscape Research (ZALF) in Müncheberg, Germany. He completed his Ph.D. at the Brandenburg University of Technology Cottbus-Senftenberg, Germany. As a biologist he is generally interested in all forms of life. However, in his research Daniel is focused on organisms that form biogenic silica structures (e.g., plant phytoliths, testate amoeba shells, or sponge spicules) and the role of biogenic silica for biogeochemical silicon cycling in terrestrial ecosystems and agricultural landscapes. With his research Daniel aims to find environmentally friendly and sustainable strategies in order to deal with the challenges that global change poses to (agro)ecosystem functioning.

Wajid Zaman

Wajid Zaman completed his PhD from the Institute of Botany, Chinese Academy of Sciences, Beijing, China. Currently, he is working as a Research Professor in the Department of Life Sciences, Yeungnam University, Republic of Korea. He is interested in developing a career that combines teaching and research, while maintaining his interest in Plant Sciences with advanced technologies.

Preface

Biogeochemical element cycling in plant–soil systems is fundamental for ecosystem dynamics and services. The articles published in this Reprint provide a comprehensive overview of the complex interactions that govern ecosystem functioning and offer important insights for sustainable ecosystem management in a changing world. We hope that the findings presented here will inspire future research and contribute to the development of policies and practices that promote healthier, more resilient ecosystems for the generations to come.

We are grateful to all authors for their contributions (a short summary is provided in the Editorial).

"All is flux, nothing is stationary."

(based on Heraclitus of Ephesus)

Daniel Puppe and Wajid Zaman

Guest Editors

biology

MDPI

Editorial

Biogeochemical Element Cycling in Plant–Soil Systems: Implications for Ecosystem Dynamics and Services

Daniel Puppe [1,*] and Wajid Zaman [2,*]

[1] Leibniz Centre for Agricultural Landscape Research (ZALF), 15374 Müncheberg, Germany
[2] Department of Life Sciences, Yeungnam University, Gyeongsan 38541, Republic of Korea
* Correspondence: daniel.puppe@zalf.de (D.P.); wajidzaman@yu.ac.kr (W.Z.)

Biogeochemical element cycling in plant–soil systems is fundamental for ecosystem dynamics and services [1,2]. In general, the term 'ecosystem dynamics' describes changes in ecosystem structures caused by interactions of organisms with the external environment in an ecosystem [3]. Ecosystem services include all ecosystem functions that are useful to humans like pollination, climate stabilization, or biomass production. In terrestrial ecosystems, plants represent the main primary producers of biomass (organic compounds) and oxygen through photosynthesis. In soils, micro- and macro-organisms recycle elements like carbon (C), nitrogen (N), and silicon (Si), maintaining soil fertility, which is essential for plant growth and corresponding ecosystem services [4,5]. Knowledge of how these organisms and plants interact is crucial to understand ecosystem dynamics and services. Nowadays, terrestrial ecosystems and their services are strongly affected by global change, which represents a grand challenge for ecosystem management and policy [6,7]. The research presented in this Special Issue seeks to address these aspects by examining the complex relationships between biogeochemical cycling, ecosystem dynamics, and the services that ecosystems provide. The 11 articles in this Special Issue provide valuable insights into how nutrient/element cycling influences ecosystem functions and how various land management practices such as afforestation or the use of fire affect microbial diversity and soil health, for example.

In two articles of this Special Issue, the impact of different agricultural practices, i.e., the use of fire [8] and Si amendment/crop straw recycling [9], on biogeochemical nutrient/element cycling in agricultural plant–soil systems was studied. Arunrat et al. [8] examined the impact of fire on soil properties and bacterial communities under rotational shifting cultivation in the tropical zone (northern Thailand). While they found positive fire effects on soil properties (e.g., increased nutrient availability), bacterial communities were negatively affected by fire (decrease in richness and diversity), but recovered relatively fast after burning, i.e., within some months. This knowledge is helpful to derive practice-related recommendations for post-fire management in traditional farming systems like rotational shifting cultivation. Puppe et al. [9] analyzed silica accumulation in potato plants and the effect of Si plant availability on long-term potato yield performance in the temperate zone (Germany). They found relatively low Si contents in the dry mass of potato leaves (up to 0.08%) and roots (up to 0.3%) and negligible Si contents in potato tuber skin and tuber flesh for plants grown in soils with different concentrations of plant-available Si (Si amendment, field experiment 1). Moreover, potato yield was not correlated to plant-available Si concentrations in soils in the long term (1965–2015, crop straw recycling, field experiment 2). Based on their results, Puppe et al. [9] ascribed the beneficial effects of Si fertilization on potato growth and yield performance reported in previous studies mainly to

 https://doi.org/10.3390/biology15020203

antifungal/osmotic effects of foliar-applied Si fertilizers and to changes in physicochemical soil properties (e.g., enhanced phosphorus (P) availability and water-holding capacity) caused by soil-applied Si fertilizers.

Other authors examined the distribution of (toxic) soil elements in areas in China, which are strongly affected by human activities. Wu et al. [10] studied the distribution of several minor and trace elements in the soil of five different forests in the core area of the Winter Olympics in the Beijing–Tianjin–Hebei region. They reported soil minor and trace elements to be mainly influenced by climatic factors and soil properties, while no direct effect of vegetation type on soil element distribution was observable. The results of Wu et al. [10] are valuable to better understand how climate change affects the distribution of minor and trace elements in soil of boreal forests in general. Xu et al. [11] analyzed soil and crop samples to assess the distribution of toxic metals ("heavy metals") in farmlands located around smelting facilities in Jinchang City. Based on their results, they identified main areas of nickel (Ni), copper (Cu), and cobalt (Co) contamination, where Ni, Cu, and Co levels in crop samples exceeded regulatory limits. The results of Xu et al. [11] provide a valuable foundation for future establishments of remediation strategies in the investigated area.

Three studies of this Special Issue deal with N dynamics in agricultural contexts. Ahmed et al. [12] performed laboratory experiments to examine how different concentrations of the two microbial inhibitors streptomycin and cycloheximide affect nitrous oxide (N_2O) emissions from strongly acidic soil. These authors found that high concentrations of the applied microbial inhibitors effectively reduced N_2O emissions, while lower microbial inhibitor concentrations resulted in increased N_2O production. The results of Ahmed et al. [12] illustrate the complexity of microbial interactions in acidic soils and highlight the importance of considering the broader ecological context when applying microbial inhibitors in agricultural plant–soil systems. Xiang et al. [13] used an automated incubation system to examine the effects of liming on N_2O emissions from acidic soils. Their results showed that N_2O emissions increased following liming, especially when combined with urea addition, by stimulating nitrification. These findings are useful for developing an optimal liming strategy, which alleviates soil acidity by increasing soil pH, while mitigating N_2O oxide emissions from agricultural soils. Xu et al. [14] investigated the effects of soil N, i.e., nitrate (NO_3^-) and ammonium (NH_4^+), availability on the grass *Hemarthria altissima*. They observed that *H. altissima* specimens grown under high NH_4^+ levels used N mainly for carboxylation processes, while *H. altissima* specimens grown under high NO_3^- levels allocated N mainly to leaf light-capturing proteins. The findings of Xu et al. [14] are useful for developing targeted N fertilization strategies to avoid over-fertilization and improve crop yields.

In three articles of this Special Issue, the results of different afforestation approaches in China are discussed in detail. Chen et al. [15] evaluated how different vegetation restoration strategies (barren land/control, disturbed short-rotation and undisturbed long-term *Eucalyptus* monocultures, a mixed native-species plantation, and a natural forest) affect soil quality and microbial communities in tropical ecosystems. While they found soil quality in undisturbed long-term *Eucalyptus* monoculture and mixed native-species plantations to be comparable to natural forest soil quality, soil microbial biomass remained lower in all plantations compared to the natural forest. The results of Chen et al. [15] represent a valuable basis for assessing the suitability of different afforestation approaches in the tropics. Du et al. [16] examined the effects of four different afforestation strategies, i.e., plantations of (i) *Salix cheilophila* and *Populus simonii*, (ii) *S. cheilophila*, (iii) *P. simonii*, and (iv) *Caragana korshinskii*, on biological soil crust properties and microbial communities in an alpine sandy land in the Gonghe Basin. Their results showed the *P. simonii* plantation to represent the most promising afforestation strategy in this area to enhance biological

https://doi.org/10.3390/biology15020203

soil crust nutrient contents and optimize bacterial community structures. Zuo et al. [17] studied the effects of five different afforestation strategies, i.e., plantations of (i) pure *Eucalyptus*, (ii) *Eucalyptus* and *Cupressus* (2:1 ratio), (iii) *Eucalyptus* and *Cupressus* (1:1 ratio), (iv) *Eucalyptus* and *Cupressus* (1:2 ratio), and (v) pure *Cupressus*, on soil physicochemical properties, microbial communities, and metabolomes in a subtropical mountain area. They found mixed *Eucalyptus–Cupressus* plantations to represent the most promising afforestation strategy, which improved soil conditions and promoted a more diverse and functionally enriched soil microbiome, particularly at a 1:2 *Eucalyptus–Cupressus* ratio.

Last but not least, Zaman et al. [18] summarized current knowledge of biogeochemical C, N, P, sulfur (S), and Si cycles, emphasizing their roles in nutrient/element cycling, plant growth, and soil health, especially in agricultural plant–soil systems. Based on their literature review, they identified research gaps that should be filled in future studies to better understand the interlinkages between biogeochemical cycles and their responses to global change. Moreover, they called for integrated approaches that combine interdisciplinary research, technological innovation, and sustainable land-use strategies to mitigate human-induced disruptions and enhance ecosystem resilience in the face of environmental change.

The articles published in this Special Issue provide a comprehensive overview of the complex interactions that govern ecosystem functioning and offer important insights for sustainable ecosystem management in a changing world. We hope that the findings presented here will inspire future research and contribute to the development of policies and practices that promote healthier, more resilient ecosystems for the generations to come.

Author Contributions: Conceptualization, writing—original draft preparation, writing—review and editing, D.P. and W.Z.; supervision, D.P. All authors have read and agreed to the published version of the manuscript.

Funding: The authors declare that no funds, grants, or other support were received during the preparation of this manuscript.

Data Availability Statement: No data was generated or analyzed during the writing of this article.

Conflicts of Interest: The authors declare that they have no known competing financial interests or personal relationships that could have appeared to influence the work reported in this paper.

References

1. Duckworth, O.W.; Franzluebbers, A.J.; Gardner, T.G. Biogeochemical processes underpin ecosystem services. *Soil Ecosyst. Serv.* **2020**, 79–99. [CrossRef]
2. Smith, P.; Cotrufo, M.F.; Rumpel, C.; Paustian, K.; Kuikman, P.J.; Elliott, J.A.; McDowell, R.; Griffiths, R.I.; Asakawa, S.; Bustamante, M. Biogeochemical cycles and biodiversity as key drivers of ecosystem services provided by soils. *Soil* **2015**, *1*, 665–685. [CrossRef]
3. Bradshaw, R.H.; Sykes, M.T. *Ecosystem Dynamics: From the Past to the Future*; John Wiley & Sons: Hoboken, NJ, USA, 2014.
4. Bardgett, R.D.; Van Der Putten, W.H. Belowground biodiversity and ecosystem functioning. *Nature* **2014**, *515*, 505–511. [CrossRef] [PubMed]
5. Meunier, J.-D.; Cornu, S.; Keller, C.; Barboni, D. The role of silicon in the supply of terrestrial ecosystem services. *Environ. Chem. Lett.* **2022**, *20*, 2109–2121. [CrossRef]
6. Scholes, R.J. Climate change and ecosystem services. *Wiley Interdiscip. Rev. Clim. Change* **2016**, *7*, 537–550. [CrossRef]
7. Malhi, Y.; Franklin, J.; Seddon, N.; Solan, M.; Turner, M.G.; Field, C.B.; Knowlton, N. Climate change and ecosystems: Threats, opportunities and solutions. *Philos. Trans. R. Soc. B Biol. Sci.* **2020**, *375*, 20190104. [CrossRef] [PubMed]
8. Arunrat, N.; Sansupa, C.; Sereenonchai, S.; Hatano, R.; Lal, R. Fire-induced changes in soil properties and bacterial communities in rotational shifting cultivation fields in Northern Thailand. *Biology* **2024**, *13*, 383. [CrossRef]
9. Puppe, D.; Busse, J.; Stein, M.; Kaczorek, D.; Buhtz, C.; Schaller, J. Silica accumulation in potato (*Solanum tuberosum* L.) plants and implications for potato yield performance—Results from field experiments in Northeast Germany. *Biology* **2024**, *13*, 828. [CrossRef]
10. Wu, X.; Zhang, H.; Wang, Z.; Tian, W.; Liu, Z. Climate and Soil Properties Drive the Distribution of Minor and Trace Elements in Forest Soils of the Winter Olympic Core Area. *Biology* **2025**, *14*, 82. [CrossRef] [PubMed]

https://doi.org/10.3390/biology15020203

11. Xu, S.; Yun, M.; Wang, Y.; Liu, K.; Wu, A.; Li, S.; Su, Y.; Wang, S.; Kang, H. Heavy Metal Contamination and Risk Assessment in Soil–Wheat/Corn Systems near Metal Mining Areas in Northwestern China. *Biology* **2025**, *14*, 1475. [CrossRef] [PubMed]
12. Ahmed, W.; Gong, H.; Xiang, X.; Chen, R.; Xu, Y.; Shi, W.; Li, B.; Yin, J.; Chen, Q. Microbial Community Imbalance Drives Nitrous Oxide Emissions from Strongly Acidic Soil—Insights from a Laboratory Experiment with Microbial Inhibitors. *Biology* **2025**, *14*, 621. [CrossRef] [PubMed]
13. Xiang, X.; Gong, H.; Ahmed, W.; Thompson, R.B.; Shi, W.; Yin, J.; Chen, Q. Liming-induced nitrous oxide emissions from acidic soils dominated by stimulative nitrification. *Biology* **2025**, *14*, 1110. [CrossRef] [PubMed]
14. Xu, N.; Wei, X.; Zhang, J.; Sun, M.; Zhang, J.; Zhao, Z.; Yang, X. Leaf Nitrogen Allocation Trade-Offs Promote Efficient Utilization of Different Nitrogen Forms in *Hemarthria altissima*. *Biology* **2025**, *14*, 1260. [CrossRef] [PubMed]
15. Chen, Y.; Zhang, F.; Cao, J.; Liu, T.; Zhang, Y. Responses of Soil Quality and Microbial Community Composition to Vegetation Restoration in Tropical Coastal Forests. *Biology* **2025**, *14*, 1120. [CrossRef] [PubMed]
16. Du, S.; Xie, H.; Zhang, G.; Qiao, F.; Geng, G.; Cao, E. Effects of Different Afforestation Measures on Biological Soil Crust Properties and Microbial Communities in an Alpine Sandy Land. *Biology* **2025**, *14*, 1530. [CrossRef] [PubMed]
17. Zuo, Y.-W.; Liu, Y.-Y.; Jiang, Y.-X.; Li, W.-Q.; Peng, Y.; Zhou, S.-M.; You, S.-Q.; Liu, S.-Q.; Deng, H.-P. Improving Soil Properties and Microbiomes by Mixed Eucalyptus–Cupressus Afforestation. *Biology* **2025**, *14*, 1667. [CrossRef] [PubMed]
18. Zaman, W.; Ayaz, A.; Puppe, D. Biogeochemical Cycles in Plant-Soil Systems: Significance for Agriculture, Interconnections, and Anthropogenic Disruptions. *Biology* **2025**, *14*, 433. [CrossRef] [PubMed]

https://doi.org/10.3390/biology15020203

biology

MDPI

Article

Fire-Induced Changes in Soil Properties and Bacterial Communities in Rotational Shifting Cultivation Fields in Northern Thailand

Noppol Arunrat [1,*], Chakriya Sansupa [2], Sukanya Sereenonchai [1], Ryusuke Hatano [3] and Rattan Lal [4]

[1] Faculty of Environment and Resource Studies, Mahidol University, Nakhon Pathom 73170, Thailand; sukanya.ser@mahidol.ac.th

[2] Department of Biology, Faculty of Science, Chiang Mai University, Chiang Mai 50200, Thailand; chakriya.sansupa@gmail.com

[3] Laboratory of Soil Science, Research Faculty of Agriculture, Hokkaido University, Sapporo 060-8589, Japan; hatano@agr.hokudai.ac.jp

[4] CFAES Rattan Lal Center for Carbon Management and Sequestration, The Ohio State University, 2021 Coffey Rd, Columbus, OH 43210, USA; lal.1@osu.edu

* Correspondence: noppol.aru@mahidol.ac.th

Simple Summary: The dynamics of the soil bacterial communities in rotational shifting cultivation fields were investigated. A six-year interval of fallow years did not result in any differences in the soil bacterial communities. A recovery in the abundance of the soil bacterial communities was observed during the rainy season. An increase in bacterial richness occurred during the year after a burn. After one year, the diversity of the bacterial communities reverted to pre-burning levels.

Abstract: Fire is a common practice in rotational shifting cultivation (RSC), but little is known about the dynamics of bacterial populations and the impact of fire disturbance in northern Thailand. To fill the research gap, this study aims to investigate the dynamics of soil bacterial communities and examine how the soil's physicochemical properties influence the bacterial communities in RSC fields over a period of one year following a fire. Surface soil samples (0–2 cm depth) were collected from sites with 6 (RSC-6Y) and 12 (RSC-12Y) years of fallow in Chiang Mai Province, northern Thailand at six different time points: before burning, 5 min after burning (summer), 3 months after burning (rainy season), 6 months after burning (rainy season), 9 months after burning (winter), and 12 months after burning (summer). The results revealed a reduction in the soil bacterial communities' diversity and an increase in soil nutrient levels immediately after the fire. The fire significantly influenced the abundance of Firmicutes, Proteobacteria, Acidobacteria, and Planctomycetes, but not that of Actinobacteria. At the genus level, Bacillus, Conexibacter, and Chthoniobacter showed increased abundance following the fire. During the rainy season, a recovery in the abundance of the soil bacterial communities was observed, although soil nutrient availability declined. Soil physicochemical properties such as pH, organic matter, organic carbon, electrical conductivity, cation exchange capacity, nitrate-nitrogen, available phosphorus, exchangeable potassium, total nitrogen, bulk density, sand, and silt contents significantly influenced the composition of bacterial communities. Alpha diversity indices indicated a decrease in diversity immediately after burning, followed by an increase from the early rainy season until the summer season, indicating that seasonal variation affected the composition of the soil bacterial communities. After one year of burning, an increase in bacterial richness was observed, while the diversity of the bacterial communities reverted to pre-burning levels.

Keywords: fire; rotational shifting cultivation; soil bacterial community; soil physicochemical properties; terrestrial microbiology

1. Introduction

Shifting cultivation is a type of rotational agricultural technique, commonly referred to as slash-and-burn or swidden. Moreover, it is a culture that has been transmitted by local indigenous populations [1]. Shifting cultivation is a relatively common practice, particularly in the highlands of south and southeast Asia [2]. It involves clearing a patch of vegetation using the slash-and-burn method and cultivating assorted crops in the cleared land [3]. The fields are left fallow after harvesting to facilitate the recovery of soil nutrients and promote the growth of plants through secondary succession [4,5]. The plant diversity may completely recover between 20 and 40 years after a fire [6]. In general, 14–15 months after the combustion phase, the soil microbial population can be the same as before [7]. In the mountainous regions of northern Thailand, shifting cultivation has been a long-standing traditional farming practice among hill tribes. However, shifting cultivation areas in Thailand are referred to as rotational shifting cultivation (RSC), because cutting native forests and building new settlements in protected forests are not allowed [3].

Fire is an element of an ecosystem that is utilized to encourage the germination or spread of plant species, especially in areas with deciduous forests [8]. However, wildfires have the apparent disadvantage of being difficult to control. One of the fundamental components of shifting agriculture is fire, which serves as a tool for land preparation [9]. However, burning vegetation can indeed alter the characteristics of the soil and its microbial community, impacting the soil's texture, structure, fertility, and physicochemical properties [10,11]. Burning of cultivated areas can alter the soil's characteristics, depending on the fire's severity and intensity. Low-intensity fires are shown to increase soil organic carbon (SOC), whereas high-intensity fires reduce SOC [12]. Ash deposition from biomass burning increases pH, calcium (Ca), phosphorus (P), potassium (K), magnesium (Mg), and the sum of basic cations [13]. The highest losses of eroded soil carbon (C) and nitrogen (N) were recorded in the first 2 years following severe wildfires in the western USA [14]. Hence, prescribed fires, specifically low-intensity fires, can be used to manage fuel and mitigate negative effects on soil properties [15].

Soil bacteria play a crucial role in biogeochemical cycles, supplying essential nutrients to the soil, and promoting plant growth [16]. Aeromicrobium, Agromyces, and Bacillus are among the common bacteria found in soil free from plant diseases [17]. Bacillus is crucial for plant development [18]. However, fire affects soil bacteria through heat and alteration of soil properties [19]. Smith et al. [20] reported a widespread presence of the Bacillus genus in the boreal forest near Chisholm, Alberta, Canada. In Bangladesh's Chittagong Hill Tracts, the Cocci, Bacillus, and Streptococcus species were found in soil under shifting cultivation [21]. Arunrat et al. [22] conducted the first study in Thailand to identify soil bacterial communities in a continuous 5-year fallow of a RSC field. They observed that Candidatus Udaeobacter dominated during the summer and winter seasons, while Bacillus was the dominant genus during the rainy season. Arunrat et al. [23] reported that the phylum Firmicutes presented a substantial increase of around 95% after a fire, with Bacillus being dominant. In the fungal community, Ascomycota experienced a significant increase, and Penicillium a dominant increase, after the fire in a RSC field in Thailand.

Hamman et al. [7] reported that the microbial communities in the burnt sites exhibited structural differences compared to the unburnt sites, while microbial biomass did not change after burning. They also indicated temperature, C content, and soil pH are the main factors that influence soil bacterial development. Some prior studies have revealed that the temporal dynamics of microbial communities are primarily influenced by seasonal periodicity [24–26], which can be attributed to the complex interactions among climate factors and variations in soil physicochemical properties. However, the dynamics of bacterial populations in shifting agricultural areas and the impact of fire disturbance remain unknown in northern Thailand. To gain a better understanding of these changes, this study was aimed at assessing the dynamics of soil bacterial communities and assessing the influence of soil physicochemical properties on bacterial communities for one year after a fire. This study was designed to test the following hypotheses: (i) burning would result in an immediate reduction in soil bacterial community diversity and an increase in soil nutrient levels, (ii) the abundance of the soil bacterial communities would begin to increase after 3 months of burning due to seasonal changes, while the availability of soil nutrients would decline, and (iii) seasonal variation and soil physicochemical properties would influence the composition of the soil microbial communities. The findings of this study can serve as a valuable scientific reference for understanding the recovery of soil properties and bacterial communities following a fire in RSC fields.

2. Materials and Methods

2.1. Study Areas

The study sites were located in Ban Mae Pok, Ban Thab Subdistrict, Mae Chaem District, Chiang Mai Province, northern Thailand (Figure 1A) [27]. Based on the data recorded by the Thai Meteorological Department in Chiang Mai Province (Doi Ang Khang and Mueang Chiang Mai stations) during 2021–2022, the annual rainfall varied between 1105–2688 mm (rainy season: May to October). The winter season (October and February) had minimum temperatures ranging from 3.2–22.1 °C, while the summer season (February to April) reached maximum temperatures ranging from 35–40 °C (Figure 1B). The soils in the highlands of Thailand with a slope greater than 35% were classified as slope complex soil series [28].

Two RSC fields were chosen for their similarity in microclimate and prior land use for the cultivation of upland rice. RSC-12Y (18°23′17″ N, 98°11′41″ E; elevation 692 m a.s.l; slope gradient 28%) was left fallow for 12 years after a harvest of upland rice. The field was subsequently cleared, burned, and used for upland rice cultivation in 2022. RSC-6Y (18°23′16″ N, 98°11′32″ E; elevation 729 m a.s.l; slope gradient 31%) was left fallow for 6 years after a harvest of upland rice. The RSC-12Y and RSC-6Y fields were cleared, the residues were burned, and the fields were used for upland rice cultivation in 2022 [27].

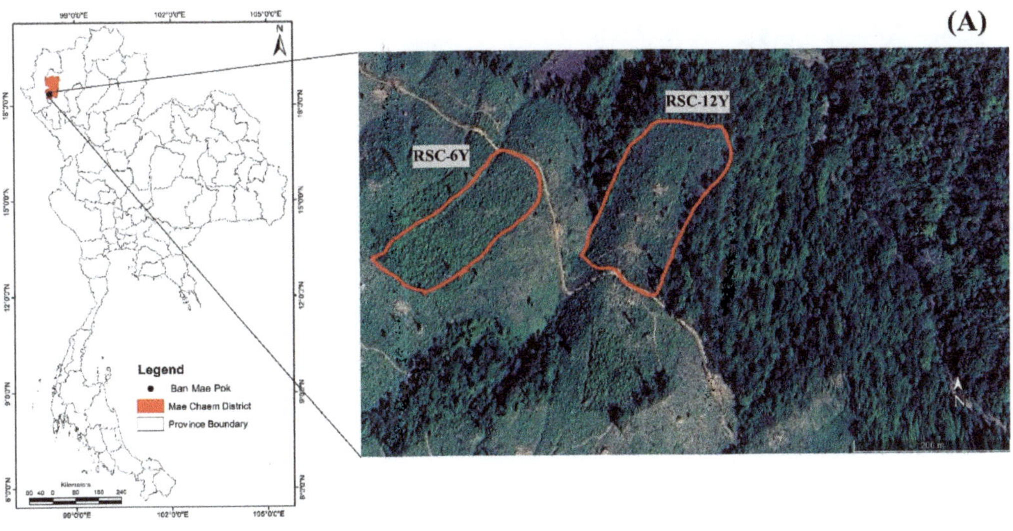

Figure 1. *Cont.*

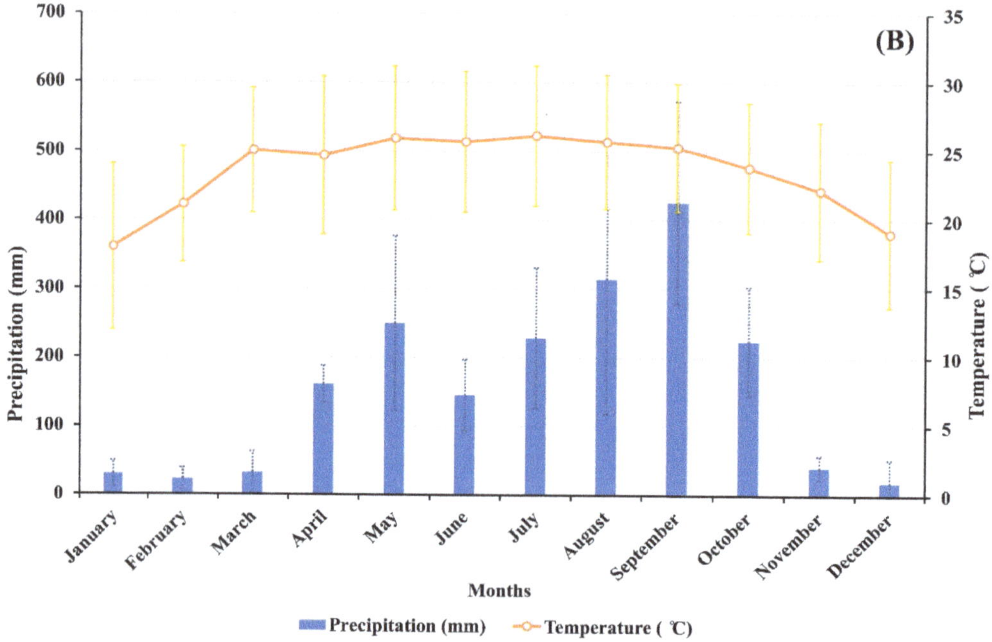

Figure 1. *Cont.*

Figure 1. Study area. (**A**) Location of RSC-6Y and RSC-12Y sites, (**B**) temperature and precipitation during 2021–2022, (**C**) fire behaviors, and (**D**) RSC fields over six time points. Photos were taken by Noppol Arunrat.

2.2. Experimental Design, Fire Measurements and Soil Sampling

Three sampling lines were established at intervals of 25 m for RSC-12Y and 15 m for RSC-6Y, ensuring statistical independence between lines. The sampling lines, with lengths of 170 m for RSC-12Y and 150 m for RSC-6Y, began at the upper slope and concluded at the lowest slope. Burning was conducted on 30 March 2022 at 15.00–17.00 h. The fire temperature during burning was measured using an infrared thermometer (PONPE 470IR). Fire measurements, including flame length, flame residence time, and spread rate, were conducted using drones for top-view image and video recordings, as well as direct observation by personnel on the ground (Figure 1C). Flame length was measured from the average flame tip to the middle of the flaming zone at the base of the fire [29]. Flame residence time was determined from video recordings, beginning when the flame front reached a specific point along the fuel bed and ending when flaming combustion was completed at that location [29]. The rate of fire spread was calculated using the cumulative spread rate method, which is derived by dividing the total distance traveled by a fire by the total time of travel [30].

In each sampling line of both the RSC-12Y and RSC-6Y fields, soil samples were collected from the surface layer (0–2 cm depth) at three sample plots, each with an approximate area of 1 × 1 m. While collecting soil samples, a measuring tape was used to ensure that samples were taken at precisely 0–2 cm depth. These sample plots were spaced approximately 50 m apart from each other. Soil samples from each plot of each sampling line were composited to obtain one composite sample per sampling line. This approach could reduce the variability in the richness and diversity of soil bacteria along the slopes, which might occur due to variations in soil nutrients caused by erosion processes in RSC fields. Collecting soil samples at a depth of 0–2 cm was considered appropriate for capturing the dynamics of soil bacteria, as this surface layer is more sensitive to changes in environmental conditions [22,31–33]. Stones, grasses, roots, ash, charcoal, and residues were removed manually. Steel knives were used to collect soil samples, following which approximately 1 kg of soil was transferred into a plastic bag for the analysis of the soil's physical and chemical properties. Meanwhile, around 100 g of soil was placed into zip-lock plastic bags and cooled at −20 °C for DNA extraction. Soil sampling was conducted at six different time points: before burning (March 2022), after burning at 5 min (March 2022), 3 months (June 2022), 6 months (September 2022), 9 months (harvest, December 2022), and 12 months (March 2023) (Figure 1D). A total of 36 soil samples were taken from the two RSC fields, comprising six time points. For each plot from where soil samples were obtained, soil temperature and soil moisture content were also recorded at 2 cm depth by using a Type K Thermocouple (PONPE 422 PR) and soil moisture meter (PONPE 301SM), respectively.

2.3. *Analysis of Soil Physical and Chemical Properties*

The soil bulk density (BD) at 0–5 cm was measured using a steel core (5.0 cm diameter and 5.5 cm length) and dried at 105 °C for 24 h [23]. Bulk samples were air-dried, grinded and sieved through a 2-mm sieve. Soil texture was determined on sieved samples by the hydrometer method [23]. Soil pH was measured by a pH meter (1:1 solids in water) [34], electrical conductivity (ECe) by using an EC meter [34], the cation exchange capacity (CEC) by the NH_4OAc method (pH 7.0) [32], total N (TN) by the micro-Kjeldahl method [34], and ammonium N (NH_4-N) and nitrate–N (NO_3-N) by the KCl extraction method [34]. In addition, exchangeable calcium (exch. Ca), magnesium (exch. Mg), and potassium (exch. K) were measured using atomic absorption spectrometry with NH_4OAc pH 7.0 extraction [35,36]. Available phosphorus (avail. P) was measured using the Bray II extraction method [35,37] and soil organic carbon (SOC) content was determined by the potassium dichromate ($K_2Cr_2O_7$) in sulfuric acid method [35,38], and converted to SOM by multiplying by 1.724.

2.4. *DNA Extraction, Bacterial 16s Amplification, and Sequencing*

DNA was extracted from the soil using a DNeasy PowerSoil Pro DNA Kit (Qiagen, Hilden, Germany), following the manufacturer's protocols. The extracted DNA was amplified using primers 341F (5′-CCTAYGG-GDBGCWSCAG) and 805R (5′-GGACTAC-NVGGGTHTCTAAT-3′), which amplified the V3–V4 region of the 16S rRNA gene [39]. The PCR product was subjected to sequencing using the paired-end Illumina Miseq platform at the Omics Sciences and Bioinformatics Center of Chulalongkorn University (Bangkok, Thailand).

2.5. *Bioinformatics Analysis*

Bioinformatics analysis was performed on QIIME2 to identify bacterial taxonomy and estimate the abundance of each taxon [40]. Briefly, raw sequence reads were imported to QIIME2. Forward and reverse primers were cut by cutadapt [41]. The DADA2 plugin was used for quality-filtering, merging, and chimera removal with the following parameters: forward read length = 230 bp, reverse read length = 210 bp, maxEE = 5, and overlap = 12 bp [42]. Amplicon sequence variants (ASVs) with less than two sequence reads (singletons) were removed. Taxonomy was assigned using the Silva v.138 database [43,44] and ASVs that were assigned to mitochondria or chloroplasts were filtered out. The remaining ASVs were then normalized to the smallest number of sequences from each sample using the rarefy plugin. Predictive enzyme abundance was assigned to the dataset using PICRUSt2 [45]. PICRUSt2 predicts potential functions based on reference genes. While this tool provides valuable insights into microbial functions, it is important to note that the results from these predictions still need to be validated by experimental data.

2.6. *Statistical Analysis*

Statistical analyses were performed on PAST [46] and the R program. The soil's physical and chemical properties at six time points were compared with one-way repeated measures ANOVA and post hoc Tukey's HSD tests. The alpha diversity indices, including observed richness, ACE, Simpson, and Shannon indices, were computed by the microeco package [47]. The alpha diversity indices in the two study sites and at the 6 sampling times were analyzed by two-way analysis of variance (two-way ANOVA). Differences of the indices between each of the sampling times were tested by ANOVA with repeated measures. Bacterial community composition was analyzed and visualized by non-metric multidimensional scaling (NMDS). The differences between each composition were tested by two-way permutational multivariate analyses of variance (two-way PERMANOVA). A redundancy analysis (RDA) was used to determine the influence of soil properties on soil bacterial community compositions, and the significance of the correlations between soil properties and bacterial communities was confirmed using the Mantel test. Differences between the abundances of bacteria at the phyla and genus levels and predictive enzymes were tested using ANOVA with repeated measures. Spearman's rank correlation was used to test the correlations between the abundances of genera, predictive enzymes, and soil properties.

3. Results

3.1. Soil Moisture, Soil Temperature, Fire Behaviors, and Soil Physicochemical Properties

No significant differences in soil moisture content were observed between pre-burning and 5 min after burning, likely due to the short flame residence time (Table 1) resulting in less moisture evaporation. However, significant differences were observed at 3 and 6 months after burning due to high precipitation (Figure 1B). At a depth of 0–2 cm, a significant increase in soil temperature was observed even at 5 min after burning, reaching an average of 46.0 °C and 47.8 °C for the RSC-6Y and RSC-12Y sites, respectively. During burning, the fire temperature in the litter layer ranged from 253 to 612 °C and 315 to 754 °C for the RSC-6Y and RSC-12Y sites, respectively. This is consistent with the fire temperatures in the topsoil of the slash and burn agricultural system in the rural community of Tijuco Preto in the Prudentópolis municipality of southern Brazil, which ranged from 534 to 777 °C [48]. Flame length and flame residence time during burning were measured in the range of 3.6 to 5.5 m and 22.0 to 33.0 s, respectively, while the spread rates varied from 12.5 to 15.5 m/min (Table 1).

Five minutes after burning, the soils exhibited significantly lower levels of SOM, SOC, TN, NO3-N, and clay content. Conversely, burned soils showed higher pH, ECe, NH_4-N, sand content, and soil nutrients (Avail. P, Exch. K, Exch. Ca, and Exch. Mg). However, bulk density and CEC remained unchanged at 5 min after burning. At 1 year after burning, the soil pH, ECe, SOM, SOC, Avail. P, and NH_4-N were higher than the initial values (pre-burning), whereas the levels of TN, Exch. K, Exch. Ca, and Exch. Mg declined over time points (Tables 2 and 3). However, it may be crucial to measure the bulk of plant litter in future studies to enhance our understanding of how soil temperatures impact the soil surface, especially in relation to SOM.

3.2. Bacterial Richness and Diversity

The bacterial diversity in two rotational shifting cultivations at six time points was examined in this study. A total of 427,788 high-quality sequence reads (11,883 sequences per sample) and 19,297 ASVs were obtained from the sequencing analysis. Alpha diversity, which represents the richness (observed richness and ACE) and diversity of bacteria (Shannon's and Simpson's indices), was observed. The two-way ANOVA analysis (Table 4) showed that sampling time had a significant impact on all diversity indices, whereas site and the interaction between both factors did not. Furthermore, the richness and diversity decreased immediately after burning (5 min) and increased three months later. The observed richness and ACE in the wet or rainy season (3 and 6 months) were comparable to those observed 5 min after burning. However, they significantly increased in the dry season (winter and summer, 9 and 12 months after burning) (Figure 2a,b). On the other hand, the Shannon's and Simpson's indices increased significantly in the wet or rainy season (3 months after burning). The Shannon's index was low during the rainy season and increased during the dry season (winter and summer) (Figure 2c). However, no significant changes were observed between the Simpson's diversity indices for different sampling times after 3 months of burning (Figure 2d).

Table 1. Soil moisture, soil temperature, and fire behaviors of study sites (mean ± SD).

Sites	Variables	Time Points						Fire Temperature in the Litter Top (°C) (min–max)	Flame Length (m)	Flame Residence Time (s)	Spread Rate (m/min)
		Pre-Burning	5 min after Burning	3 Months after Burning	6 Months after Burning	9 Months after Burning	12 Months after Burning				
RSC-6Y	Soil moisture (%)	34.2 ± 5.5 ^{b}	31.7 ± 6.9 ^{b}	41.3 ± 7.5 ^{a}	43.5 ± 8.0 ^{a}	33.8 ± 5.5 ^{b}	32.5 ± 5.5 ^{b}	253–612	3.6 ± 1.0	22.0 ± 19.0	12.5 ± 8.5
	Soil temperature (°C)	23.3 ± 3.1 ^{a}	46.0 ± 2.5 ^{b}	25.2 ± 3.5 ^{a}	23.4 ± 3.0 ^{a}	26.5 ± 3.0 ^{a}	27.5 ± 2.5 ^{a}				
RSC-12Y	Soil moisture (%)	35.8 ± 4.4 ^{b}	30.4 ± 5.8 ^{b}	40.1 ± 6.0 ^{a}	45.4 ± 5.8 ^{a}	32.5 ± 6.0 ^{b}	32.1 ± 6.2 ^{b}	315–754	5.5 ± 2.0	33.0 ± 21.0	15.5 ± 9.0
	Soil temperature (°C)	23.6 ± 2.2 ^{a}	47.8 ± 1.0 ^{b}	25.1 ± 2.0 ^{a}	23.5 ± 2.5 ^{a}	26.1 ± 3.5 ^{a}	27.5 ± 3.0 ^{a}				

^{a,b} significant statistical differences ($p < 0.05$), as determined by using one-way repeated measures ANOVA with post-hoc Tukey's HSD.

Table 2. Soil physical and chemical properties: bulk density (BD), electrical conductivity (ECe), soil organic matter (SOM), soil organic carbon (SOC), total nitrogen (TN), and proportion of sand, silt and clay.

Site	Time Point	pH (1:1)		EC_e (dS m^{-1})		BD (Mg m^{-3})		SOM (%)		SOC (%)		TN (%)		%Sand		%Silt		%Clay	
		Mean	SD	Mean	SD	Mean	SD	Mean	SD	Mean	SD	Mean	SD	Mean	SD	Mean	SD	Mean	SD
RSC-6Y	pre-burning	5.69 ^{a}	0.03	0.14 ^{a}	0.01	1.35 ^{a}	0.02	7.03 ^{a}	0.31	4.08 ^{a}	0.18	0.26 ^{a}	0.02	24.18 ^{a}	0.71	48.30 ^{a}	0.57	27.52 ^{a}	0.74
	5 min after burning	6.72 ^{b}	0.05	1.22 ^{b}	0.01	1.29 ^{a}	0.01	5.74 ^{b}	0.11	3.33 ^{b}	0.06	0.19 ^{b}	0.01	34.46 ^{b}	0.48	45.47 ^{a}	0.04	20.07 ^{b}	0.52
	3 months after burning	6.93 ^{b}	0.13	1.60 ^{b}	0.01	1.31 ^{a}	0.02	5.21 ^{b}	0.11	3.02 ^{b}	0.06	0.14 ^{b}	0.01	22.16 ^{a}	0.05	54.12 ^{b}	0.05	23.72 ^{b}	0.09
	6 months after burning	6.97 ^{b}	0.12	1.11 ^{b}	0.02	1.32 ^{a}	0.01	5.33 ^{b}	0.09	3.09 ^{b}	0.06	0.19 ^{b}	0.02	22.38 ^{a}	0.05	53.04 ^{b}	0.05	24.58 ^{a}	0.09
	9 months after burning	7.02 ^{b}	0.11	1.14 ^{b}	0.01	1.39 ^{a}	0.01	5.38 ^{b}	0.06	3.12 ^{b}	0.04	0.13 ^{b}	0.01	22.83 ^{a}	0.05	50.91 ^{a}	0.04	26.26 ^{a}	0.08
	12 months after burning	6.37 ^{b}	0.15	1.11 ^{b}	0.01	1.42 ^{a}	0.01	5.40 ^{b}	0.02	3.13 ^{b}	0.01	0.21 ^{a}	0.02	27.06 ^{a}	1.09	45.62 ^{a}	0.75	27.32 ^{a}	0.84
RSC-12Y	pre-burning	5.31 ^{a}	0.03	0.13 ^{a}	0.01	1.28 ^{a}	0.02	7.45 ^{a}	0.21	4.32 ^{a}	0.12	0.31 ^{a}	0.02	35.19 ^{a}	0.90	45.66 ^{a}	1.14	19.15 ^{a}	1.99
	5 min after burning	6.26 ^{b}	0.03	0.56 ^{b}	0.02	1.25 ^{a}	0.01	6.14 ^{b}	0.21	3.56 ^{a}	0.12	0.21 ^{b}	0.03	33.43 ^{a}	2.77	51.29 ^{b}	0.36	15.28 ^{a}	3.00
	3 months after burning	6.23 ^{b}	0.02	0.67 ^{b}	0.01	1.27 ^{a}	0.02	5.48 ^{b}	0.08	3.18 ^{a}	0.05	0.23 ^{b}	0.02	22.31 ^{b}	0.16	54.19 ^{b}	0.09	23.50 ^{b}	0.18
	6 months after burning	6.03 ^{b}	0.07	1.03 ^{b}	0.07	1.28 ^{a}	0.02	6.03 ^{b}	0.07	3.50 ^{a}	0.04	0.24 ^{b}	0.02	22.53 ^{b}	0.16	53.10 ^{b}	0.08	24.37 ^{b}	0.18
	9 months after burning	6.12 ^{b}	0.03	1.20 ^{b}	0.03	1.35 ^{a}	0.02	6.22 ^{b}	0.09	3.61 ^{a}	0.05	0.16 ^{b}	0.02	22.98 ^{b}	0.17	50.98 ^{b}	0.08	26.04 ^{b}	0.19
	12 months after burning	6.04 ^{b}	0.09	1.20 ^{b}	0.06	1.36 ^{a}	0.02	6.28 ^{b}	0.11	3.64 ^{a}	0.06	0.20 ^{b}	0.02	26.70 ^{b}	0.99	50.64 ^{b}	2.35	22.66 ^{b}	1.53

^{a,b} significant statistical differences ($p < 0.05$), as determined by using one-way repeated measures ANOVA with post-hoc Tukey's HSD.

Table 3. Soil chemical properties: cation exchange capacity (CEC); available phosphorus (avail.P); exchangeable K, Ca, and Mg, NH4-N, and NO3-N.

Site	Time Point	CEC (cmol kg^{-1})		Avail. P (mg kg^{-1})		Exch. K (mg kg^{-1})		Exch. Ca (mg kg^{-1})		Exch. Mg (mg kg^{-1})		NH_4-N (mg kg^{-1})		NO_3-N (mg kg^{-1})	
		Mean	SD	Mean	SD	Mean	SD	Mean	SD	Mean	SD	Mean	SD	Mean	SD
RSC-6Y	pre-burning	12.52 ^{a}	0.00	2.49 ^{a}	0.09	147.66 ^{a}	3.38	455.21 ^{a}	16.26	175.99 ^{a}	4.40	7.11 ^{a}	0.00	14.21 ^{a}	0.00
	5 min after burning	13.99 ^{a}	0.36	50.31 ^{b}	1.92	398.92 ^{b}	9.72	1024.87 ^{b}	46.60	278.51 ^{b}	9.27	68.68 ^{b}	4.10	0.00 ^{b}	0.00
	3 months after burning	11.27 ^{a}	0.00	30.75 ^{b}	0.37	256.20 ^{b}	7.70	1244.45 ^{b}	70.04	265.38 ^{b}	5.77	28.42 ^{c}	0.00	14.21 ^{a}	0.00
	6 months after burning	10.39 ^{a}	0.07	25.82 ^{b}	0.31	215.20 ^{b}	6.47	1045.33 ^{b}	58.83	222.92 ^{b}	4.84	35.53 ^{c}	0.00	28.42 ^{b}	0.00
	9 months after burning	10.72 ^{a}	0.37	16.52 ^{c}	0.20	137.72 ^{a}	4.14	669.01 ^{a}	37.65	142.66 ^{a}	3.10	48.35 ^{c}	0.00	35.53 ^{b}	0.00
	12 months after burning	11.68 ^{a}	0.16	11.01 ^{c}	0.29	135.38 ^{a}	23.61	634.73 ^{a}	23.90	139.54 ^{a}	10.51	35.53 ^{c}	0.00	14.21 ^{a}	0.00
RSC-12Y	pre-burning	12.52 ^{a}	0.00	2.62 ^{a}	0.12	93.94 ^{a}	1.02	174.54 ^{a}	6.77	87.43 ^{a}	0.60	35.53 ^{a}	0.00	132.63 ^{a}	4.10
	5 min after burning	10.65 ^{a}	0.63	50.31 ^{b}	6.74	334.23 ^{b}	4.93	546.43 ^{b}	11.81	165.03 ^{b}	8.39	90.00 ^{b}	4.10	7.11 ^{b}	0.00
	3 months after burning	14.40 ^{a}	0.00	19.79 ^{c}	0.91	248.57 ^{b}	2.45	311.13 ^{b}	11.51	174.38 ^{b}	10.82	78.16 ^{c}	0.00	35.53 ^{c}	0.00
	6 months after burning	11.26 ^{a}	0.33	16.62 ^{c}	0.76	208.80 ^{b}	2.06	261.34 ^{b}	9.67	146.47 ^{b}	9.09	78.16 ^{c}	0.00	59.21 ^{d}	4.10
	9 months after burning	11.92 ^{a}	0.42	10.63 ^{c}	0.49	133.63 ^{a}	1.32	167.25 ^{a}	6.18	93.73 ^{a}	5.82	78.72 ^{c}	0.15	60.89 ^{d}	0.19
	12 months after burning	12.30 ^{a}	0.15	8.39 ^{c}	0.27	106.27 ^{a}	6.63	139.66 ^{a}	1.40	83.16 ^{a}	8.17	78.16 ^{c}	0.00	35.53 ^{c}	0.00

$^{a-d}$ significant statistical differences ($p < 0.05$), as determined by using one-way repeated measures ANOVA with post-hoc Tukey's HSD.

Table 4. Alpha and Beta diversity indices of bacteria.

Factors	Alpha Diversity				Beta Diversity
	Richness		Diversity Index		Bray–Curtis
	Observed	ACE	Shannon	Simpson	
Site	0.44	0.54	0.13	0.41	3.10 *
Sampling time	14.28 *	14.03 *	15.98 *	7.96 *	4.10 *
Site:Sampling time	0.22	0.17	0.13	0.3	2.16 *

* denote statistically significant ($p < 0.05$), as determined by using two-way ANOVA and PERMANOVA analysis.

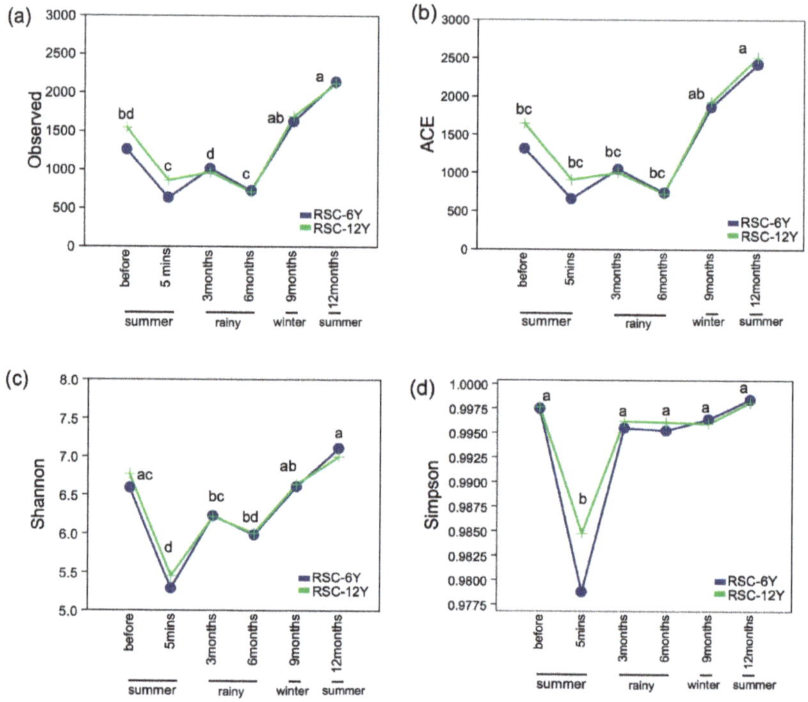

Figure 2. Alpha diversity indices across all samples. (**a**) Observed richness, (**b**) ACE richness, (**c**) Shannon diversity index, and (**d**) Simpson diversity index. Different letters in the plots indicate statistically different ($p < 0.05$).

3.3. *Bacterial Taxonomic Distribution*

Overall, we observed 35 phyla, 91 classes, 184 orders, 258 families, and 502 genera. The most abundant phyla across all samples belonged to Actinobacteria, Proteobacteria, Firmicutes, Chloroflexi, and Planctomycetes (Figure 3). Burning significantly affected the abundance of several phyla, such as Firmicutes, Proteobacteria, Acidobacteria, and Planctomycetes, while it did not have a significant effect on certain phyla, such as Actinobacteria. The abundance of Actinobacteria at 5 min after burning was similar to that before burning, but increased during the wet or rainy season (3 months after burning) and decreased during the dry season (winter and summer; 9 and 12 months after burning) (Figure 3). At the genus level, the most abundant genera across all samples were *Bacillus*, followed by *Geodermatophilus*. Burning significantly affected the abundance of *Bacillus*, *Conexibacter*, and *Chthoniobacter*. The abundance of each genus changed at each time point. One year after burning, the proportions of bacteria were not similar to those before burning.

3.3.1. Taxonomic Distribution in RSC-6Y

Chloroflexi (19.44%), Actinobacteria (18.61%), and Proteobacteria (18.39%) dominated the community before burning. However, the most fluctuating phyla after burning were Fermicute, Proteobacteria, Acidobacteria, and Planctomycetes. The abundance of Firmicutes increased from 9% to 30% after burning (5 min), then decreased to 10% during the wet or rainy season (3 and 6 months after burning), to 4% during winter (9 months after burning), and 7% in summer (12 months after burning). Proteobacteria and Acidobacteria decreased from 18% to 7% and 12% to 2%, respectively, after burning (5 min). The abundance of these two phyla increased during the rainy season, after 3 months (Proteobacteria: 22% and Acidobacteria: 4%) and 6 months (Proteobacteria: 16% and Acidobacteria: 8%). Then, during summer (12 months after burning), Proteobacteria totaled 11.67% and Acidobacteria totaled 10.53% (Figure 3a).

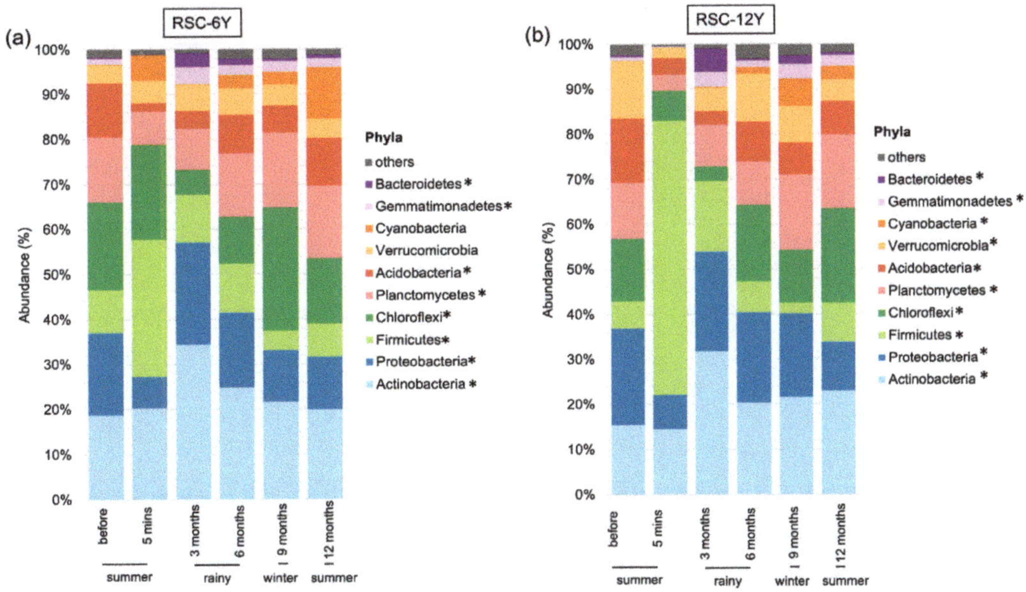

Figure 3. Stacked bar plot showing the most abundant bacterial phyla in (**a**) rotational shifting cultivation—6 years (RSC-6Y) and (**b**) rotational shifting cultivation—12 years (RSC-12Y). * denote statistically significant ($p < 0.05$).

The most abundant genus before burning was *Bacillus* (4.65%), followed by *Conexibacter* (3.80%) and *Gemmata* (2.29%). The abundance of *Bacillus* and *Nocardioides* increased after burning (5 min), while that of *Gemmata*, *Conexibacter*, Candidatus *Udaeobacter*, and *Sphingomonas* decreased. However, significant change was observed only in *Conexibacter*. The data from this study also showed that *Sphingomonas* significantly changed with the changes of the seasons. Their abundance was high during the wet or rainy season (3 and 6 months after burning) and significantly decreased during the dry season (winter and summer) (Figure 4). One year after burning, the most prevalent genera belonged to *Bacillus* (5.60%), followed by *Geodermatophilus* (3.30%) and *Gemmata* (2.67%).

Spearman's rank analysis revealed positive correlations between *Bacillus* and Avail. P and Exch. K, as well as a negative correlation between *Bacillus* and BD, albeit these correlations were not statistically significant. Contrarily, significant correlations were observed between the genus *Geodermatophillus* and pH (positive) and TN (negative) (see in Supplementary Figure S1).

3.3.2. Taxonomic Distribution in RSC-12Y

Before burning, the most abundant phyla were Proteobacteria (21.57%), Actinobacteria (15.43%), and Acidobacteria (14.17%). Burning significantly changed the abundance of some phyla, including Firmicutes, Proteobacteria, Acidobacteria, Planctomycetes, and Verrucomicrobia. The abundance of Firmicutes increased from 6% to 61% after burning (5 min), then decreased to 16% at 3 months (rainy season) and 8% after 12 months (summer). Proteobacteria decreased from 22% to 8% after burning (5 min), then increased to 22% after 3 months and declined to 10% after 12 months. Similarly, Acidobacteria also decreased from 14% to 4% after burning and increased to 7% after 12 months. However, Actinobacteria did not change significantly immediately after burning (Figure 3b).

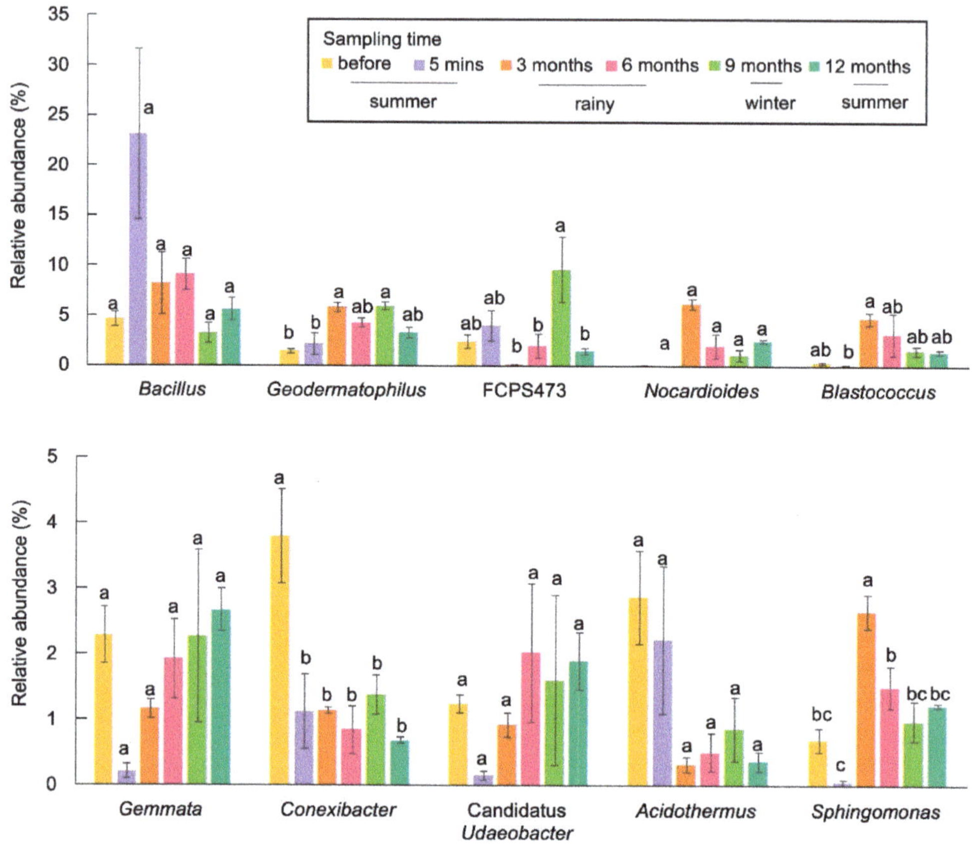

Figure 4. Bar plot showing the abundance of the most abundant genera in RSC-6Y. Different letters in the plots indicate statistically different ($p < 0.05$).

At the genus level, *Bacillus* (1.43%), Candidatus *Udaeobacter* (1.37%), and HSB OF53-F07 (1.49%) were the most prevalent taxa in samples before burning. After burning (5 min), the abundance of *Bacillus* increased significantly, whereas that of *Chthoniobacter* decreased (Figure 5). There were some taxa which increased significantly during the wet season and decreased during the dry season. These included *Conexibacter*, HSB OF53-F07, and *Acidothemus*. One year after burning, the most prevalent taxa belonged to FCPS473 (1.44%), *Bacillus* (1.43%), and *Geodermatophilus* (0.67%).

Spearman's rank correlation analysis showed positive correlations between Avail. P, Exch. K, Exch. Mg, and Exch. Ca with *Bacillus*, while these nutrients showed negative correlations with other genera. Soil BD and clay showed negative correlations with *Bacillus* but positive with *Geodermatophillus*, FCPS473, *Chthoniobacter*, and *Gemmata*. However, no significant differences were observed in any of these correlations (see in Supplementary Figure S2).

3.4. Community Composition and Correlation with Soil Properties

Two-way PERMANOVA analysis revealed that both sampling time and site significantly affected beta diversity or bacterial community composition (Table 4). Bacteria in each time point and site were separate from each other, and this indicated the difference in the community compositions (see in Supplementary Figure S3). One year after burning, the community compositions of bacteria were not similar to those before burning. Thus, redundancy analyses (RDAs) in the two study sites were performed separately. RDA showed that soil properties explained 62.7% and 61.8% of the total variations in the community compositions in RSC-6Y and RSC-12Y, respectively (Figure 6). According to the Mantel test, several soil parameters (i.e., pH, SOM, SOC, ECe, CEC, NO_3-N, Avail. P, Exch. K, TN, BD, sand, and silt) significantly impacted the bacterial communities. Specifically, pH and NO_3-N were significant only in RSC-12Y (Figure 6, Table 5).

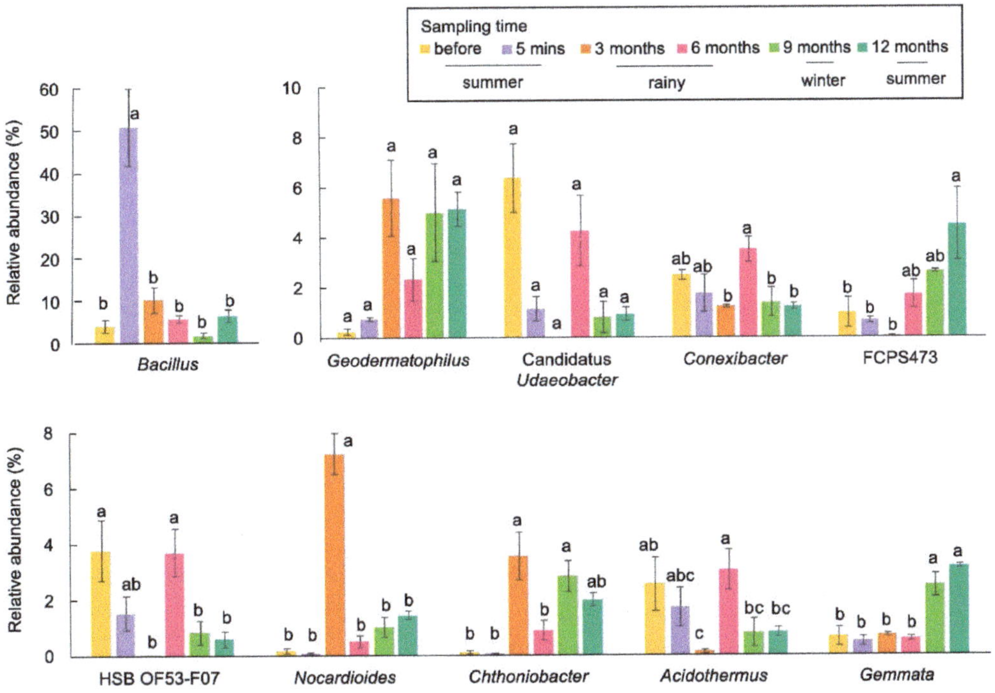

Figure 5. Bar plot showing the abundance of the most abundant genera in RSC-12Y. Different letters in the plots indicate statistically different ($p < 0.05$).

Table 5. Correlation coefficient between bacterial community and soil properties.

Soil Properties	RSC-6Y		RSC-12Y	
	Correlation Coefficient	*p*-Value	Correlation Coefficient	*p*-Value
pH	0.184	0.054	0.257	0.008 *
SOM	0.339	0.004 *	0.302	0.009 *
SOC	0.341	0.004 *	0.300	0.010 *
ECe	0.296	0.005 *	0.404	0.001 *
CEC	0.354	0.002 *	0.298	0.003 *
NH_4-N	0.410	0.001 *	0.304	0.001 *
NO_3-N	0.131	0.119	0.442	0.001 *
Avail. P	0.424	0.001 *	0.320	0.002 *
Exch. K	0.340	0.001 *	0.303	0.003 *
Exch. Ca	0.235	0.013 *	0.296	0.006 *
Exch. Mg	0.183	0.030 *	0.218	0.026 *
Total. N	0.138	0.100	0.155	0.086
BD	0.237	0.012 *	0.228	0.012 *
Sand	0.301	0.011 *	0.360	0.001 *
Silt	0.218	0.019 *	0.221	0.034 *
Clay	0.333	0.004 *	0.255	0.023 *

* denote statistically significant ($p < 0.05$), as determined by using Mantel test.

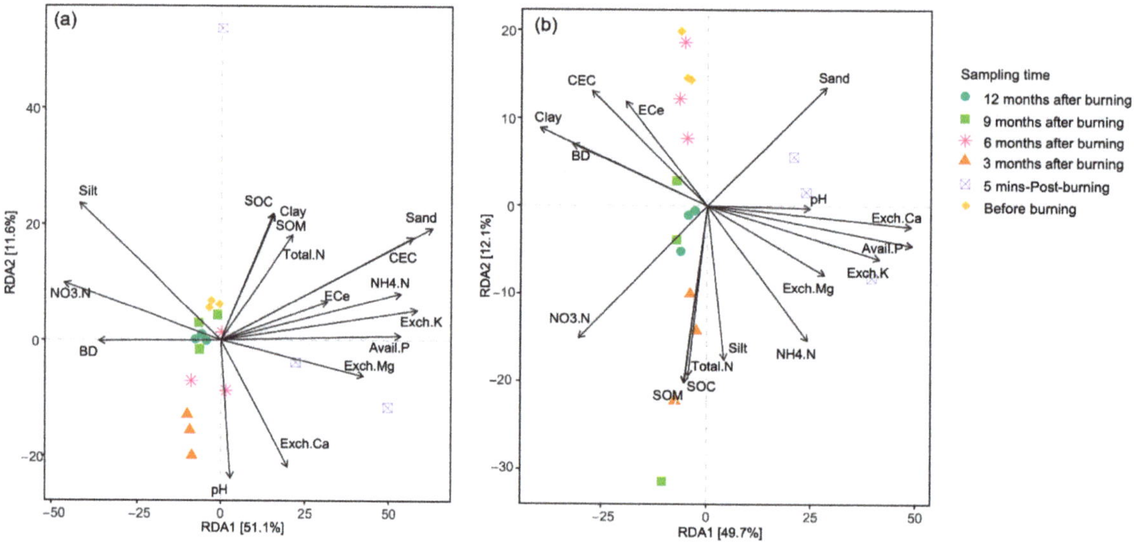

Figure 6. RDA ordination illustrating soil properties significantly correlated with bacterial community composition in (**a**) RSC-6Y and (**b**) RSC-12Y. Significant parameters indicated with the Mantel test ($p < 0.05$).

3.5. Predictive Function

A total of 19,265 (99.85%) ASVs were assessed for their impact on various soil functions. Specifically, 10 enzymes were highlighted in relation to soil systems, including nitrogenase, nitrate reductase, alkaline phosphatase, acid phosphatase, arylsulfatase, chitinase, β-glucosidase, cellulase, amidase, and urease. In both study sites, acid phosphatase levels decreased significantly immediately after burning and increased after a few months. Similar trends were observed for nitrate reductase and urease in the RSC-12Y site. In contrast, cellulase and amidase levels in both study sites increased significantly immediately after burning and then decreased after some months (Figure 7).

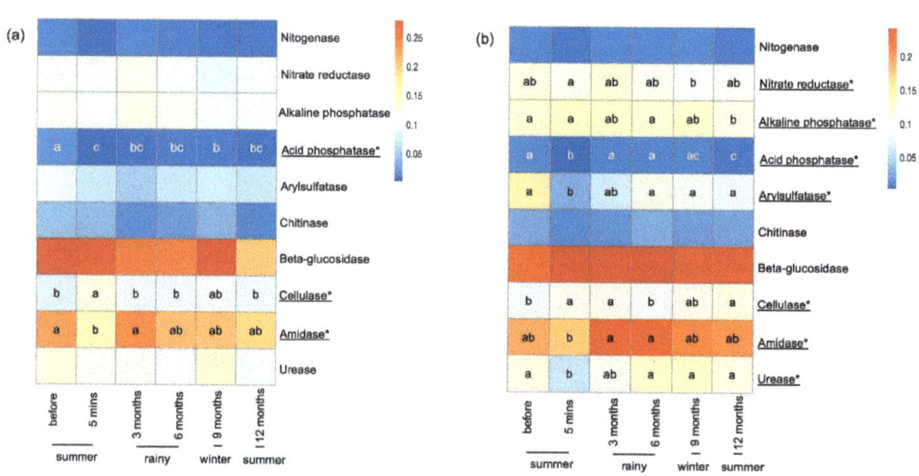

Figure 7. The abundance of predictive enzymes in RSC-6Y (**a**) and RSC-12Y (**b**). Color represents relative abundance of each enzyme. * denote statistically significant ($p < 0.05$).

In terms of correlations, a significant positive correlation was observed between cellulase and ECe (electrical conductivity of the saturation extract) in RSC-6Y (Figure 8a). In the case of RSC-12Y, significant positive correlations were observed between acid phosphatase and arysulfatase as well as between acid phosphatase and NO_3-N (nitrate nitrogen) (Figure 8b).

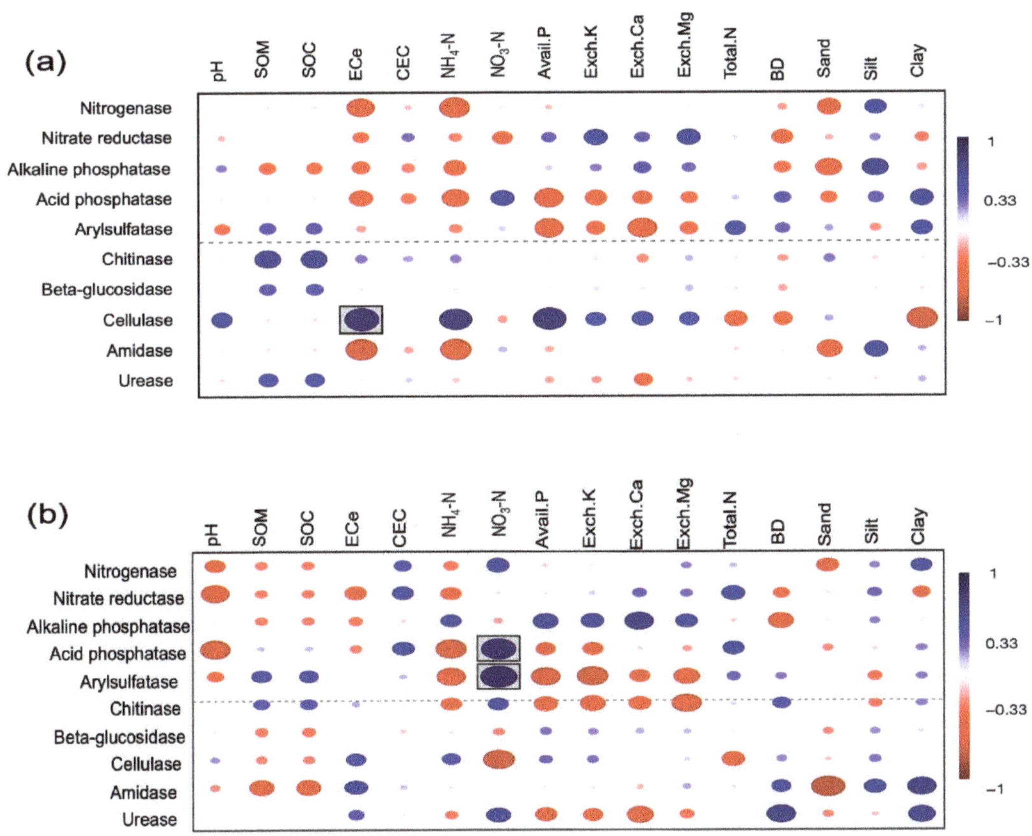

Figure 8. Correlations between the predictive enzymes and soil properties in RSC-6Y (**a**) and RSC-12Y (**b**). Color represents correlation coefficient. Blue = positive correlation, red = negative correlation, and circles within boxes are significant correlations. Circle size corresponds with the correlation coefficient.

4. Discussion

4.1. Effects of Seasonal Variations on Soil Microbial Communities

Soil microbial composition changes with the seasons, influenced by variations in soil temperature, moisture, and nutrients content [49–51]. Furthermore, seasonal variations lead to changes in environmental conditions, including photosynthesis, root exudates, and the accumulation of litter, which can have an impact on soil microbial communities [52,53]. The data in Table 4 showed that sampling time significantly impacted all diversity indices. Alpha diversity indices across all samples showed a decline immediately after burning and increased with the onset of the early rainy season until the summer season (Figure 2). This trend may be attributed to the increase in soil moisture resulting from precipitation during the rainy season, along with the release of available nutrients from ash and charcoal following the fire (Tables 2 and 3). Additionally, the increase in temperature from late winter to the summer season contributes to an increase in the composition of soil bacterial communities. The data presented herein are in accord with those reported by Luo et al. [54], who also demonstrated that the microbial community in an orchard in southeastern China exhibited a peak during the summer, likely attributed to the increases in temperature and rainfall. Similar trends have also been observed in diverse ecosystems, including an evergreen broadleaf forest in southwestern China [55], temperate grasslands [56], an apple orchard in the temperate monsoon climate zone [57], and the Mongolian oak-dominant Gwangneung forest in Korea [58].

Seasonal changes in climate parameters can have a significant impact on the growth of vegetation, which directly affects the microbial community [59,60]. This trend may be due to the variations in the decomposition rate of litter [61] and substances released through root exudates [62]. It is also in accord with the trend observed in the alpha diversity indices shown in Figure 2. Soil moisture content plays a crucial role in plant growth and the microbial community by influencing osmotic potential, nutrient and energy transport, and cellular metabolism [63]. Griffths et al. [64] observed that increased availability of carbon sources and root growth during the growing season led to an increase in soil microbial biomass and activity. Jefferies et al. [65], Zhang et al. [66], and Li et al. [67] also reported that soil microbial growth was limited during winter but increased during summer due to an increase in temperature.

In RSC fields, fire is an anthropogenic disturbance that significantly alters soil microbial communities (Figures 1 and 2), which play a crucial role in supporting the recovery process following biomass burning, particularly in response to seasonal variations (Figures 3–5). The abundance of Firmicutes, Proteobacteria, Acidobacteria, and Planctomycetes increased after the occurrence of fire. After the fire, there was also a significant increase in the abundance of Firmicutes, making it the most prominent group. However, the abundance of Actinobacteria remained unchanged immediately after burning and exhibited variation throughout the seasons. It showed an increase during the rainy season but a decrease during the winter and summer seasons, which could be attributed to intense competition and interactions between microbial species (Figure 3). The phylum Firmicutes includes diverse thermophilic, antibiotic-producing, and endospore-forming organisms that exhibit high resistance to desiccation, heat, and radiation [68]. A study by Nelson et al. [69] reported that Actinobacteria possess genes associated with heat resistance, rapid growth, and the ability to utilize pyrogenic carbon, which enables their survival following a fire. At the genus level, the abundance of *Sphingomonas* decreased significantly after the fire event in RSC-6Y. However, its abundance showed a significant increase during the rainy season and decreased during the dry season (winter and summer) (Figure 4). This trend suggests that *Sphingomonas* may be a species that exhibits a rapid response to changes in precipitation and temperature, possibly indicating its adaptability to fluctuating environmental conditions. In the RSC-12Y site, the genera *Conexibacter*, HSB OF53-F07, and *Acidothemus* exhibited a positive response during the rainy season and a decrease during the dry season (Figure 4). In contrast to the other genera, the abundance of FCPS473 and *Gemmata* remained unchanged after the occurrence of fire and exhibited a significant increase during the dry season (Figures 4 and 5).

4.2. Relationship between Soil Physical and Chemical Properties, and Soil Bacterial Communities

The combustion of biomass and SOM during the fire resulted in the accumulation of ash on the soil surface. This ash accumulation plays a crucial role in altering the physical and chemical properties of the soil [70], ultimately influencing the dynamics of soil microbial communities [71]. The occurrence of fire significantly increased the soil pH, ECe, NH_4-N, and nutrients (Avail. P, Exch. K, Exch. Ca, and Exch. Mg) (Tables 2 and 3). This increase was attributed to the deposition of ash during the fire event. However, over time, these values declined due to runoff, leaching, and the movement of nutrients into deeper soil layers through infiltration and percolation. These findings are also in accord with those reported by previous studies [3,12,19,72–75].

Fire behavior directly impacts the extent of combustion and the maximum temperature reached during the fire event, which in turn influence the soil's physical, chemical, and biological properties. High-intensity fires result in the high volatilization of SOM and oxidation processes in the topsoil layer [76]. On the other hand, incomplete combustion during low-intensity fires can produce semi-pyrolyzed ash [77]. The accumulation of ash affects soil microbial communities by increasing nutrient availability in the short term [73]. Additionally, the darkening of the soil due to ash deposition reduces the albedo, resulting in an increase in soil surface temperature [78]. The data presented herein show significant correlations between bacterial communities and several soil properties, including pH, SOM, SOC, ECe, CEC, NO_3-N, Avail. P, Exch. K, TN, BD, and the contents of sand and silt (Table 5). The data presented herein also indicate an increase NO_3-N and a decrease in NH_4-N during the rainy and winter seasons (3, 6, and 9 months after burning, corresponding to June, September, and December) (Table 3). This observation suggests a rapid growth of soil microorganisms in the tropical zone, which can be attributed to the high soil moisture conditions during these seasons. These findings are also consistent with the trend observed in the alpha diversity indices (Figure 2), implying that soil bacteria, particularly *Bacillus*, play a role in the post-fire ecosystem by oxidizing NH_4-N to NO_2-N and subsequently converting it to NO_3-N, which is important for plant growth. A study conducted in northern Sweden by DeLuca et al. [79] reported findings similar to those presented herein. The data in Table 5 specifically show that pH and NO_3-N played a significant role in changing the bacterial communities in RSC-12Y, where fire intensity and soil temperature were higher compared to RSC-6Y (Table 1). These findings may be attributed to the significant impact of increased soil temperature and soil pH on nitrogen cycle processes. Jones et al. [80] and Verma and Jayakumar [81] have revealed that an increase in soil temperature and higher pH have an impact on the dynamics of inorganic nitrogen (NH_4-N and NO_3-N) through mineralization and nitrification processes, as well as the ammonification process [12].

The short-term increase in available nutrients and soil pH following a fire can facilitate the rapid regrowth of plants, leading to an increased release of plant root exudates. These exudates play a crucial role in changing the composition of soil bacteria [82]. Meharg and Killham [83] and Ren et al. [84] observed that a high pH value promotes bacterial growth by increasing the exudation of organic compounds from plants. This provides another possible explanation for the observed increase in bacterial richness and diversity three months after the fire in our study (Figure 2). Fire alters the functional diversity of microbial communities, which is essential for the recovery of ecosystem processes, influencing decomposition and nutrient cycling [85]. D'Ascoli et al. [86] observed short-term changes in functional diversity after fire in the Mediterranean region, suggesting rapid recovery despite persistent reductions in microbial community activity and structural changes. Fontúrbel et al. [71] observed higher values of diversity indices and pH immediately after fire, indicating more favorable soil conditions for stimulating bacterial populations. However, fire-induced changes in the functional diversity of microbial communities in RSC fields need to be further studied in the future.

5. Conclusions

Understanding the impact of fire on soil physicochemical properties and bacterial communities in RSC fields is crucial for effective post-fire management and regeneration of burnt areas. The data presented herein show that immediately after the fire, there was a decrease in soil bacterial community diversity and an increase in soil nutrient levels at the 0–2 cm soil depth. However, 3 months after burning (during the rainy season), there was a recovery in the abundance of the soil bacterial communities, despite the decline in soil nutrient availability. The composition of soil microbial communities was influenced by seasonal variation and the soil's physicochemical properties. After one year of burning, there was an increase in soil nutrient levels, pH, and ECe compared to those of the initial values (pre-burning), along with an increase in bacterial richness. However, the diversity of the bacterial communities reverted to pre-burning levels.

Supplementary Materials: The following supporting information can be downloaded at: https://www.mdpi.com/article/10.3390/biology13060383/s1. Figure S1: Correlation between the abundant genera and soil properties in RSC-6Y. Color represents correlation coefficient. Blue = positive correlation, red = negative correlation, and circles within boxes are significant correlations; Figure S2: Correlation between the abundant genera and soil properties in RSC-12Y. Color represents correlation coefficient. Blue = positive correlation, red = negative correlation, and circles within boxes are significant correlations; Figure S3: NMDS ordination based on Bray–Curtis distances showing bacterial community compositions across all samples.

Author Contributions: Conceptualization, N.A., C.S., S.S., R.H. and R.L.; Methodology, N.A. and C.S.; Formal analysis, N.A. and C.S.; Investigation, N.A. and S.S.; Data curation, N.A.; Writing—original draft, N.A., C.S., S.S., R.H. and R.L.; Writing—review and editing, N.A. and C.S.; Supervision, R.H. and editing and reviewing of data on soil physical and chemical properties, R.L.; Funding acquisition, N.A. All authors have read and agreed to the published version of the manuscript.

Funding: This work was supported by the Office of the Permanent Secretary, Ministry of Higher Education, Science, Research, and Innovation (OPS MHESI), Thailand Science Research and Innovation (TSRI), and Mahidol University (Grant No. RGNS 64-153).

Institutional Review Board Statement: This study was conducted according to the guidelines of the Declaration of Helsinki and approved by the Institutional Review Board of the Institute for Population and Social Research, Mahidol University (IPSR-IRB) (COA. No. 2022/03-059).

Informed Consent Statement: Not applicable.

Data Availability Statement: The original contributions presented in the study are publicly available. The data can be found at the National Center for Biotechnology Information (NCBI) under the BioProject accession number PRJNA1003674.

Acknowledgments: The authors extend their appreciation to the Office of the Permanent Secretary, Ministry of Higher Education, Science, Research, and Innovation (OPS MHESI), Thailand Science Research and Innovation (TSRI), and Mahidol University.

Conflicts of Interest: The authors declare that they have no known competing financial interests or personal relationships that could have appeared to influence the work reported in this paper.

References

1. Cherrier, J. Shifting Cultivation: Misconception of the Asian Governments. *J. Int. Dev. Coop.* **2018**, *24*, 71–82.
2. Rasul, G.; Thapa, G.B. Shifting cultivation in the mountains of South and Southeast Asia: Regional patterns and factors influencing the change. *Land Degrad. Dev.* **2003**, *14*, 495–508. [CrossRef]
3. Arunrat, N.; Sereenonchai, S.; Hatano, R. Effects of fire on soil organic carbon, soil total nitrogen, and soil properties under rotational shifting cultivation in northern Thailand. *J. Environ. Manag.* **2022**, *302*, 113978. [CrossRef]
4. Stygler, E.; Harivelo, M.R.; Max, J.; Pfefer, M.J.; Fernandes, E.C.M.; Bates, D.M. Infuence of slash and burn farming practices on fallow succession and land degradation in the rainforest region of Madagascar. *Agric. Ecosyst. Environ.* **2007**, *19*, 257–269. [CrossRef]
5. Mertz, O.; Padoch, C.; Fox, J.; Cramb, R.A.; Leisz, S.J.; Lam, N.I.; Vien, T.D. Swidden change in Southeast Asia: Understanding causes and consequences. *Hum. Ecol.* **2009**, *37*, 259–264. [CrossRef]
6. Finegan, B. Pattern and process in neotropical secondary rain forests: The first 100 years of succession. *Trends Ecol.* **1996**, *1*, 119–124. [CrossRef] [PubMed]
7. Hamman, S.T.; Burke, I.C.; Stromberger, M.E. Relationships between microbial community structure and soil environmental conditions in a recently burned system. *Soil Biol. Biochem.* **2007**, *39*, 1703–1711. [CrossRef]
8. Emery, S.M.; Uwimbabazi, J.; Flory, S.L. Fire intensity effects on seed germination of native and invasive Eastern deciduous forest understory plants. *For. Ecol. Manag.* **2011**, *261*, 1401–1408. [CrossRef]
9. Ketterings, Q.M.; Tri Wibowo, T.; van Noordwijk, M.; Penot, E. Farmers' perspectives on slash-and-burn as a land clearing method for small-scale rubber producers in Sepunggur, Jambi Province, Sumatra, Indonesia. *For. Ecol. Manag.* **1999**, *120*, 157–169. [CrossRef]
10. Barreiro, A.; Díaz-Raviña, M. Fire impacts on soil microorganisms: Mass, activity, and diversity. *Curr. Opin. Environ. Sci. Health* **2021**, *2*, 100264. [CrossRef]
11. Agbeshie, A.A.; Abugre, S.; Atta-Darkwa, T.; Awuah, R. A review of the effects of forest fire on soil properties. *J. For. Res.* **2022**, *33*, 1419–1441. [CrossRef]
12. Caon, L.; Vallejo, V.R.; Ritsema, C.J.; Geissen, V. Effects of wildfire on soil nutrients in Mediterranean ecosystems. *Earth-Sci. Rev.* **2014**, *139*, 47–58. [CrossRef]

13. Dzwonko, Z.; Loster, S.; Gawroński, S. Impact of fire severity on soil properties and the development of tree and shrub species in a Scots pine moist forest site in southern Poland. *For. Ecol. Manag.* **2015**, *342*, 56–63. [CrossRef]
14. Pierson, D.N.; Robichaud, P.R.; Rhoades, C.C.; Brown, R.E. Soil carbon nitrogen eroded after severe wildfire erosion mitigation treatments. *Int. J. Wildland Fire* **2019**, *28*, 814. [CrossRef]
15. Alcañiz, M.; Outeiro, L.; Francos, M.; Úbeda, X. Effects of prescribed fires on soil properties: A review. *Sci. Total Environ.* **2018**, *613–614*, 944–957. [CrossRef] [PubMed]
16. Hayat, R.; Ali, S.; Amara, U.; Khalid, R.; Ahmed, I. Soil beneficial bacteria and their role in plant growth promotion: A review. *Ann. Microbiol.* **2010**, *60*, 579–598. [CrossRef]
17. Wang, R.; Zhang, H.; Sun, L.; Qi, G.; Chen, S.; Zhao, X. Microbial community composition is related to soil biological and chemical properties and bacterial wilt outbreak. *Sci. Rep.* **2017**, *7*, 343. [CrossRef] [PubMed]
18. Saxena, A.K.; Kumar, M.; Chakdar, H.; Anuroopa, N.; Bagyaraj, D.J. Bacillus species in soil as a natural resource for plant health nutrition. *J. Appl. Microbiol.* **2020**, *128*, 1583–1594. [CrossRef] [PubMed]
19. Certini, G. Effects of fire on properties of forest soils: A review. *Oecologia* **2005**, *143*, 1–10. [CrossRef] [PubMed]
20. Smith Nancy, R.; Kishchuk Barbara, E.; Mohn William, W. Effects of Wildfire and Harvest Disturbances on Forest Soil Bacterial Communities. *Appl. Environ. Microbiol.* **2008**, *74*, 216–224. [CrossRef]
21. Miah, S.; Dey, S.; Sirajul Haque, S.M. Shifting cultivation effects on soil fungi and bacterial population in Chittagong Hill Tracts, Bangladesh. *J. For. Res.* **2010**, *21*, 311–318. [CrossRef]
22. Arunrat, N.; Sansupa, C.; Sereenonchai, S.; Hatano, R. Stability of soil bacteria in undisturbed soil and continuous maize cultivation in Northern Thailand. *Front. Microbiol.* **2023**, *14*, 1285445. [CrossRef] [PubMed]
23. Arunrat, N.; Sansupa, C.; Sereenonchai, S.; Hatano, R. Short-term response of soil bacterial and fungal communities to fire in rotational shifting cultivation, Northern Thailand. *Appl. Soil Ecol.* **2024**, *196*, 105303. [CrossRef]
24. Bardgett, R.D.; Freeman, C.; Ostle, N.J. Microbial contributions to climate change through carbon cycle feedbacks. *ISME J.* **2008**, *2*, 805–814. [CrossRef] [PubMed]
25. Allison, S.D.; Treseder, K.K. Warming and drying suppress microbial activity and carbon cycling in boreal forest soils. *Glob. Chang. Biol.* **2008**, *14*, 2898–2909. [CrossRef]
26. Gilbert, J.A.; Steele, J.A.; Caporaso, J.G.; Steinbrück, L.; Reeder, J.; Temperton, B.; Huse, S.; McHardy, A.C.; Knight, R.; Joint, I.; et al. Defining seasonal marine microbial community dynamics. *ISME J.* **2012**, *6*, 298–308. [CrossRef]
27. Arunrat, N.; Sereenonchai, S.; Kongsurakan, P.; Yuttitham, M.; Hatano, R. Variations of soil properties and soil surface loss after fire in rotational shifting cultivation in Northern Thailand. *Front. Environ. Sci.* **2023**, *11*, 1213181. [CrossRef]
28. LDD (Land Development Department). *Preliminary Study on Highland Development Project in Northern Thailand*; Soil Survey Division, Land Development Department: Bangkok, Thailand, 1992.
29. Butler, B.; Quarles, S.; Standohar-Alfano, C.; Morrison, M.; Jimenez, D.; Sopko, P.; Wold, C.; Bradshaw, L.; Atwood, L.; Landon, J.; et al. Exploring fire response to high wind speeds: Fire rate of spread, energy release and flame residence time from fires burned in pine needle beds under winds up to 27 m s^{-1}. *Int. J. Wildland Fire* **2019**, *29*, 81–92. [CrossRef]
30. Gould, J.S.; Sullivan, A.L. Two methods for calculating wildland fire rate of forward spread. *Int. J. Wildland Fire* **2020**, *29*, 272–281. [CrossRef]
31. Mehring, M.; Glaser, B.; de Camargo, P.B.; Zech, W. Impact of forest organic farming change on soil microbial C turnover using ^{13}C of phospholipid fatty acids. *Agron. Sustain. Dev.* **2011**, *31*, 719–731. [CrossRef]
32. Liu, X.; Liu, Y.; Zhang, L.; Yin, R.; Wu, G.-L. Bacterial contributions of bio-crusts and litter crusts to nutrient cycling in the mu us Sandy land. *Catena* **2021**, *199*, 105090. [CrossRef]
33. Barbour, K.M.; Weihe, C.; Allison, S.D.; Martiny, J.B.H. Bacterial community response to environmental change varies with depth in the surface soil. *Soil Biol. Biochem.* **2022**, *172*, 108761. [CrossRef]
34. Soil Survey Staff. *Soil Survey Field and Laboratory Methods Manual*; Burt, R., Soil Survey Staff, Eds.; Soil Survey Investigations Report No. 51, Version 2.0; U.S. Department of Agriculture, Natural Resources Conservation Service: Washington, DC, USA, 2014.
35. Jones, J.B., Jr. *Laboratory Guide for Conducting Soil Tests and Plant Analysis*; CRC Press: Boca Raton, FL, USA, 2001.
36. Thomas, G.W. Exchangeable cations. In *Methods of Soil Analysis, Part 2 Chemical and Microbiological Properties*, 2nd ed.; Page, A.L., Ed.; ASA; SSSA: Madison, WI, USA, 1982; Chapter 9; pp. 159–165.
37. Bray, R.A.; Kurtz, L.T. Determination of total organic and available form of phosphorus in soil. *Soil Sci.* **1945**, *59*, 39–45. [CrossRef]
38. Walkley, A.; Black, J.A. An examination of the dichormate method for determining soil organic matter and a proposed modification of the chromic acid titration method. *Soil Sci.* **1934**, *37*, 29–32. [CrossRef]
39. Klindworth, A.; Pruesse, E.; Schweer, T.; Peplies, J.; Quast, C.; Horn, M.; Glöckner, F.O. Evaluation of general 16S ribosomal RNA gene PCR primers for classical and next-generation sequencing-based diversity studies. *Nucleic Acids Res.* **2013**, *41*, e1. [CrossRef] [PubMed]
40. Estaki, M.; Jiang, L.; Bokulich, N.A.; McDonald, D.; González, A.; Kosciolek, T.; Martino, C.; Zhu, Q.; Birmingham, A.; Vázquez-Baeza, Y.; et al. QIIME 2 Enables Comprehensive End-to-End Analysis of Diverse Microbiome Data and Comparative Studies with Publicly Available Data. *Curr. Protoc. Bioinform.* **2020**, *70*, e100. [CrossRef] [PubMed]
41. Martin, M. Cutadapt removes adapter sequences from high-throughput sequencing reads. *EMBnet J.* **2011**, *17*, 10–12. [CrossRef]

42. Callahan, B.J.; McMurdie, P.J.; Rosen, M.J.; Han, A.W.; Johnson, A.J.A.; Holmes, S.P. DADA2: High-Resolution Sample Inference from Illumina Amplicon Data. *Nat. Methods* **2016**, *13*, 581–583. [CrossRef] [PubMed]
43. Quast, C.; Pruesse, E.; Yilmaz, P.; Gerken, J.; Schweer, T.; Yarza, P.; Peplies, J.; Glöckner, F.O. The SILVA ribosomal RNA gene database project: Improved data processing and web-based tools. *Nucleic Acids Res.* **2013**, *41*, D590–D596. [CrossRef]
44. Glöckner, F.O.; Yilmaz, P.; Quast, C.; Gerken, J.; Beccati, A.; Ciuprina, A.; Bruns, G.; Yarza, P.; Peplies, J.; Westram, R.; et al. 25 years of serving the community with ribosomal RNA gene reference databases and tools. *J. Biotechnol.* **2017**, *261*, 169–176. [CrossRef]
45. Douglas, G.M.; Maffei, V.J.; Zaneveld, J.R.; Yurgel, S.N.; Brown, J.R.; Taylor, C.M.; Huttenhower, C.; Langille, M.G.I. PICRUSt2 for Prediction of Metagenome Functions. *Nat. Biotechnol.* **2020**, *38*, 685–688. [CrossRef]
46. Hammer, O.; Harper, D.; Ryan, P. PAST: Paleontological Statistics Software Package for Education and Data Analysis. *Palaeontol. Electron.* **2001**, *4*, 1–9.
47. Liu, C.; Cui, Y.; Li, X.; Yao, M. *microeco*: An R package for data mining in microbial community ecology. *FEMS Microbiol. Ecol.* **2021**, *97*, fiaa255. [CrossRef]
48. Thomaz, E.L. Fire changes the larger aggregate size classes in slash-and-burn agricultural systems. *Soil Till. Res.* **2017**, *165*, 210–217. [CrossRef]
49. Habekost, M.; Eisenhauer, N.; Scheu, S.; Steinbeiss, S.; Weigelt, A.; Gleixner, G. Seasonal changes in the soil microbial community in a grassland plant diversity gradient four years after establishment. *Soil Biol. Biochem.* **2008**, *40*, 2588–2595. [CrossRef]
50. Landesman, W.J.; Freedman, Z.B.; Nelson, D.M. Seasonal, sub-seasonal and diurnal variation of soil bacterial community composition in a temperate deciduous forest. *FEMS Microbiol. Ecol.* **2019**, *95*, fiz002. [CrossRef]
51. Arunrat, N.; Sereenonchai, S.; Sansupa, C.; Kongsurakan, P.; Hatano, R. Effect of rice straw and stubble burning on soil physicochemical properties and bacterial communities in Central Thailand. *Biology* **2023**, *12*, 501. [CrossRef] [PubMed]
52. Koranda, M.; Kaiser, C.; Fuchslueger, L.; Kitzler, B.; Sessitsch, A.; Zechmeister-Boltenstern, S.; Richter, A. Seasonal variation in functional properties of microbial communities in beech forest soil. *Soil Biol. Biochem.* **2013**, *60*, 95–104. [CrossRef]
53. Yao, M.; Rui, J.; Niu, H.; Heděnec, P.; Li, J.; He, Z.; Wang, J.; Cao, W.; Li, X. The differentiation of soil bacterial communities along a precipitation and temperature gradient in the eastern Inner Mongolia steppe. *Catena* **2017**, *152*, 47–56. [CrossRef]
54. Luo, X.; Wang, M.K.; Hu, G.; Weng, B. Seasonal Change in Microbial Diversity and Its Relationship with Soil Chemical Properties in an Orchard. *PLoS ONE* **2019**, *14*, e0215556. [CrossRef]
55. Zhu, W.Z.; Cai, X.H.; Liu, X.L.; Wang, J.X.; Cheng, J.X.; Cheng, S.; Zhang, X.; Li, D.; Li, M. Soil microbial population dynamics along a chronosequence of moist evergreen broad-leaved forest succession in southwestern China. *J. Mount. Sci.* **2010**, *7*, 327–338. [CrossRef]
56. Bardgett, R.D.; Kandeler, E.; Tscherko, D.; Hobbs, P.J.; Bezemer, T.M.; Jones, T.H.; Thompson, L.J. Below-ground microbial community development in a high temperature world. *Oikos* **1999**, *85*, 193. [CrossRef]
57. Shishido, M.; Sakamoto, K.; Yokoyama, H.; Momma, N.; Miyashita, S. Changes in microbial communities in an apple orchard and its adjacent bush soil in response to season, land-use, and violet root rot infestation. *Soil Biol. Biochem.* **2008**, *40*, 1460–1473. [CrossRef]
58. Kim, C.S.; Nam, J.W.; Jo, J.W.; Kim, S.Y.; Han, J.G.; Hyun, M.W.; Sung, G.H.; Han, S.K. Studies on seasonal dynamics of soilhigher fungal communities in Mongolian oak-dominant Gwangneung forest in Korea. *J. Microbiol.* **2016**, *54*, 14–22. [CrossRef] [PubMed]
59. Mitchell, R.J.; Hester, A.J.; Campbell, C.D.; Chapman, S.J.; Cameron, C.M.; Hewison, R.L.; Potts, J.M. Is vegetation composition or soil chemistry the best predictor of the soil microbial community? *Plant Soil* **2010**, *3*, 417–430. [CrossRef]
60. Bargali, K.; Manral, V.; Padalia, K.; Bargali, S.S.; Upadhyay, V.P. Effect of vegetation type and season on microbial biomass carbon in Central Himalayan forest soils, India. *Catena* **2018**, *171*, 125–135. [CrossRef]
61. Thoms, C.; Gleixner, G. Seasonal differences in tree species' influence on soil microbial communities. *Soil Biol. Biochem.* **2013**, *6*, 239–248. [CrossRef]
62. Huang, X.F.; Chaparro, J.M.; Reardon, K.F.; Zhang, R.; Shen, Q.; Vivanco, J.M. Rhizosphere interactions: Root exudates, microbes, and microbial communities. *Botany* **2014**, *92*, 267–275. [CrossRef]
63. Kempf, B.; Bremer, E. A novel amidohydrolase gene from *Bacillus subtilis* cloning: DNA-sequence analysis and map position of *amhX*. *FEMS Microbiol. Lett.* **1996**, *141*, 129–137. [CrossRef]
64. Griffiths, R.I.; Whitely, A.S.; O'Donnell, A.G.; Bailey, M.J. Influence of depth and sampling time on bacterial community structure in an upland grassland soil. *FEMS Microbiol. Ecol.* **2003**, *43*, 35–43. [CrossRef]
65. Jefferies, R.L.; Walker, N.A.; Edwards, K.A.; Dainty, J. Is the decline of soil microbial biomass in late winter coupled to changes in the physical state of cold soils? *Soil Biol. Biochem.* **2010**, *42*, 129–135. [CrossRef]
66. Zhang, B.; Chen, S.Y.; Zhang, J.F.; He, X.Y.; Liu, W.J.; Zhao, Q.; Zhao, L.; Tian, C.J. Depth-related responses of soil microbial communities to experimental warming in an alpine meadow on the Qinghai-Tibet Plateau. *Eur. J. Soil Sci.* **2015**, *6*, 496–504. [CrossRef]
67. Li, G.; Kim, S.; Han, S.H.; Chang, H.; Du, D.; Son, Y. Precipitation affects soil microbial and extracellular enzymatic responses to warming. *Soil Biol. Biochem.* **2018**, *120*, 212–221. [CrossRef]
68. Bukar, M.; Sodipo, O.; Dawkins, K.; Ramirez, R.; Kaldapa, J.T.; Tarfa, M.; Esiobu, N. Microbiomes of Top and Sub-Layers of Semi-Arid Soils in North-Eastern Nigeria Are Rich in Firmicutes and Proteobacteria with Surprisingly High Diversity of Rare Species. *Adv. Microbiol.* **2019**, *9*, 102–118. [CrossRef]

69. Nelson, A.R.; Narrowe, A.B.; Rhoades, C.C.; Fegel, T.S.; Daly, R.A.; Roth, H.K.; Chu, R.K.; Amundson, K.K.; Young, R.B.; Steindorff, A.S.; et al. Wildfire-dependent changes in soil microbiome diversity and function. *Nat. Microbiol.* **2022**, *7*, 1419–1430. [CrossRef] [PubMed]
70. Rodríguez-Berbel, N.; Ortega, R.; Lucas-Borja, M.E.; Solé-Benet, A.; Miralles, I. Long-Term Effects of Two Organic Amendments on Bacterial Communities of Calcareous Mediterranean Soils Degraded by Mining. *J. Environ. Manag.* **2020**, *271*, 110920. [CrossRef] [PubMed]
71. Fontúrbel, M.T.; Barreiro, A.; Vega, J.A.; Martín, A.; Jiménez, E.; Carballas, T.; Fernández, C.; Díaz-Raviña, M. Effects of an Experimental Fire and Post-Fire Stabilization Treatments on Soil Microbial Communities. *Geoderma* **2012**, *191*, 51–60. [CrossRef]
72. Arocena, J.M.; Opio, C. Prescribed fire-induced changes in properties of sub-boreal forest soils. *Geoderma* **2003**, *113*, 1–16. [CrossRef]
73. Badía, D.; Martí, C.; Aguirre, A.J.; Aznar, J.M.; González-Pérez, J.A.; De la Rosa, J.M.; León, J.; Ibarra, P.; Echeverría, T. Wildfire effects on nutrients and organic carbon of a Rendzic Phaeozem in NE Spain: Changes at cm-scale topsoil. *CATENA* **2014**, *113*, 267–275. [CrossRef]
74. Alcañiz, M.; Outeiro, L.; Francos, M.; Farguell, J.; Úbeda, X. Long-term dynamics of soil chemical properties after a prescribed fire in a Mediterranean forest (Montgrí massif, Catalonia, Spain). *Sci. Total Environ.* **2016**, *572*, 1329–1335. [CrossRef]
75. Arunrat, N.; Sereenonchai, S.; Kongsurakan, P.; Iwai, C.B.; Yuttitham, M.; Hatano, R. Post-fire recovery of soil organic carbon, soil total nitrogen, soil nutrients, and soil erodibility in rotational shifting cultivation in Northern Thailand. *Front. Environ. Sci.* **2023**, *1*, 1117427. [CrossRef]
76. Giorgis, M.A.; Zeballos, S.R.; Carbone, L.; Zimmermann, H.; von Wehrden, H.; Aguilar, R.; Ferreras, A.E.; Tecco, P.A.; Kowaljow, E.; Barri, F.; et al. A review of fire effects across South American ecosystems: The role of climate and time since fire. *Fire Ecol.* **2021**, *17*, 11. [CrossRef]
77. Doerr, S.H.; Santín, C.; Merino, A.; Belcher, C.M.; Baxter, G. Fire as a removal mechanism of pyrogenic carbon from the environment: Effects of fire and pyrogenic carbon characteristics. *Front. Earth Sci.* **2018**, *6*, 1–13. [CrossRef]
78. Ulery, A.L.; Graham, R.C. Forest Fire Effects on Soil Color and Texture. *Soil Sci. Soc. Am. J.* **1993**, *57*, 135–140. [CrossRef]
79. DeLuca, T.; Nilsson, M.-C.; Zackrisson, O. Nitrogen mineralization and phenol accumulation along a fire chronosequence in northern Sweden. *Oecologia* **2002**, *133*, 206–214. [CrossRef] [PubMed]
80. Jones, R.; Chambers, J.C.; Johnson, D.W.; Blank, R.R.; Board, D.I. Effect of repeated burning on plant and soil carbon and nitrogen in cheatgrass (*Bromus tectorum*) dominated ecosystems. *Plant Soil* **2015**, *386*, 47–64. [CrossRef]
81. Verma, S.; Jayakumar, S. Effect of recurrent fires on soil nutrient dynamics in a tropical dry deciduous forest of Western Ghats, India. *J. Sustain. For.* **2018**, *37*, 678–690. [CrossRef]
82. Lynch, J.M.; Whipps, J.M. Substrate flow in the rhizosphere. *Plant Soil* **1990**, *129*, 1–10. [CrossRef]
83. Meharg, A.; Killham, K. Carbon distribution within the plant and rhizosphere in laboratory and field-grown Lolium perenne at different stages of development. *Soil Biol. Biochem.* **1990**, *2*, 471–477. [CrossRef]
84. Ren, B.H.; Hu, Y.M.; Chen, B.D.; Zhang, Y.; Thiele, J.; Shi, R.J.; Liu, M.; Bu, R.C. Soil pH and plant diversity shape soil bacterial community structure in the active layer across the latitudinal gradients in continuous permafrost region of northeastern china. *Sci. Rep.* **2018**, *8*, 5619. [CrossRef]
85. Giller, K.E.; Beare, M.H.; Lavalle, P.; Izac, A.M.N.; Swift, M.J. Agricultural intensification, soil biodiversity and agroecosystem function. *Appl. Soil Ecol.* **1997**, *6*, 3–16. [CrossRef]
86. D'Ascoli, R.; Rutigliano, F.A.; De Pascale, R.A.; Gentile, A.; Virzo De Santo, A. Functional diversity of the microbial community in Mediterranean maquis soils as affected by fires. *Int. J. Wildland Fire* **2005**, *14*, 355–363. [CrossRef]

biology

MDPI

Article

Silica Accumulation in Potato (*Solanum tuberosum* L.) Plants and Implications for Potato Yield Performance—Results from Field Experiments in Northeast Germany

Daniel Puppe [1,*], Jacqueline Busse [1], Mathias Stein [2], Danuta Kaczorek [1,3], Christian Buhtz [1] and Jörg Schaller [1]

1 Leibniz Centre for Agricultural Landscape Research (ZALF), 15374 Müncheberg, Germany
2 Soil Science and Soil Protection, Martin Luther University Halle-Wittenberg, 06120 Halle (Saale), Germany
3 Department of Soil Science, Warsaw University of Life Sciences (SGGW), 02-776 Warsaw, Poland
* Correspondence: daniel.puppe@zalf.de

Simple Summary: The potato is the most important non-cereal food crop worldwide. Silicon (Si) fertilizers have been reported to improve potato growth and yield. We used results from two field experiments in the temperate zone to gain insight into silica accumulation in potato plants as well as corresponding long-term potato yield performance. We found relatively low Si contents in potato plants grown in soils with different concentrations of plant-available Si (field experiment 1). Moreover, potato yield was not correlated to plant-available Si concentrations in soils in the long term (1965–2015, field experiment 2). Based on our results, we ascribe the reported positive effects of Si fertilization on potatoes rather to effects of the used Si fertilizers than to silica accumulation in potato plants. While Si fertilizers applied directly to the leaves can prevent fungal infections, soil-applied Si fertilizers can enhance phosphorus and water availability in agricultural soils. With our study, we aim to inspire further research on Si fertilization–potato relationships. The corresponding results will help to derive practice-oriented recommendations for potato growers worldwide to cope with the challenges of climate change.

Abstract: The potato is the most important non-cereal food crop, and thus improving potato growth and yield is the focus of agricultural researchers and practitioners worldwide. Several studies reported beneficial effects of silicon (Si) fertilization on potato performance, although plant species from the family Solanaceae are generally considered to be non-Si-accumulating. We used results from two field experiments in the temperate zone to gain insight into silica accumulation in potato plants, as well as corresponding long-term potato yield performance. We found relatively low Si contents in potato leaves and roots (up to 0.08% and 0.3% in the dry mass, respectively) and negligible Si contents in potato tuber skin and tuber flesh for plants grown in soils with different concentrations of plant-available Si (field experiment 1). Moreover, potato yield was not correlated to plant-available Si concentrations in soils in the long term (1965–2015, field experiment 2). Based on our results, we ascribe the beneficial effects of Si fertilization on potato growth and yield performance reported in previous studies mainly to antifungal/osmotic effects of foliar-applied Si fertilizers and to changes in physicochemical soil properties (e.g., enhanced phosphorus availability and water-holding capacity) caused by soil-applied Si fertilizers.

Keywords: phytogenic silica; crop production; phytoliths; sustainability; biogenic silica; stress resilience; plant-available silicon; silica amendment; long-term field experiment

1. Introduction

As silicon (Si) is the second most abundant element in the Earth's crust, Si can be found virtually everywhere. Thus, it is not surprising that Si is also an important component in many organisms like protists, sponges, and plants, which use dissolved monomeric silicic acid ($Si(OH)_4$) for the formation of biogenic silica (i.e., amorphous hydrated silica, $SiO_2 \cdot nH_2O$). This process of biosilicification has been found to represent a key factor in the global Si cycle [1–3]. In terrestrial ecosystems, Si cycling by vegetation has been the focus of research [4–6], although the role of protists (i.e., testate amoebae) has been highlighted since the beginning of the 21st century [7].

Precipitated biogenic silica in plants is called phytogenic silica, which can be found within cells (i.e., in the cell wall and the cell lumen) and in intercellular spaces and extracellular (cuticular) layers. While intercellular and extracellular phytogenic silica structures are quite delicate/fragile, cell wall and lumen silica precipitates are quite resilient and can persist in soils as microfossils (phytoliths) up to hundreds and thousands of years [8,9]. These phytoliths are routinely used in many scientific fields like archaeology, (paleo)botany, (evolutionary) biology, plant taxonomy, or climatology, and thus a phytolith nomenclature and classification system has evolved [10]. As phytoliths can also contain various elements like carbon, aluminum, calcium, iron, manganese, phosphorus, lead, copper, cadmium, or arsenic, their potential for carbon and metal(loid) long-term sequestration in soils has been recognized recently [11–13].

In general, Si accumulation in plants has been shown to enhance plants' resistance to abiotic and biotic stress with implications for plant performance and ecosystem functioning [14]. In agricultural plant–soil systems, Si fertilizers are widely used to increase yields of Si-accumulating crops like rice, maize, wheat, and sugarcane, especially in the (sub)tropics, where soils usually are desilicated much stronger than in the temperate zone [15–17]. In this context, Si-rich materials used for fertilization comprise industrial waste matter (i.e., slags or silica fume), manufactured fertilizers (e.g., fused magnesium phosphate or potassium silicate), and minerals mined from the earth's surface (e.g., wollastonite or diatomaceous earth) [15,18]. However, the production of these fertilizers is quite energy-consuming, and fertilization with some of these products can cause environmental problems (e.g., metal(loid) contamination of soils). Biochar has been discussed as a comparatively environmentally friendly Si source in agriculture [19], but it has to be considered that its production by pyrolysis of crop residues and manures is relatively CO_2-intensive [20]. In the long term, the maximum restoration of the Si cycle in agricultural plant–soil systems by crop straw recycling might represent the most promising and environmentally friendly approach for the sustainable agricultural production of resilient crops [21,22].

The potato (*Solanum tuberosum* L., family Solanaceae) represents one of the most important crops worldwide. According to the Food and Agriculture Organization (FAO) of the United Nations, approximately 375 million tons of potatoes were produced worldwide in the year 2022 [23]. Despite the fact that plant species from the family Solanaceae are considered to be non-Si-accumulating [24], some studies reported beneficial effects of Si fertilization on potato production. Crusciol et al. [25], for example, found that Si application in a greenhouse pot experiment significantly increased potato tuber yield and Si concentrations in potato leaves. While some other greenhouse experiments corroborated the beneficial effects of Si (soil and foliar) fertilization on potato growth [26–30], Vulavala et al. [31] found no significantly changed silica accumulation in potato roots or leaves after Si fertilization, although they observed an upregulated expression of a gene (called *StLsi1*), encoding a corresponding Si-influx protein in these plant organs.

Notably, field experiments on Si fertilization of potatoes were performed only in a few studies, which were mostly limited to the foliar application of Si in the temperate zone [32–34]. One of the rare studies that analyzed soil Si fertilization of potatoes under field conditions was conducted in the tropics by Nyawade et al. [35], who reported synergistic effects of soil Si fertilization and potato–legume intercropping in Kenya. Moreover, the previous studies mainly focused on the effects of Si (soil/foliar) fertilization on potato production using specific plant growth indicators like leaf numbers/areas, protein/saccharide concentrations in leaves, or tuber dry weights. However, the accumulation of silica in specific plant organs on a cellular level has not been the focus of research until now, although

Si concentrations in potato plant shoots/organs or tubers were also reported in some of the previous studies (e.g., [25,27,28]).

In our study, we used a scanning electron microscope (SEM) equipped with an energy-dispersive X-ray spectroscopy (EDX) instrument to analyze silica accumulation in potato plants (leaves, roots, tuber flesh, and tuber skin) on a cellular level. The potato plants were taken from a field experiment with control (no Si addition) and Si (addition of artificial silica to the soil) plots and Si concentrations in the microscopically examined plant materials were also determined spectroscopically in corresponding plant extracts. To gain further insights into the effects of Si on potato growth, we additionally used potato yield data from an ongoing long-term field experiment (LTFE) with plots where plant-available Si in soils has been increased via crop straw recycling. The combination of microscopical and spectroscopical techniques, as well as the combined analysis of results from two different field experiments in our study, will help us to evaluate the effects of Si supply on potato cultivation in detail. The corresponding results will not only be interesting for agricultural scientists, but also for potato growers worldwide.

2. Materials and Methods

2.1. Study Sites and Sampling

The two field experiments, "V434" (silica amendment experiment) and "V140" (LTFE), are located in the experimental area of the Leibniz Centre for Agricultural Landscape Research (ZALF). The experimental sites are managed according to "Good Agricultural Practice". The climate is characterized by a mean annual precipitation of 535 mm and a mean annual temperature of 9.3 °C based on the reference period of 1991–2020 (measured by a weather station of the German Meteorological Service installed on the ZALF area).

The silica amendment experiment was established in 2020 [36]. This experimental site consists of 12 plots measuring 3 m × 4 m each. Six of these plots serve as control and received no Si addition. The soil at six other plots was amended with artificial amorphous silica (ASi; Aerosil 300, Evonik Industries, Essen, Germany) in the upper 25 cm (Ap horizon) in different amounts. While the soil at three of the Si plots was mixed with 1.8 kg ASi m^{-2}, resulting in a mass percentage of 0.5%, the soil at the three other Si plots was mixed with 3.6 kg ASi m^{-2}, resulting in a mass percentage of 1.0%. A block design was used for practical purposes, with buffer strips of equal size between the Si-treated and control plots to prevent cross-treatment interferences (Figure 1). The ASi was carefully mixed into the soil by hand in the first step and, subsequently, a cultivator was used to homogenously distribute the ASi in the topsoil. To ensure comparable soil conditions, the soil at the control plots was also treated with the cultivator, but without mixing in ASi. Finally, an overhead sprinkler system was used to irrigate all plots with 60 mm m^{-2} of water, ensuring uniform soil moisture across the entire field. In the first two years, wheat (*Triticum aestivum*) was cultivated on the different plots of the silica amendment experiment [36,37]. In June 2020, soil samples were taken and prepared (i.e., air-dried and passed through a 2 mm sieve) for further analyses, i.e., the extraction of plant-available Si (see Section 2.2).

Seed potatoes (cultivar "Talent") were planted on 22 April 2022. In May 2022, a mineral fertilizer ("Piasan 25/6"; 120 kg nitrogen ha^{-1} and 29 kg sulfur ha^{-1}) was applied. Pesticides were applied in May (herbicide) and June (insecticide, fungicide) 2022. Potato plant samples (shoots, roots, and tubers) were carefully taken on two dates in 2022, i.e., on the 30 June and the 28 July. On the first date, several specimens of the ten-lined potato beetle (*Leptinotarsa decemlineata* Say, 1824) were already observed. Despite a three-time application of insecticides, the beetle infestation increased and, on the second date, the shoots of the potato plants were heavily damaged. Plant samples were thoroughly washed, dried, and, finally, used for Si analyses (see Section 2.2).

The LTFE (52°31′01″ N, 14°07′19″ E) at ZALF was established in a full randomized block design in 1963 to analyze the effects of different mineral and organic fertilizers on yields and soil fertility [21,38]. The sand-dominated soil is classified as Albic Luvisol (Arenic, Neocambic [39]), with two argic horizons in depths of 80–120 cm (Bt1) and

120–160 cm (Bt2). The experimental setup includes different treatments, i.e., (i) NPK fertilization in steps of 5 rates related to N (plots NPK 1–5), (ii) organic fertilization (manure or straw plots), and (iii) control plots, with 8 field repetitions per treatment (168 single plots in total). Soil samples have been regularly taken by the staff of the Experimental Station of ZALF and analyzed (e.g., pH, phosphate concentrations) in the Central Laboratory of ZALF. Plant biomass (yield) per plot has been determined every year. The effect of crop straw recycling on anthropogenic desilication was analyzed in detail in a previous study using selected plots of the LTFE (i.e., control, NPK 1, NPK 1 + Straw, NPK 3, NPK 3 + Straw, NPK 5, and NPK 5 + Straw plots; see Figure 2) [21]. For our study, we used published (plant-available Si concentrations in soils of the different treatments; samples from 1976, 1998, and 2018 [21]) and unpublished (potato yields for the corresponding treatments; stated for all years, in which potatoes were grown at the LTFE, i.e., for 1965, 1967, 1973, 1983, 1987, 1991, 1999, 2007, and 2015) data to examine the long-term effects of Si supply on potato yield performance. Monthly temperature and precipitation data (1965–2015) were used to analyze the potential effects of climate change on potato yields.

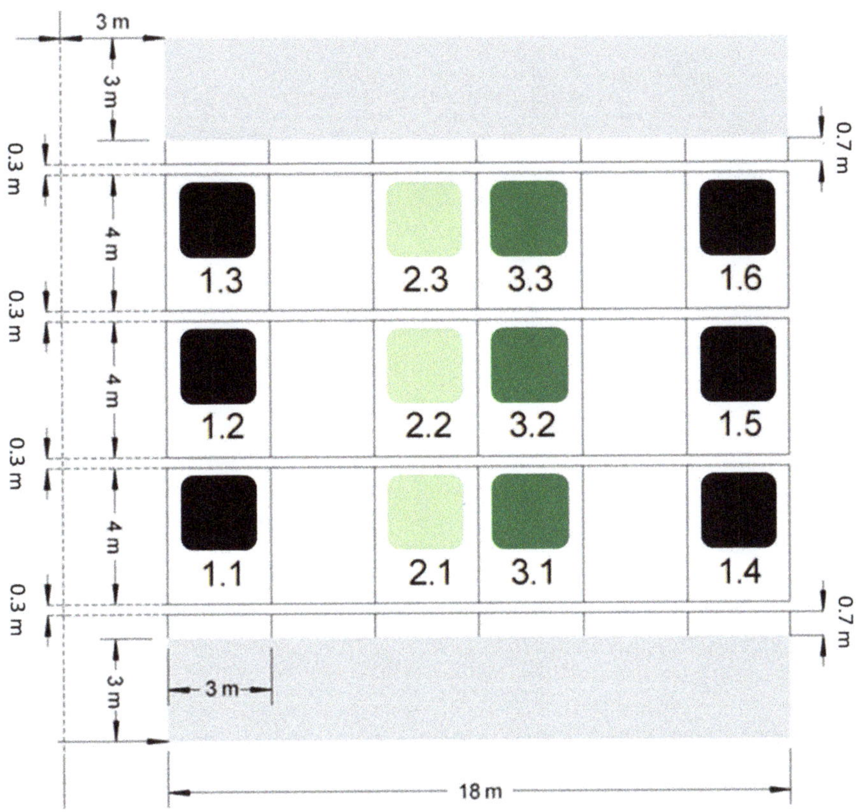

Figure 1. Overview of the plots at the silica amendment experiment. While six plots serve as control (i.e., plot numbers 1.1–1.6, marked by black squares), six plots represent Si treatments with 0.5% (i.e., plot numbers 2.1–2.3, marked by light green squares) or 1.0% (i.e., plot numbers 3.1–3.3, marked by dark green squares) amorphous silica mass percentage.

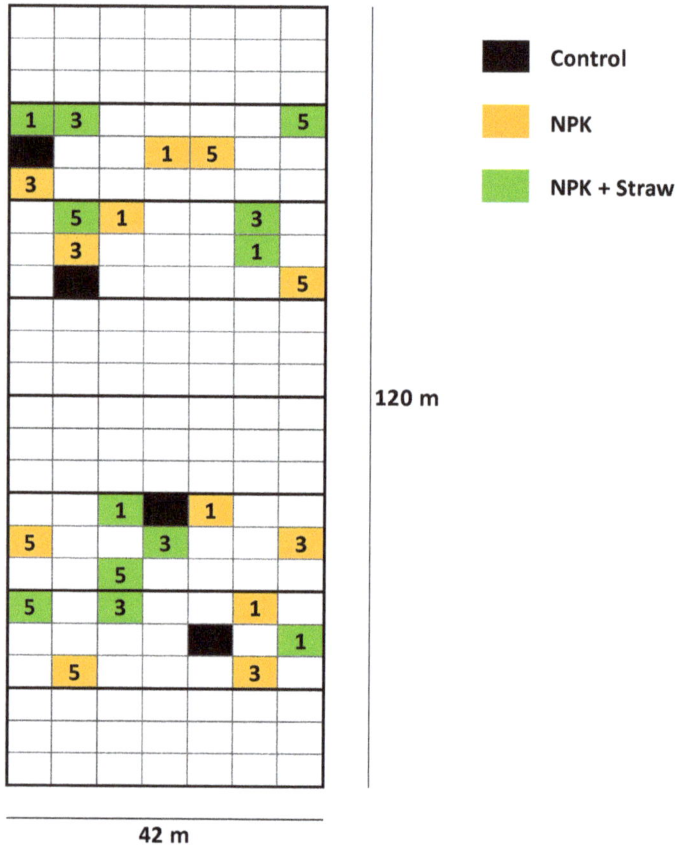

Figure 2. Overview of the plots at the LTFE (modified from Puppe et al. [21]). The plots used in the study by Puppe et al. [21] are highlighted in color (see legend). 1 = low fertilization rate (NPK 1, ~30 kg N ha^{-1} y^{-1}), 3 = medium (i.e., common) fertilization rate (NPK 3, ~98 kg N ha^{-1} y^{-1}), and 5 = high fertilization rate (NPK 5, ~166 kg N ha^{-1} y^{-1}). At plots with crop straw recycling (NPK + Straw), NPK fertilization has been supplemented by incorporation of 4.0 t (dry mass) straw ha^{-1} every second year using chopped straw of the recently harvested cereal crop. At the control plots, neither NPK fertilization, nor crop straw recycling has been performed.

2.2. *Soil and Plant Analyses*

Plant-available Si in soil samples of the silica amendment experiment was extracted following the procedures described by Haysom and Chapman [40] and de Lima Rodrigues et al. [41]. In short, two-gram samples of soil were placed in 50 mL plastic centrifuge tubes, mixed with 20 mL of a 0.01 M calcium chloride ($CaCl_2$) solution, and agitated continuously on a swivel roller mixer for 16 h. Finally, the extracted solutions were centrifuged at 4000 revolutions per minute (equal to a relative centrifugal force (RCF) of ≈1700) for 30 min and filtrated using 0.45 µm polyamide membrane filters (Whatman NL 17). Si concentrations in the $CaCl_2$ extracts were measured via inductively coupled plasma optical emission spectrometry (ICP-OES; iCAP 6300 Duo, Thermo Scientific, Waltham, MA, USA) in the ZALF Central Laboratory.

Si was extracted from plant materials of the silica amendment experiment following the procedure described by Puppe et al. [42]. In short, 30 mg of plant samples were weighed into 50 mL centrifuge tubes, and a 30 mL aliquot of the Tiron solution (pH 10.5) was added. The tubes were then heated at 80 °C in a water bath for 1 h. The samples were gently shaken by hand twice, one time directly before heating and one time after 30 min in the heated water bath. Finally, the extracted solutions were centrifuged at 1700 RCF for 30 min and filtrated (0.45 μm polyamide membrane filters, Whatman NL 17). Si concentrations in the Tiron extracts were measured via microwave plasma atomic emission spectroscopy (MP-AES; 4210 MP-AES instrument, Agilent Technologies Inc., Waldbronn, Germany) following the procedure described by Puppe et al. [43].

All analyses were performed in two lab replicates and three single ICP-OES or MP-AES measurements per replicate resulting in six (n = 6) measured data per sample. Blank sample Si concentrations were subtracted from sample Si concentrations and Si contents in plant samples were calculated considering the weighed portion (2 g or 30 mg), the extractant volume (20 mL or 30 mL), and the degree of dilution (1:10). To avoid any potential Si contamination only plastic equipment was used during the entire laboratory work.

Si analyses on a cellular level were performed using a SEM (ZEISS EVO MA10) equipped with an element detector for EDX (Bruker QUANTAX EDS). We used plant materials from both sampling dates (i.e., 30 June and 28 July 2022) to analyze potential plant growth-related changes in Si accumulation. However, for the first sampling date, only leaf samples were analyzed to check if Si is translocated from the roots to the shoots. For the second date, potato leaves, tubers (tuber skin and tuber flesh), and roots were analyzed. At each sample, several regions of interest were analyzed via SEM-EDX (EDX spectra and compositional maps for Si) to obtain a reliable data set. All SEM-EDX scans were performed using samples sputter-coated with gold (coating thickness approx. 5 nm) and the relative abundances of detected elements were displayed as normalized mass percent.

2.3. Statistical Analyses

Linear and monotonic relationships in the data set were analyzed via Pearson's (r) and Spearman's rank (r_s) correlations (α level of 0.05), respectively. Differences between means were tested with the Mann–Whitney U test or the Kruskal–Wallis analysis of variance (ANOVA) followed by pairwise multiple comparisons (Dunn's post hoc test). All statistical analyses were performed using the software package SPSS Statistics (version 22.0.0.0, IBM Corp., Armonk, NY, USA).

3. Results

3.1. Silica Accumulation in Potato Plants—Results from the Silica Amendment Experiment

Concentrations of plant-available Si in soils of Si plots (mean for 0.5% Asi, 11.0 mg kg^{-1}; mean for 1.0% Asi, 13.5 mg kg^{-1}) were significantly higher than in soils of control plots (mean: 4.6 mg kg^{-1}). However, differences in concentrations of plant-available Si in soils of 0.5% ASi and 1.0% ASi plots were not statistically significant (Figure 3). The differences in Si availability were directly reflected in the potato leaves collected at the first sampling date: In leaves from plants grown at control plots, we found a relative Si abundance of about 0.2%, while relative Si abundances in leaves of 0.5% ASi and 1.0% ASi plots were about 0.5% and 0.7%, respectively (SEM-EDX measurements, see Figure 4). In leaves of the 1.0% ASi plots collected at the second sampling date, relative Si abundances were even higher (1.1%), indicating a plant growth-related Si accumulation. Relative Si abundances in tuber skin (0.16%), tuber flesh (0.04%), and roots (0.14%) were relatively low in the plants collected at the second sampling date. Selected micrographs and EDX spectra, as well as an exemplary compositional map of our SEM-EDX analyses, can be found in Figure 5. In general, we found only slight silicification on a cellular level in all cross-sections of all analyzed potato plant samples, which was directly related to Si availability (cf. Figure 4). However, no recognizable phytoliths were observed at all. The results of our SEM-EDX analyses were generally corroborated by our Si extraction results: Leaves and roots showed a plant growth-related Si accumulation, i.e., Si contents at sampling date 1 were lower compared to the ones at sampling date 2 with only one exception (i.e., for root samples from 1.0% ASi plots, see Table 1). Moreover, Si accumulation again reflected the Si availability in the soil: Lowest Si contents were detected in plant samples collected at control plots, while Si contents in plant materials collected at Si plots were higher (control < 0.5% ASi < 1.0% ASi). However, as Si contents were relatively inhomogeneous in the analyzed samples (reflected in relatively high standard deviations), we found neither statistical significance for the growth-related differences (Mann–Whitney U test, $p > 0.05$ for $\text{control}_{\text{sampling date 1}}$ vs. $\text{control}_{\text{sampling date 2}}$, 0.5% $\text{ASi}_{\text{sampling date 1}}$ vs. 0.5% $\text{ASi}_{\text{sampling date 2}}$, and 1.0% $\text{ASi}_{\text{sampling date 1}}$ vs. 1.0% $\text{ASi}_{\text{sampling date 2}}$) nor between the different treatments at the two sampling dates (Kruskal–Wallis ANOVA, $p > 0.05$ for $\text{control}_{\text{sampling date 1}}$ vs. 0.5% $\text{ASi}_{\text{sampling date 1}}$ vs. 1.0% $\text{ASi}_{\text{sampling date 1}}$ and $\text{control}_{\text{sampling date 2}}$ vs. 0.5% $\text{ASi}_{\text{sampling date 2}}$ vs. 1.0% $\text{ASi}_{\text{sampling date 2}}$). For tuber skin and tuber flesh samples Si contents were all below the detection limit.

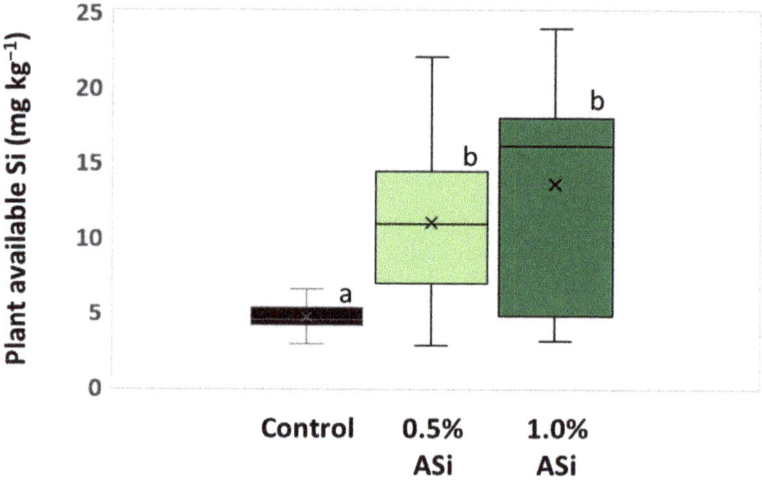

Figure 3. Plant-available Si in soils of control and Si plots of the silica amendment experiment. Means are marked by "x" in the boxplots each. Different letters indicate statistically significant differences (Kruskal–Wallis ANOVA, $p < 0.05$) between the plots.

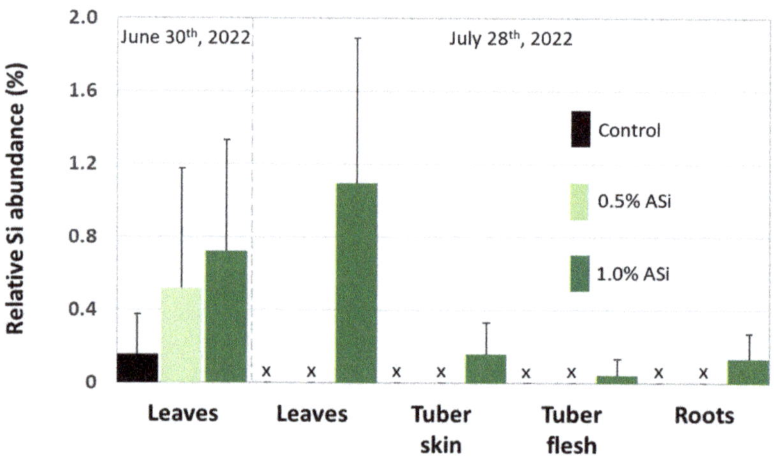

Figure 4. Relative Si abundance (SEM-EDX) in leaves, tubers (i.e., tuber skin and tuber flesh), and roots of potato plants taken at control and Si plots of the silica amendment experiment. Black and green bars represent means of normalized mass percent, error bars represent corresponding standard deviations. x = no data available.

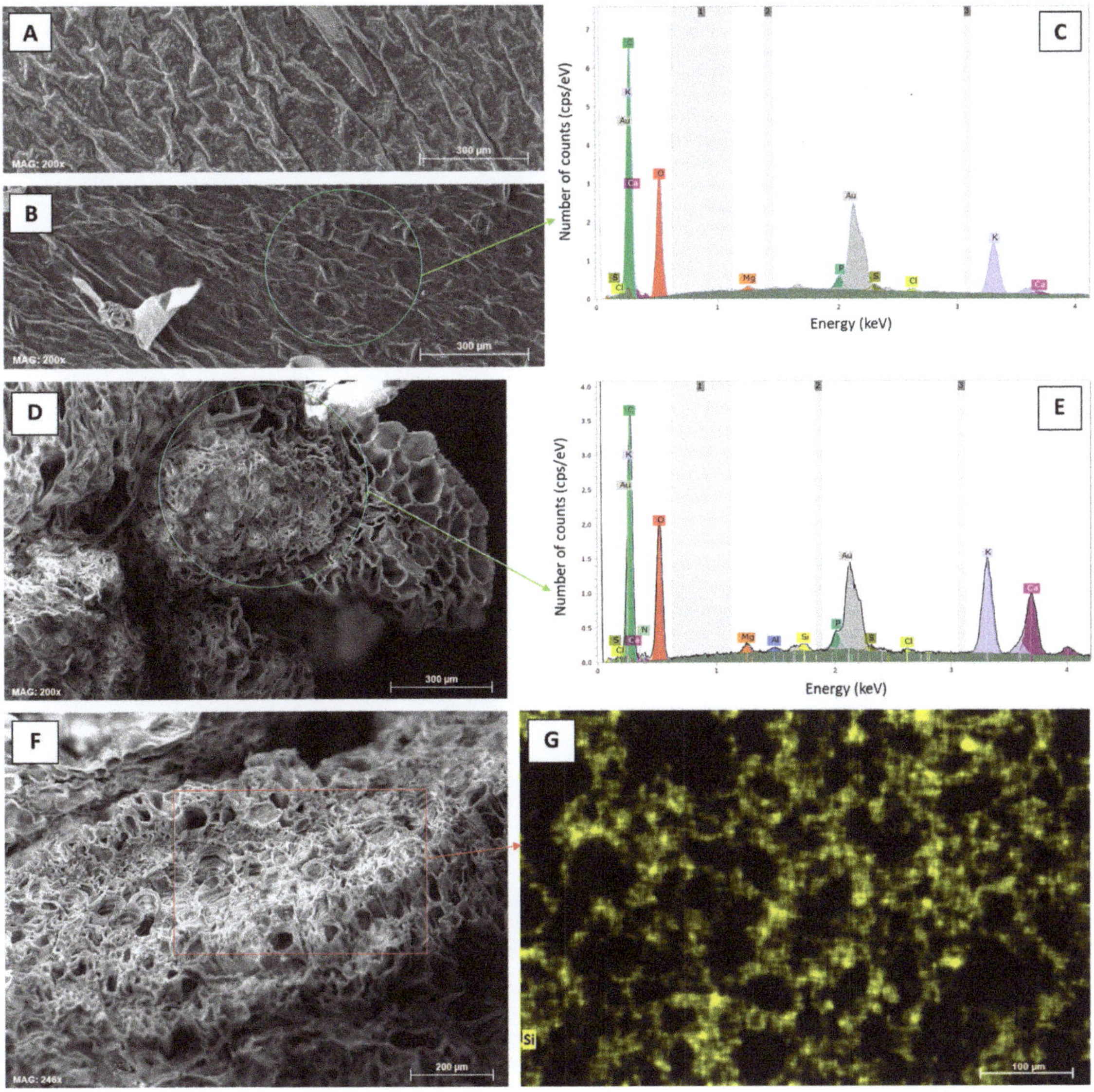

Figure 5. Elemental analyses (SEM-EDX) of leaf samples from potato plants taken at the first sampling date (30 June 2022) at control and Si plots of the silica amendment experiment. (**A**) Micrograph of the leaf top epidermis (control), (**B**) micrograph of the leaf undersurface epidermis (control), (**C**) corresponding exemplary EDX spectra derived from SEM-EDX measurements performed in a specific region of interest in (**B**) (green circle), (**D**) micrograph of a leafstalk cross-section (1.0% ASi), (**E**) corresponding exemplary EDX spectra derived from SEM-EDX measurements performed in a specific region of interest in (**D**) (green circle), (**F**) micrograph of a leafstalk cross-section (1.0% ASi), and (**G**) corresponding compositional map for Si in a specific region of interest in (**F**) (red rectangle).

Table 1. Si contents (Tiron extraction) in leaves, tubers (i.e., tuber skin and tuber flesh), and roots of potato plants taken at control and Si plots of the silica amendment experiment. SD = standard deviation.

		Si Content (mg kg^{-1})			
		30 June 2022		28 July 2022	
Treatment	Plant Material	Mean	SD	Mean	SD
Control	Leaves	0	--	50	0.2
0.5% ASi	Leaves	0	--	646	--
1.0% ASi	Leaves	12	263	789	--
Control	Tuber skin	0	--	0	--
0.5% ASi	Tuber skin	0	--	0	--
1.0% ASi	Tuber skin	0	--	0	--
Control	Tuber flesh	0	--	0	--
0.5% ASi	Tuber flesh	0	--	0	--
1.0% ASi	Tuber flesh	0	--	0	--
Control	Roots	316	405	860	929
0.5% ASi	Roots	936	762	1669	2361
1.0% ASi	Roots	3198	2081	2401	3326

--: no data available.

3.2. *Si Effects on Potato Yields—Results from the Long-Term Field Experiment*

In general, potato yields showed a decreasing trend within the analyzed 50-year period at low, medium (i.e., common), and high fertilization plots. At low fertilization plots, yields at NPK + Straw plots were statistically significantly higher than yields at control plots in 6 out of 9 years (Figure 6A). At medium (i.e., common) fertilization plots, yields at NPK + Straw plots were statistically significantly higher than yields at control plots in 4 out of 9 years (Figure 6B). At high fertilization plots, yields at NPK + Straw plots were statistically significantly higher than yields at control plots in 8 out of 9 years (Figure 6C). Moreover, yields at NPK + Straw plots were slightly higher than at NPK plots in 6, 8, and 5 out of 9 years at low, medium (i.e., common), and high fertilization plots, respectively. However, these differences were not statistically significant.

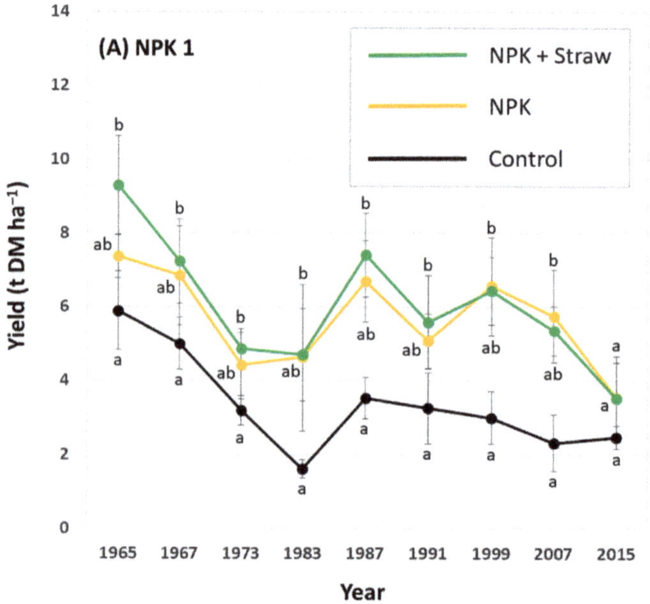

Figure 6. *Cont.*

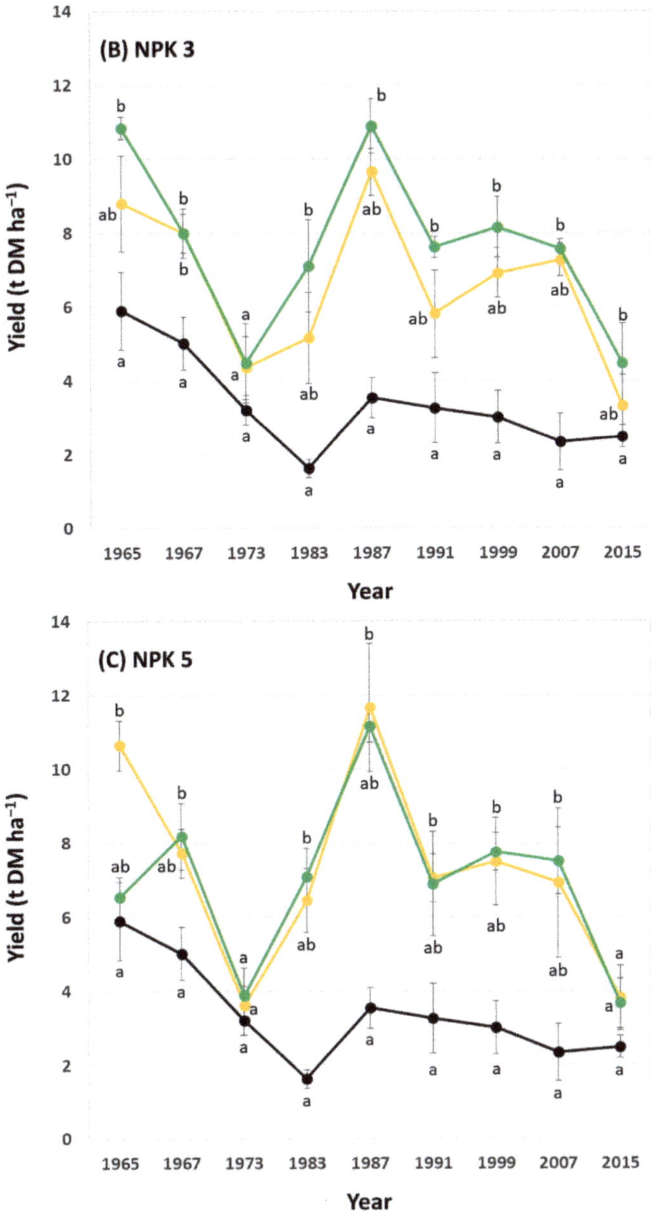

Figure 6. Potato yields for low (**A**), medium (i.e., common) (**B**), and high (**C**) fertilization plots (NPK 1, NPK 3, and NPK 5, respectively) at the long-term field experiment. Yields are stated for all years in which potatoes were grown during the ongoing long-term field experiment. Different letters indicate statistically significant differences (Kruskal–Wallis ANOVA, $p < 0.05$) between control, NPK, and NPK + Straw plots in a specific year.

Plant-available Si in soils increased at all plots with experiment duration, especially at NPK + Straw plots. Compared to a mean of 6.3 mg plant-available Si kg^{-1} soil in the year 1976 (range: 5.1–7.6 mg Si kg^{-1}, see Puppe et al. [21]), means of plant-available Si in soils increased to 7.2 mg Si kg^{-1} (range: 5.9–8.4 mg Si kg^{-1}) and 9.2 mg Si kg^{-1} (range: 8.3–9.9 mg Si kg^{-1}) in the years 1998 and 2018, respectively. However, this time-related increase in plant-available Si in soils was statistically significant only at NPK 3 + Straw and NPK 5 + Straw plots (Figure 7). Plant-available Si in soils (data from the years 1976, 1998, and 2018; see Figure 7) and potato yields (data from the years 1973, 1999, and 2015; see Figure 6B,C) at these plots showed low to moderate negative correlations, which were not statistically significant (for NPK 3 + Straw: $r = -0.436$, $p = 0.713$ and $r_s = -0.500$, $p = 0.667$; for NPK 5 + Straw: $r = -0.371$, $p = 0.758$ and $r_s = -0.500$, $p = 0.667$).

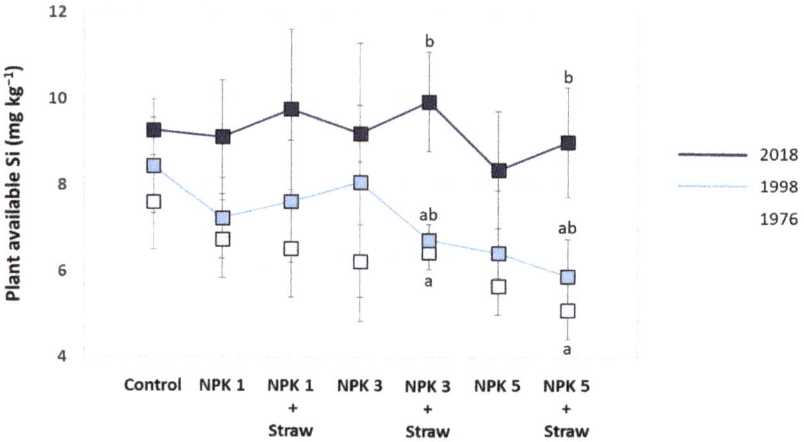

Figure 7. Plant-available Si in soils of the different plots at the long-term field experiment for the years 1976, 1998, and 2018 (data taken from Puppe et al. [21]). Different letters indicate statistically significant differences (Kruskal–Wallis ANOVA, $p < 0.05$) between the three years for specific plots. If no statistical significances were found for a specific plot, no letters were stated.

In the region, where our study sites are located, mean annual temperatures increased from 8.0 °C in the year 1965 to 10.8 °C in the year 2015 (1967: 9.9 °C, 1973: 9.0 °C, 1983: 10.0 °C, 1987: 7.7 °C, 1991: 9.2 °C, 1999: 10.2 °C, 2007: 10.7 °C). This increase was also reflected in elevated temperatures in the potato growing season (April–September) in Brandenburg, Germany (Figure 8). Mean growing season temperatures (1965: 14.1 °C, 1967: 15.4 °C, 1973: 15.0 °C, 1983: 16.2 °C, 1987: 14.2 °C, 1991: 15.0 °C, 1999: 16.5 °C, 2007: 16.5 °C, 2015: 16.2 °C) were moderately highly ($r = 0.60$–0.79) negatively correlated with corresponding potato yields for most plots (i.e., control: $r = -0.680$, $p = 0.044$; NPK 1 + Straw: $r = -0.728$, $p = 0.026$; NPK 5: $r = -0.687$, $p = 0.041$). For NPK 1, NPK 3, and NPK 3 + Straw plots, we found moderate ($r = 0.40$–0.59) to moderately high negative correlations at an α level of 0.10 (i.e., NPK 1: $r = -0.617$, $p = 0.077$; NPK 3: $r = -0.608$, $p = 0.082$; NPK 3 + Straw: $r = -0.585$, $p = 0.098$). Yields at NPK 5 + Straw plots were not statistically significantly correlated with the mean growing season temperatures ($r = -0.360$, $p = 0.342$).

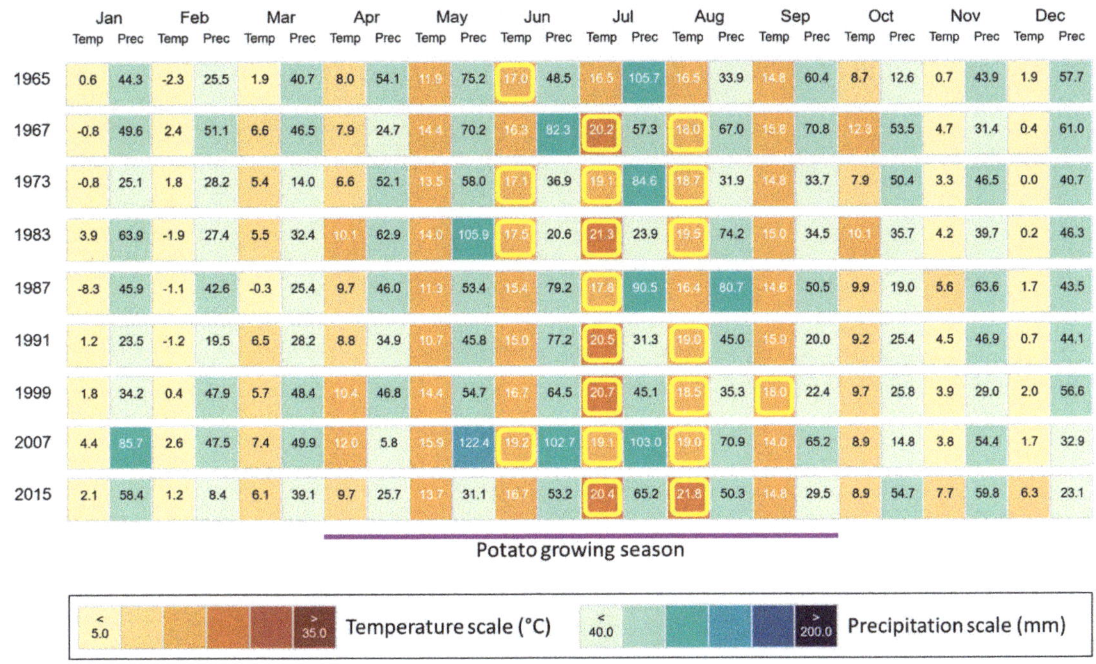

Figure 8. Monthly climate data (temperature and precipitation) for the region, where our study sites are located. Climate data are stated for all years in which potatoes were grown during the ongoing long-term field experiment at ZALF. Temperatures ≥ 17 °C (diminishment of potato tuberization) in the potato growing season (April–September) in Brandenburg, Germany, are highlighted in yellow. Figure created using "ClimateCharts.net" [44], modified.

4. Discussion

Due to the fact that plant species from the family Solanaceae are considered non-Si-accumulating in general [24], our results showing only quite low Si accumulation in potato plant samples are not surprising at all. Compared to strong Si-accumulating crops like wheat (*Triticum aestivum*, mean shoot Si concentration of about 2.5% in the dry mass) or rice (*Oryza sativa*, mean shoot Si concentration of about 4.2% in the dry mass) [45], we found Si accumulation in potato leaves (max. Si concentration of about 0.08% in the dry mass) and roots (max. Si concentration of about 0.3% in the dry mass) to be about 30–50 or 8–14 times lower, respectively. For tuber skin and tuber flesh samples, Si contents were even below the detection limit of the used MP-AES (i.e., 7.9 µg L^{-1} for Si) [43]. In general, the Si contents we found are within the range of Si contents stated in previous studies [25–28,31,46]. However, this range is quite big, spanning from 0.2 to 2000 mg Si kg^{-1} dry mass in potato tubers, representing a difference of four orders of magnitude, for example (Table 2).

As the Si contents in our study were at the bottom of the reported Si content range, most previous studies showed considerably higher Si contents in potato plant materials. Vulavala et al. [31], for example, found considerably higher Si contents in potato leaves (about 0.15–0.24% Si in the dry mass), roots (about 1.6–4.4% Si in the dry mass, but results were most likely biased by contaminations with the Si-rich growth medium perlite), and tuber skin (peel, 0.1–0.4% Si in the dry mass) samples collected from control and Si treatments in a pot experiment. In another pot experiment, Crusciol et al. [25] found Si contents in potato leaves to be about 0.4% in the dry mass, which is five times higher than in our study. Soratto et al. [27] reported even higher Si contents in potato plants (for roots up to

1.2% and for shoots up to 1%), which are slightly higher than the mean Si content of maize (*Zea mays*) shoots (0.8%) [45].

Table 2. Overview of reported Si contents in potato plant materials of various potato cultivars.

Year	Potato Cultivar	Plant Material	Si Content (mg kg^{-1} DM)		Si Contents of Control and Si Treatments Statistically Significantly Different?	Reference
			Control	Si Treatment(s)		
2009	Bintje	Leaves	3700–4100	4200–4700	yes (under drought stress)/no (without stress)	Crusciol et al. [25]
2013	Agata	Leaves	4100	8300–10,000	yes	Pilon et al. [26]
		Stems	6300	7600–10,100	yes (soil Si application)/no (foliar Si application)	
		Roots	3800	4000–5900	yes (soil Si application)/no (foliar Si application)	
		Tubers	2000	2100–2200	no	
2016	Winston	Leaves	1400–2300	1500–2200	no	Vulavala et al. [31]
		Roots [a]	15,600–41,300	17,300–34,200	no	
		Tuber skin	950–2000	850–3900	no	
2018	Agria	Shoots + roots	26	27–50	ns	Soltani et al. [28]
		Tubers	37	40–46	ns	
2019	Agata	Leaves	8300	8400–8600	no	Soratto et al. [27]
		Roots	11,000	11,600–12,300	no	
		Shoots	8100	8300–9600	yes (high Si fertilization level)/no (low Si fertilization level)	
		Tubers	1200	2100–2300	yes	
2023	Catania	Tubers	0.2	0.3	no	Wadas and Kondraciuk [46]
2024	Talent	Leaves	0–50	0–790	no	This study
		Tuber skin	0	0	no	
		Tuber flesh	0	0	no	
		Roots	320–860	940–3200	no	

[a] = most likely contaminated with the Si-rich growth medium perlite; DM = dry mass; ns = not specified.

In two of the previous studies [26,27] the identical potato cultivar ("Agata") was examined showing comparable Si contents. This indicates that the reported Si contents in potato plant materials seem to be directly related to the potato cultivar (Table 2). In this context, the differences in reported potato Si contents might be mainly related to the ability of different potato cultivars to take up and transport silicic acid. This ability, in turn, is directly related to the presence of transport/channel proteins that allow silicic acid transportation in the plant [47–49]. In general, several influx (called low silicon "Lsi" 1 and Lsi 6) and efflux (Lsi2 and Lsi3) proteins for the transport of silicic acid have been described for rice (Poaceae), but also some other plants like horsetail (Equisetaceae), strawberry (Rosaceae), tomato (Solanaceae), or pumpkin (Cucurbitaceae) [48]. While Lsi1 and Lsi6 represent specific aquaporins that belong to the Nodulin-26-like Intrinsic Proteins (NIPs), Lsi2 and Lsi3 are members of the anion transporter superfamily. The localization of influx and efflux proteins in planta and the expression of corresponding protein-encoding *Lsi* genes control silicic acid transport [50].

Regarding potato plants (*Solanum tuberosum*), Vulavala et al. [31] found *Lsi1* genes (i.e., *StLsi1*) to be expressed in roots and leaves, whereby gene expression was more pronounced in Si compared to control treatments. Expression of the *StLsi2* gene was observed in all potato materials (tuber flesh and skin, stolon, root, stem, and leaf samples) analyzed by these authors, whereby no differences between gene expression in control and Si treatments were observed. However, although Vulavala et al. [31] observed an upregulated expression of *StLsi1* genes in potato roots and leaves (cultivar "Winston"), they found no significantly changed silica accumulation in these plant organs after Si fertilization. Based on their results, Vulavala et al. [31] concluded that the space of 109 amino acids between the asparagine–proline–alanine (NPA) motifs (aquaporins are characterized by two highly conserved hydrophobic NPA motifs, which form a pore or channel for water and/or small molecules like glycerol, urea, or silicic acid) in *StLsi1* explains the low Si accumulation in their potato samples.

This is underpinned by a study by Deshmukh et al. [51], who showed that the ability of plants to take up silicic acid is related to a precise distance of 108 amino acids between the NPA motifs. In total, they analyzed the genomes of 25 plant species including 2 lower plant species (*Physcomitrella patens* and *Selaginella moellendorffii*), 1 gymnosperm species (*Picea abies*), 7 monocot species (e.g., *Oryza sativa*, *Sorghum bicolor*, and *Zea mays*), and 15 dicot species (e.g., *Arabidopsis thaliana*, *Glycine max*, *Solanum tuberosum*, and *Solanum lycopersicum*). Their results showed that Si-accumulating plants had a precise distance of 108 amino acids between the NPA motifs, while plants with 107 or 109 amino acids between the NPA motifs were not able to take up silicic acid in higher amounts. For the wild tomato species *Solanum pimpinellifolium* (Solanaceae), these authors found 109 amino acids between the NPA motifs as well. From their findings, Deshmukh et al. [51] hypothesized that this distance of 109 amino acids most likely originates not from a domestication-related genome alteration in cultivated Solanaceae species, but has its origin in the genome of the wild ancestors. However, this hypothesis is derived from the analysis of only one Solanaceae species (wild tomato), and further research is necessary to draw general conclusions regarding this aspect.

Regarding potatoes, it is assumed that the more than 4000 cultivars globally known originate from a relatively small sample of South American clones only, but with a relatively large amount of genetic diversity [52]. Thus, it cannot be ruled out that there might be differences in NPA motif amino acid distances between different potato cultivars controlling their ability to take up silicic acid. What we need now are detailed genome analyses (NPA motifs) of the various potato cultivars grown worldwide to clarify this aspect. Moreover, Thorne et al. [53] recently found that different, widely cultivated rice cultivars grown under hydroponic conditions showed different, cultivar-specific shoot and root Si concentrations, which were dependent on the levels of sodium chloride (salinity stress) and Si (plant Si availability). As plant Si availability is another crucial factor for Si uptake by plants [18], the relationship between the ability of specific potato cultivars to take up silicic acid and the concentrations of plant-available Si in agricultural soils has to be considered in future studies as well. Combined potato cultivar genome and soil Si availability studies will allow us to better understand the cultivar-specific differences in the uptake of silicic acid and to derive corresponding practice-oriented recommendations for potato growers worldwide.

The potato yields at our LTFE showed a decreasing trend within the analyzed 50-year period at low, medium, and high fertilization plots. We attribute this yield decrease to climate change to a certain degree, because the yield of many potato genotypes is quite sensitive to elevated temperatures, as potatoes originate from the Andes in South America, i.e., from a region with relatively cool temperatures. In fact, temperatures above 17 °C lead to a diminishment of potato tuberization, and thus global warming has been predicted to lead to decreased potato yields on a global scale in general [54,55]. In the region, where our study sites are located, mean annual temperatures increased from 8.0 °C in the year 1965 to 10.8 °C in the year 2015. This increase was also reflected in elevated temperatures in the potato growing season (April–September) in Brandenburg, Germany, which were negatively correlated to potato yields. However, it has to be stated here that we do not know to which extent other climate-related factors (e.g., drought, pest infestation, or heavy precipitation) and/or changes in soil properties (e.g., soil moisture, soil organic matter, or soil pH) (cf. [56–59]) affected potato yields at our experimental fields. The evaluation of such interactions was outside the scope of our study, which aimed at the analysis of silica accumulation in potato plants and the relationship between plant-available Si in agricultural soils and corresponding potato yield performance in the long term.

We found no relationship between the concentration of silicic acid (plant-available Si) in soils and corresponding potato yields in our study at all. Based on our (long-term) results and because Si contents of potato plant materials from control and Si treatments often show no statistically significant differences (see Table 2), we assume that silica accumulation in potato plants has no effect on potato yield performance. Consequently, we ascribe the reported (beneficial) effects of Si fertilization on potato growth and yield performance [26–30,32,34] mainly to antifungal/osmotic effects of foliar-applied Si fertilizers [16] and to changes in physicochemical soil properties (e.g., enhanced phosphorus availability and water-holding capacity) caused by soil-applied Si fertilizers [36,60]. In fact, potato plants can suffer from numerous diseases, which are caused by fungi (e.g., *Alternaria solani* (early blight), *Rhizoctonia solani* (black scurf), *Synchytrium endobioticum* (black scab), or *Fusarium* spec. (colored rots)) or fungus-like microorganisms (e.g., *Phytophthora infestans* (late blight)) in most cases [61,62]. Moreover, phosphorus and water availability in agricultural soils represent the main controls for potato growth and yield, because potatoes are characterized by a relatively high phosphorus requirement and susceptibility to even mild water stress [63–66]. However, as research on the effects of Si fertilization on potato performance is still limited to a few potato cultivars (cf. Table 2), we are calling for more studies dealing with the aspects discussed above.

5. Conclusions

Based on our results, we assume that the beneficial effects of Si fertilization on potato growth and yield performance reported in previous studies are related to the effects of the used Si fertilizers, rather than to silica accumulation in potato plants. In this context, antifungal/osmotic effects of foliar-applied Si fertilizers and changes in physicochemical soil properties (e.g., enhanced phosphorus availability and water-holding capacity) caused by soil-applied Si fertilizers seem to be the strongest candidates to explain the phenomena observed. To derive practice-oriented recommendations for potato growers worldwide, future research should aim at elucidating the complex relationships between the cultivated potato cultivar, the used Si fertilizer, and the prevalent soil properties as well as climate conditions. In this context, the following questions might be of particular interest:

(i) How big is the range of Si contents in potato plants considering the numerous cultivars worldwide? Recently, published data show that Si contents in potato tubers represent a difference of four orders of magnitude, for example (Table 2).
(ii) Which foliar Si fertilizer formula, at which dose, is most effective against which disease caused by fungi or fungus-like microorganisms?
(iii) How do different soil Si fertilizers (e.g., slags, fused magnesium phosphate, wollastonite, or biochar) affect soil properties in different soils under different climate conditions?

Author Contributions: Conceptualization, D.P. and J.S.; Formal Analysis, D.P.; Investigation, J.B., M.S., D.K. and C.B.; Visualization, D.P., J.B. and M.S.; Writing—Original Draft, D.P. and M.S.; Writing—Review and Editing, D.P., J.B., M.S., D.K., C.B. and J.S. All authors have read and agreed to the published version of the manuscript.

Funding: This research received no external funding.

Institutional Review Board Statement: Not applicable.

Informed Consent Statement: Not applicable.

Data Availability Statement: All relevant data are presented within the paper. Underlying data will be made available by the corresponding author upon reasonable request.

Acknowledgments: We are grateful to Dietmar Barkusky (Experimental Station Müncheberg, ZALF) and his team for providing potato yield data and managing the field experiments. We would like to thank Kristina Holz (ZALF Central Laboratory) and her team, especially Ellen Janiszewski, for the ICP-OES analyses. Finally, many thanks to the three anonymous reviewers, whose insightful comments on our manuscript helped to improve its quality.

Conflicts of Interest: The authors declare no conflicts of interest.

References

1. Laruelle, G.G.; Roubeix, V.; Sferratore, A.; Brodherr, B.; Ciuffa, D.; Conley, D.; Dürr, H.; Garnier, J.; Lancelot, C.; Le Thi Phuong, Q. Anthropogenic perturbations of the silicon cycle at the global scale: Key role of the land-ocean transition. *Glob. Biogeochem. Cycles* **2009**, *23*. [CrossRef]
2. Struyf, E.; Smis, A.; Van Damme, S.; Meire, P.; Conley, D.J. The Global Biogeochemical Silicon Cycle. *Silicon* **2009**, *1*, 207–213. [CrossRef]

3. Tréguer, P.J.; Sutton, J.N.; Brzezinski, M.; Charette, M.A.; Devries, T.; Dutkiewicz, S.; Ehlert, C.; Hawkings, J.; Leynaert, A.; Liu, S.M. Reviews and syntheses: The biogeochemical cycle of silicon in the modern ocean. *Biogeosciences* **2021**, *18*, 1269–1289. [CrossRef]
4. Carey, J.C.; Fulweiler, R.W. The terrestrial silica pump. *PLoS ONE* **2012**, *7*, e52932. [CrossRef] [PubMed]
5. Street-Perrott, F.A.; Barker, P.A. Biogenic silica: A neglected component of the coupled global continental biogeochemical cycles of carbon and silicon. *Earth Surf. Process. Landf.* **2008**, *33*, 1436–1457. [CrossRef]
6. Struyf, E.; Conley, D.J. Emerging understanding of the ecosystem silica filter. *Biogeochemistry* **2012**, *107*, 9–18. [CrossRef]
7. Puppe, D. Review on protozoic silica and its role in silicon cycling. *Geoderma* **2020**, *365*, 114224. [CrossRef]
8. Hodson, M.J. The development of phytoliths in plants and its influence on their chemistry and isotopic composition. Implications for palaeoecology and archaeology. *J. Archaeol. Sci.* **2016**, *68*, 62–69. [CrossRef]
9. Sangster, A.; Hodson, M.; Tubb, H. Silicon deposition in higher plants. In *Studies in Plant Science*; Elsevier: Amsterdam, The Netherlands, 2001; Volume 8, pp. 85–113.
10. International Committee for Phytolith Taxonomy. International Code for Phytolith Nomenclature (ICPN) 2.0. *Ann. Bot.* **2019**, *124*, 189–199. [CrossRef]
11. Hodson, M.J. The Relative Importance of Cell Wall and Lumen Phytoliths in Carbon Sequestration in Soil: A Hypothesis. *Front. Earth Sci.* **2019**, *7*, 167. [CrossRef]
12. Hodson, M.J.; Guppy, C.N. Special issue on silicon at the root-soil interface. *Plant Soil* **2022**, *477*, 1–8. [CrossRef]
13. Puppe, D.; Kaczorek, D.; Stein, M.; Schaller, J. Silicon in Plants: Alleviation of Metal (loid) Toxicity and Consequential Perspectives for Phytoremediation. *Plants* **2023**, *12*, 2407. [CrossRef] [PubMed]
14. Katz, O.; Puppe, D.; Kaczorek, D.; Prakash, N.B.; Schaller, J. Silicon in the Soil-Plant Continuum: Intricate Feedback Mechanisms within Ecosystems. *Plants* **2021**, *10*, 652. [CrossRef] [PubMed]
15. Haynes, R.J. Significance and Role of Si in Crop Production. *Adv. Agron.* **2017**, *146*, 83–166.
16. Puppe, D.; Sommer, M. Experiments, Uptake Mechanisms, and Functioning of Silicon Foliar Fertilization—A Review Focusing on Maize, Rice, and Wheat. *Adv. Agron.* **2018**, *152*, 1–49.
17. Savant, N.K.; Korndörfer, G.H.; Datnoff, L.E.; Snyder, G.H. Silicon nutrition and sugarcane production: A review1. *J. Plant Nutr.* **1999**, *22*, 1853–1903. [CrossRef]
18. Haynes, R.J. A contemporary overview of silicon availability in agricultural soils. *J. Plant Nutr. Soil Sci.* **2014**, *177*, 831–844. [CrossRef]
19. Li, Z.; Delvaux, B. Phytolith-rich biochar: A potential Si fertilizer in desilicated soils. *GCB Bioenergy* **2019**, *11*, 1264–1282. [CrossRef]
20. Gustafsson, O.; Krusa, M.; Zencak, Z.; Sheesley, R.J.; Granat, L.; Engstrom, E.; Praveen, P.; Rao, P.; Leck, C.; Rodhe, H. Brown clouds over South Asia: Biomass or fossil fuel combustion? *Science* **2009**, *323*, 495–498. [CrossRef]
21. Puppe, D.; Kaczorek, D.; Schaller, J.; Barkusky, D.; Sommer, M. Crop straw recycling prevents anthropogenic desilication of agricultural soil–plant systems in the temperate zone—Results from a long-term field experiment in NE Germany. *Geoderma* **2021**, *403*, 115187. [CrossRef]
22. Yang, X.; Song, Z.; Qin, Z.; Wu, L.; Yin, L.; Van Zwieten, L.; Song, A.; Ran, X.; Yu, C.; Wang, H. Phytolith-rich straw application and groundwater table management over 36 years affect the soil-plant silicon cycle of a paddy field. *Plant Soil* **2020**, *454*, 343–358. [CrossRef]
23. FAOSTAT: Food and Agriculture Data of the Statistics Division of the Food and Agriculture Organization (FAO) of the United Nations. Available online: https://www.fao.org/faostat/en/#home (accessed on 31 May 2024).
24. Hodson, M.J.; White, P.J.; Mead, A.; Broadley, M.R. Phylogenetic variation in the silicon composition of plants. *Ann. Bot.* **2005**, *96*, 1027–1046. [CrossRef] [PubMed]
25. Crusciol, C.A.; Pulz, A.L.; Lemos, L.B.; Soratto, R.P.; Lima, G.P. Effects of silicon and drought stress on tuber yield and leaf biochemical characteristics in potato. *Crop Sci.* **2009**, *49*, 949–954. [CrossRef]
26. Pilon, C.; Soratto, R.P.; Moreno, L.A. Effects of soil and foliar application of soluble silicon on mineral nutrition, gas exchange, and growth of potato plants. *Crop Sci.* **2013**, *53*, 1605–1614. [CrossRef]
27. Soratto, R.P.; Fernandes, A.M.; Pilon, C.; Souza, M.R. Phosphorus and silicon effects on growth, yield, and phosphorus forms in potato plants. *J. Plant Nutr.* **2019**, *42*, 218–233. [CrossRef]
28. Soltani, M.; Kafi, M.; Nezami, A.; Taghiyari, H. Effects of silicon application at nano and micro scales on the growth and nutrient uptake of potato minitubers (*Solanum tuberosum* var. Agria) in greenhouse conditions. *BioNanoScience* **2018**, *8*, 218–228. [CrossRef]
29. Dorneles, A.O.S.; Pereira, A.S.; Possebom, G.; Sasso, V.M.; Rossato, L.V.; Tabaldi, L.A. Growth of potato genotypes under different silicon concentrations. *Adv. Hortic. Sci.* **2018**, *32*, 289–295.
30. Xue, X.; Geng, T.; Liu, H.; Yang, W.; Zhong, W.; Zhang, Z.; Zhu, C.; Chu, Z. Foliar application of silicon enhances resistance against Phytophthora infestans through the ET/JA-and NPR1-dependent signaling pathways in potato. *Front. Plant Sci.* **2021**, *12*, 609870. [CrossRef]
31. Vulavala, V.K.; Elbaum, R.; Yermiyahu, U.; Fogelman, E.; Kumar, A.; Ginzberg, I. Silicon fertilization of potato: Expression of putative transporters and tuber skin quality. *Planta* **2016**, *243*, 217–229. [CrossRef]
32. Artyszak, A. Effect of Silicon Fertilization on Crop Yield Quantity and Quality-A Literature Review in Europe. *Plants* **2018**, *7*, 54. [CrossRef]

33. Wadas, W. Nutritional value and sensory quality of new potatoes in response to silicon application. *Agriculture* **2023**, *13*, 542. [CrossRef]
34. Wadas, W. Possibility of increasing early potato yield with foliar application of silicon. *Agron. Sci.* **2022**, *77*, 61–75. [CrossRef]
35. Nyawade, S.; Gitari, H.I.; Karanja, N.N.; Gachene, C.K.; Schulte-Geldermann, E.; Sharma, K.; Parker, M.L. Enhancing climate resilience of rain-fed potato through legume intercropping and silicon application. *Front. Sustain. Food Syst.* **2020**, *4*, 566345. [CrossRef]
36. Schaller, J.; Scherwietes, E.; Gerber, L.; Vaidya, S.; Kaczorek, D.; Pausch, J.; Barkusky, D.; Sommer, M.; Hoffmann, M. Silica fertilization improved wheat performance and increased phosphorus concentrations during drought at the field scale. *Sci. Rep.* **2021**, *11*, 20852. [CrossRef]
37. Schaller, J.; Macagga, R.; Kaczorek, D.; Augustin, J.; Barkusky, D.; Sommer, M.; Hoffmann, M. Increased wheat yield and soil C stocks after silica fertilization at the field scale. *Sci. Total Environ.* **2023**, *887*, 163986. [CrossRef]
38. Barkusky, D. Müncheberger Nährstoffsteigerungsversuch, V140. In *Ministerium für Ländliche Entwicklung, Umwelt und Verbraucherschutz, Landesamt für Verbraucherschutz, Landwirtschaft und Flurneuordnung (ed.): Dauerfeldversuche in Brandenburg und Berlin-Beiträge für eine nachhaltige landwirtschaftliche Bodenbenutzung*; MLUL: Brandenburg, Germany, 2009; pp. 103–109.
39. WRB. *World Reference Base (WRB) for Soil Resources 2014, Update 2015. International Soil Classification System for Naming Soils and Creating Legends for Soil Maps*; FAO: Rome, Italy, 2015.
40. Haysom, M.B.C.; Chapman, L.S. Some aspects of the calcium silicate trials at Mackay. *Proc. Aust. Soc. Sugar Cane Technol.* **1975**, *42*, 117–122.
41. de Lima Rodrigues, L.; Daroub, S.H.; Rice, R.W.; Snyder, G.H. Comparison of three soil test methods for estimating plant-available silicon. *Commun. Soil Sci. Plant Anal.* **2003**, *34*, 2059–2071. [CrossRef]
42. Puppe, D.; Kaczorek, D.; Buhtz, C.; Schaller, J. The potential of sodium carbonate and Tiron extractions for the determination of silicon contents in plant samples—A method comparison using hydrofluoric acid digestion as reference. *Front. Environ. Sci.* **2023**, *11*, 1145604. [CrossRef]
43. Puppe, D.; Buhtz, C.; Kaczorek, D.; Schaller, J.; Stein, M. Microwave plasma atomic emission spectroscopy (MP-AES)—A useful tool for the determination of silicon contents in plant samples? *Front. Environ. Sci.* **2024**, *12*, 1378922. [CrossRef]
44. Zepner, L.; Karrasch, P.; Wiemann, F.; Bernard, L. ClimateCharts. net–an interactive climate analysis web platform. *Int. J. Digit. Earth* **2021**, *14*, 338–356. [CrossRef]
45. Guntzer, F.; Keller, C.; Meunier, J.-D. Benefits of plant silicon for crops: A review. *Agron. Sustain. Dev.* **2011**, *32*, 201–213. [CrossRef]
46. Wadas, W.; Kondraciuk, T. Effect of silicon on micronutrient content in new potato tubers. *Int. J. Mol. Sci.* **2023**, *24*, 10578. [CrossRef] [PubMed]
47. Ma, J.F.; Tamai, K.; Yamaji, N.; Mitani, N.; Konishi, S.; Katsuhara, M.; Ishiguro, M.; Murata, Y.; Yano, M. A silicon transporter in rice. *Nature* **2006**, *440*, 688–691. [CrossRef] [PubMed]
48. Mitani-Ueno, N.; Ma, J.F. Linking transport system of silicon with its accumulation in different plant species. *Soil Sci. Plant Nutr.* **2021**, *67*, 10–17. [CrossRef]
49. Exley, C.; Guerriero, G.; Lopez, X. How is silicic acid transported in plants? *Silicon* **2020**, *12*, 2641–2645. [CrossRef]
50. Wani, A.H.; Mir, S.H.; Kumar, S.; Malik, M.A.; Tyub, S.; Rashid, I. Silicon en route-from loam to leaf. *Plant Growth Regul.* **2023**, *99*, 465–476. [CrossRef]
51. Deshmukh, R.K.; Vivancos, J.; Ramakrishnan, G.; Guérin, V.; Carpentier, G.; Sonah, H.; Labbé, C.; Isenring, P.; Belzile, F.J.; Bélanger, R.R. A precise spacing between the NPA domains of aquaporins is essential for silicon permeability in plants. *Plant J.* **2015**, *83*, 489–500. [CrossRef]
52. Birch, P.R.; Bryan, G.; Fenton, B.; Gilroy, E.M.; Hein, I.; Jones, J.T.; Prashar, A.; Taylor, M.A.; Torrance, L.; Toth, I.K. Crops that feed the world 8: Potato: Are the trends of increased global production sustainable? *Food Secur.* **2012**, *4*, 477–508. [CrossRef]
53. Thorne, S.J.; Stirnberg, P.M.; Hartley, S.E.; Maathuis, F.J.M. The Ability of Silicon Fertilisation to Alleviate Salinity Stress in Rice is Critically Dependent on Cultivar. *Rice* **2022**, *15*, 8. [CrossRef]
54. Hijmans, R.J. The effect of climate change on global potato production. *Am. J. Potato Res.* **2003**, *80*, 271–279. [CrossRef]
55. Raymundo, R.; Asseng, S.; Robertson, R.; Petsakos, A.; Hoogenboom, G.; Quiroz, R.; Hareau, G.; Wolf, J. Climate change impact on global potato production. *Eur. J. Agron.* **2018**, *100*, 87–98. [CrossRef]
56. Bomers, S.; Ribarits, A.; Kamptner, A.; Tripolt, T.; von Gehren, P.; Prat, N.; Söllinger, J. Survey of Potato Growers' Perception of Climate Change and Its Impacts on Potato Production in Germany, Switzerland, and Austria. *Agronomy* **2024**, *14*, 1399. [CrossRef]
57. Porter, G.A.; Bradbury, W.B.; Sisson, J.A.; Opena, G.B.; McBurnie, J.C. Soil management and supplemental irrigation effects on potato: I. Soil properties, tuber yield, and quality. *Agron. J.* **1999**, *91*, 416–425. [CrossRef]
58. Po, E.A.; Snapp, S.S.; Kravchenko, A. Potato yield variability across the landscape. *Agron. J.* **2010**, *102*, 885–894. [CrossRef]
59. Cambouris, A.; Nolin, M.; Zebarth, B.; Laverdière, M. Soil management zones delineated by electrical conductivity to characterize spatial and temporal variations in potato yield and in soil properties. *Am. J. Potato Res.* **2006**, *83*, 381–395. [CrossRef]
60. Schaller, J.; Cramer, A.; Carminati, A.; Zarebanadkouki, M. Biogenic amorphous silica as main driver for plant available water in soils. *Sci. Rep.* **2020**, *10*, 2424. [CrossRef]
61. Arora, R.; Khurana, S.P. Major fungal and bacterial diseases of potato and their management. In *Fruit and Vegetable Diseases*; Springer: Berlin/Heidelberg, Germany, 2004; pp. 189–231.

62. Yuen, J. Pathogens which threaten food security: Phytophthora infestans, the potato late blight pathogen. *Food Secur.* **2021**, *13*, 247–253. [CrossRef]
63. Koch, M.; Naumann, M.; Pawelzik, E.; Gransee, A.; Thiel, H. The importance of nutrient management for potato production Part I: Plant nutrition and yield. *Potato Res.* **2020**, *63*, 97–119. [CrossRef]
64. Rosen, C.J.; Kelling, K.A.; Stark, J.C.; Porter, G.A. Optimizing phosphorus fertilizer management in potato production. *Am. J. Potato Res.* **2014**, *91*, 145–160. [CrossRef]
65. Wolfe, D.W.; Fereres, E.; Voss, R.E. Growth and yield response of two potato cultivars to various levels of applied water. *Irrig. Sci.* **1983**, *3*, 211–222. [CrossRef]
66. Vos, J.; Haverkort, A. Water availability and potato crop performance. In *Potato Biology and Biotechnology*; Elsevier: Amsterdam, The Netherlands, 2007; pp. 333–351.

biology

MDPI

Article

Climate and Soil Properties Drive the Distribution of Minor and Trace Elements in Forest Soils of the Winter Olympic Core Area

Xiaochang Wu [1], Huayong Zhang [1,2,*], Zhongyu Wang [1], Wang Tian [1] and Zhao Liu [2]

1 Research Center for Engineering Ecology and Nonlinear Science, North China Electric Power University, Beijing 102206, China
2 Theoretical Ecology and Engineering Ecology Research Group, School of Life Sciences, Shandong University, Qingdao 266237, China
* Correspondence: zhanghuayong@sdu.edu.cn

Simple Summary

Minor and trace elements are found in the environment in extremely small quantities but have powerful bioscience roles, and they may cause hazardous effects when levels exceed certain limits. Consequently, it is essential to ascertain the concentrations of these elements within the soil. This paper addresses the distribution of 12 minor and trace elements in five different forests and soil depths in the Winter Olympic core area and explores the mechanisms driving element distribution. The results of this study indicate that soil minor and trace elements are mainly influenced by climatic factors and soil properties, and no direct effect of vegetation type on soil element distribution was observed in this study. Understanding the distribution of minor and trace elements in forest soils across the core area can provide important information on soil succession patterns in the boreal forest and may help elucidate the response of elemental distributions to climate change.

Abstract

Minor and trace elements in soil play a crucial role in regulating ecological processes that sustain the functionality of forest ecosystems. In this study, we have selected three conifer forests (*Pinus sylvestris*, *Picea asperata*, *Larix principis-rupprechtii*), one broadleaf forest (*Betula Platyfilla*) and one mixed forest of *Betula Platyfilla* and *Larix principis-rupprechtii* in the Winter Olympic core area and determined the pattern of 12 typical elements (B, Fe, V, Cr, Ni, Co, Mn, As, Cu, Zn, Sn and Se) in soils and their main drivers in the three different soil layers (A, B and C horizon) in each soil profile. Our results showed that the concentrations of B, Fe, Cr, Cu, Ni and Sn were mainly enriched in the broadleaf forest and mixed broadleaf–conifer forest zones, and the average concentrations of Co, Mn, V, Zn, As and Se were mainly enriched in coniferous forest zones in contrast. We have observed that the mean concentrations of Fe, Cr, Ni, Zn, As, Sn and Co increase with soil depth in the BP forest. The concentrations of Se and Cu were higher in the A layer than the C layer. The piecewise structural equation modeling (piecewiseSEM) results visualized a direct and negative effect on B, Fe, V, Cr and Ni concentrations due to soil temperature, while the concentrations of Se is mainly influenced by soil temperature and soil properties.

Keywords: soil minor and trace elements; climate; soil properties; the core area of the Winter Olympics

https://doi.org/10.3390/biology14010082

1. Introduction

Boreal forests represent a significant element of terrestrial ecosystems [1], comprising approximately one-third of the world's forested area and contributing 20% to the global carbon sink [2]. Boreal forests have significant economic benefits and cultural value to society [3,4]. Climate change and human activities are known to affect boreal forest ecosystems [5,6], which greatly affect the structure and ecological processes of boreal forests [3]. The core area of the Winter Olympics is characterized by forest vegetation, with ecological succession predominantly involving *Betula Platyfilla*. In addition, the area is planted with a large number of coniferous forests such as *Pinus sylvestris, Picea asperata, Larix principis-rupprechtii,* which are mixed to varying degrees with *Betula Platyfilla* to form mixed broadleaf–conifer forest. As the world's largest comprehensive winter games, the Winter Olympics will cause natural forest disturbances to the local ecosystem from the planning of the competition area to the construction of the venues. These increased disturbances would pose growing pressures on boreal forests.

The most important characteristic of minor and trace elements is their very low concentration in the environment [7]. They include a variety of chemical groups—metals (e.g., Fe, Cu, and Zn), metalloids (As) and non-metals (Se)—which vary in their physicochemical properties in the environment [8]. The minor and trace elements have powerful bioscientific effects and may cause harm when levels exceed certain limits [9,10]. Some of these trace elements are required for many biological activities (e.g., Fe, Cu and Zn), for example, for the transport of oxygen, protein synthesis and enzyme activities [11]; others such as Cr, Co and B may replace essential metals, and they can disrupt the proper functioning of cofactors and enzymes, which can lead to toxic effects in organisms [12]. However, even the former metal can have toxic effects if the concentration is elevated [12,13]. The core area of the Winter Olympics also undertakes the important function of water conservation in the Beijing–Tianjin–Hebei region, Consequently, it is essential to ascertain the concentrations of these elements within the soil.

The biogeochemical cycling of elements is influenced by soil matrix characteristics and soil weathering processes [14]. These cycles are further regulated by the activities of vegetation and microorganisms [15,16] and are also affected by climatic conditions [17–19]. Climate is the most dominant factor influencing the spatial distribution of elements [20]. Temperature further influences trace element distribution [21] by regulating plant growth and microbial activity [22]. Soil moisture plays a crucial role in the formation and evolution of various soil types and the transport and transformation of trace elements [23]. An analysis of 9830 soil samples from southeastern China revealed spatial variability in the concentrations of 53 elements related to temperature, precipitation and pH [24]. Patterns of the lowest soil element (Cu, Fe, Mn, Zn and Ni) concentrations were revealed within arid and tropical ecosystems globally. In addition, a temperature threshold of 12–14 °C was identified, above which all soil minor and trace elements are abruptly reduced [25]. The Winter Olympic core area is known for its long, bitter winters, short summers, and snowfall, which makes up the majority of the yearly precipitation. The fundamental factors contributing to the fluctuations in the concentrations of various minor and trace elements remain largely unidentified. Comprehending the influence of environmental gradients on variations in elemental concentrations could provide insights into the sudden alterations in ecological responses to these elements [26,27].

We have selected five forests, *Pinus sylvestris* (PS), *Picea asperata* (PA), *Larix principis-rupprechtii* (LP), *Betula Platyfilla* (BP) and the mixed forest of *Betula Platyfilla* and *Larix principis-rupprechtii* (MF), to investigate the distribution characteristics of soil minor and trace elements in the Winter Olympic core area. We specifically focus on the forest type and soil depths. Additionally, we aim to investigate how environmental factors affect soil trace and minor elements by piecewise structural equation modeling (piecewiseSEM). Our specific objectives of this research are (1) to determine the distribution of soil minor and trace element concentrations in the Winter Olympic core area and (2) to identify the main drivers affecting trace element distribution in the Winter Olympic core area.

2. Materials and Methods

2.1. Study Area

The 118 km^2 Winter Olympic core area is located in Hebei Province, China, about 22 km northeast of Chongli District (40°47′ N to 41°17′ N and 114°17′ E to 115°34′ E). This region, which is between 1797 and 2003 m above sea level, has a typical East Asian continental monsoon climate, with cold winters and cool, humid summers. With yearly precipitation of 483.3 mm and average temperatures between 3.7 and 19 °C, the climate is categorized as continental monsoon. As of 2021, the forest coverage rate in the region is reported to be 67%, characterized by a relatively simplistic forest structure. The forests of Chongli are home to a total of 553 species of terrestrial wild plants, which are classified into 301 genera across 80 families.

2.2. Experimental Design

We conducted sampling in five forest types, including *Pinus sylvestris* (PS), *Picea asperata* (PA), *Larix principis-rupprechtii* (LP), *Betula Platyfilla* (BP) and the mixed forest of *Betula Platyfilla* and *Larix principis-rupprechtii* (MF). The sampling was carried out in July 2019. Three samples of each type of forest were selected in order from top to bottom along the direction of runoff. Each profile was categorized into three distinct layers of soil formation: A (surface horizon, characterized by a darker color), B (granular with a prismatic structure) and C (bottom layer consisting of unconsolidated earth material). In each layer, samples of ring knives were collected, and composite soil samples were subsequently analyzed to assess their physical and chemical properties. Since these forests were distributed in close proximity and there were no significant differences in mean annual temperature and annual precipitation conditions, we used mean annual soil temperature (ST) and mean annual soil humidity (SH) instead of temperature and precipitation (Table S1).

2.3. Soil Sample Collection and Chemical Analyses

Following the removal of litter and other debris from the soil surface, a soil profile measuring one meter in depth was excavated using a spade. Soil samples were then obtained utilizing a ring knife. Two samples from each plot were collected with the ring knife, and the composite samples were subsequently analyzed to assess the chemical and physical properties of the soil.

The soil was subjected to sieving using a 2 mm mesh. The bulk density (BD) of the soil was assessed by drying the sample in an oven at 105 °C for a duration of 48 h or until a consistent weight was obtained. The bulk density was subsequently calculated by dividing the weight of the oven-dried soil by its volume. Total porosity was assessed by measuring the weight of soil that had absorbed an adequate amount of water to reach saturation, which was calculated by taking the difference between the weight of the saturated soil and the weight of the oven-dried soil and then dividing this value by the volume of the soil. Soil pH was assessed utilizing a pH meter (type: PHS-3C by Shjingmi, Shanghai, China) employing a soil-to-water ratio of 2.5:1. Soil organic matter (SOM) was determined by the external heating method of potassium dichromate and concentrated sulfuric acid [28]. Purging and trapping methodologies were employed to ascertain the concentration of alkali-hydrolyzable nitrogen (AHN) utilizing an elemental analyzer (type: Elementar Vario Macro cube by Elementar, Frankfurt, Germany). The concentrations of available phosphorus (AP) [29] was quantified using inductively coupled plasma–optical emission spectrometry (type: Agilent 5110 ICP–OES by Agilent Technologies, Santa Clara, CA, USA). Soil samples were digested using aqua regia ($3HCl + HNO_3$). Two blank controls, duplicate samples and a standard reference material (GBW07404 by IGGE, Langfang, Hebei, China) in each batch were simultaneously digested to ensure quality control. The concentrations of V, Mn, Fe, Co, B and Sn were determined by inductively coupled plasma–mass spectrometry (ICP-MS, Agilent 5110 by Agilent Technologies, Santa Clara, CA, USA), while Cr, Ni, Cu, As and Zn were measured by atomic fluorescence spectrometry (240AFS AA by Agilent Technologies, Santa Clara, CA, USA). Additionally, Se concentrations were assessed using atomic fluorescence spectrometry (AFS-930, Beijing, China).

2.4. Statistical Analysis

The data presented in this study were characterized using the mean and standard deviation. The concentrations of SOM, AHN and AP were described by mass content. Normalization of data prior to analysis using Z-Score. Analysis of variance (ANOVA) and Duncan's test were employed to systematically assess variations in soil chemical and physical characteristics and the distribution of soil elements across different soil horizons and forest types. Statistical significance was established at a threshold of $p < 0.05$. Pearson correlation analysis and principal component analysis (PCA) were conducted to investigate the relationships among various concentrations of elements and the associated environmental factors. The classification of 12 elements using PCA will help to further explore the mechanisms by which climate and soil properties drive soil elemental distributions. Using the KOM spherical test, the principal components with eigenvalues greater than 1 are selected according to the eigenvalue size order, and the elemental loadings in each principal component are obtained by rotating the matrix. The initial four principal components (PC1, PC2, PC3, and PC4) derived from the principal component analysis (PCA) explained 81.2% of the variance associated with soil elements, thereby successfully representing the variability in the concentration levels of all twelve minor and trace elements analyzed.

Subsequently, the piecewise structural equation modeling (piecewiseSEM) approach was used to explore the mechanisms of soil minor and trace elements in response to environmental factors. This modeling approach effectively and clearly illustrates both the direct and indirect relationships between essential ecological factors and their impacts on soil components. Based on the existing theoretical framework research progress, several variables were selected for inclusion in the model. The piecewise SEM analysis was conducted utilizing the "piecewiseSEM" package [30]. We employed Fisher's C test to assess the adequacy of the model fitting results [31]. The models underwent iterative modifications and refinements, guided by the statistical significance of the pathways ($p < 0.05$) and the overall goodness of fit ($0 \leq$ Fisher's C/df ≤ 2 and $0.05 < p \leq 1.00$).

All statistical analyses were performed utilizing SPSS 25.0 Software (IBM, Armonk, NY, USA) and R 4.0.5 (R Development Core Team 2021 R Foundation for Statistical Vienna, Austria).

3. Results

3.1. Difference in Concentrations of Minor and Trace Elements in Five Forest Soils

The concentrations of minor and trace elements in the soils predominantly exhibited a positively skewed or bimodal distribution pattern (Figure 1). Various coefficients of variation (CV) (Table S3) for elements in different forests showed great variation, indicating significant heterogeneity. The mean concentrations of B, Fe, Cr, Cu, Ni and Sn in the five forests were highest in BP followed by LP and MF. The average concentrations of Co, Mn and Zn were highest in LP, and V was highest in PA. PS soils present higher As and Se contents; however, these levels remain within a comparable range. The average concentrations of V and Se were similar in all five forests, but the high CVs (V: 42%, Se: 42%) in PA indicated high spatial variability.

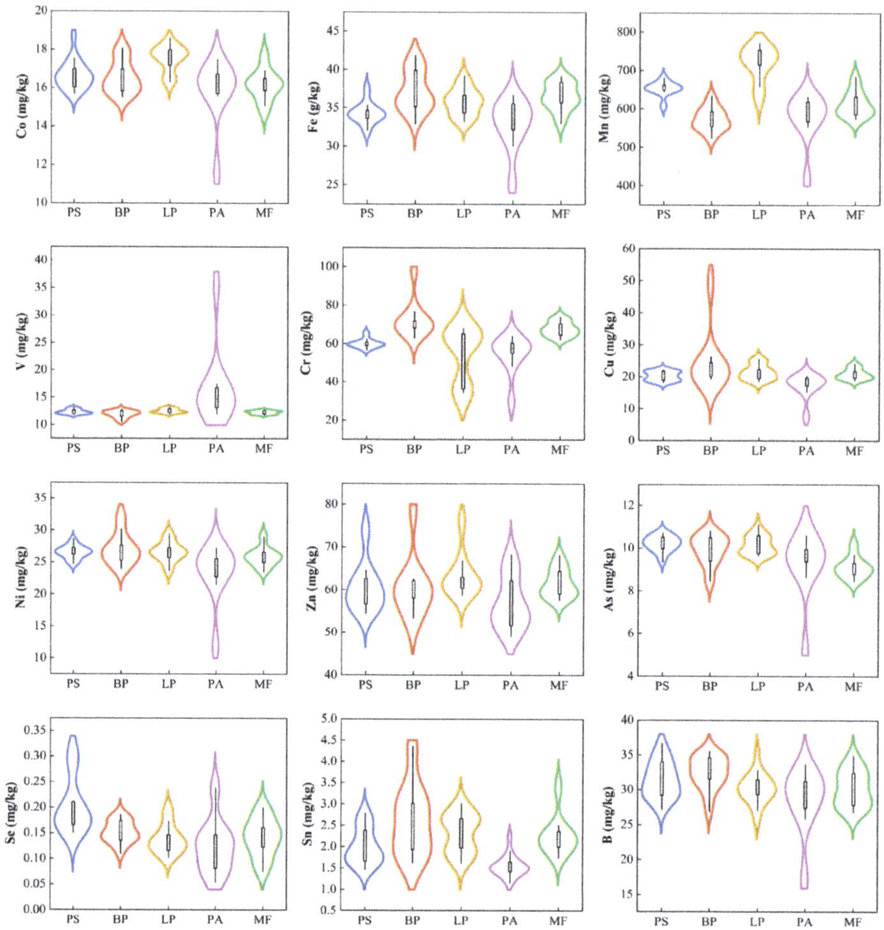

Figure 1. Violin plot of soil minor and trace elements with boxes in five forests.

3.2. Influences of Soil Depth and Forest Types on Minor and Trace Elements Concentrations

The concentrations of all tested elements in five forest soils exhibited remarkable spatial variability with soil depth (Figure 2). The mean concentrations of Fe, Cr, Ni, Zn, As, Sn and Co increase with soil depth in the BP forest. BP soils had the highest Fe, Cr, Ni and Sn concentrations. There was no significant trend in concentrations of Zn, As and Co among all five forests.

The B concentrations present a trend of increasing then decreasing with soil depth in PS, BP, PA and MF but a trend of decreasing then increasing in LP. The concentrations of Cu decrease with increasing soil depth in PS, PA and MF forests. The concentrations of Se were higher in the A layer than the C layer among all five forests, and PS soil had the highest Se concentrations. There was no significant trend between soil layers for Mn and V concentrations. LP forest soils had the highest Mn concentrations, while PA present much higher soil V concentrations than those in other forests.

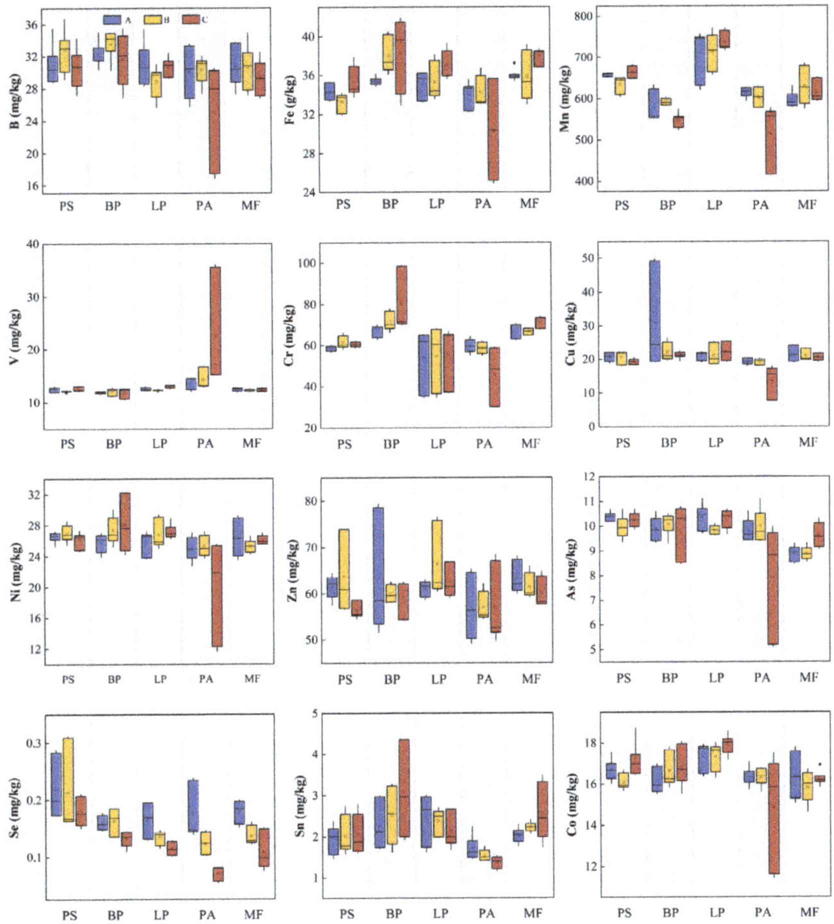

Figure 2. Vertical distribution of soil minor and trace elements in different layers among five forests.

3.3. Relationships of Soil Minor and Trace Elements with Climate Soil Proportions

Based on the findings from the principal component analysis (Figure 3a), we categorized the 12 measured elements into four principal components by analyzing the explained variance associated (Figure 3b) with each component, as well as the loadings of minor and trace elements and micronutrients (Figure 3c). PC1 explained 46.3% of the total variance and exhibited a significant correlation with the concentrations of B, Fe, V, Cr and Ni contents; these can be categorized as siderophile elements. PC2 accounted for 13% of the variance observed in the original dataset, with Co, Mn and As exhibiting the most significant loadings. PC3 accounted for 12.4% of the total variance, correlated to Cu, Zn and Sn contents. PC4 explained 9.5% of the variance, while the concentrations of Se mainly explained. The initial four principal components derived from the PCA (PC1, PC2, PC3 and PC4) collectively explained 81.2% of the total variance.

The Pearson correlation analysis indicated a significant relationship between minor and trace elements in the soil and various soil properties (Figure 4). The soil temperature showed significantly positive correlations with the concentrations of As, and pH was negatively correlated with Cr. The soil BD was significantly negatively correlated with the Fe, Cr, Ni and Sn concentrations, and porosity was correlated with V. In contrast, the soil total porosity was positively correlated with the concentrations of Fe, Cr, Ni and Sn and negatively correlated with V. Further, the soil Se contents were significantly positive correlations with the concentrations of AHN and SOM. The contents of AP were significantly positive correlations with the contents of Cu and As.

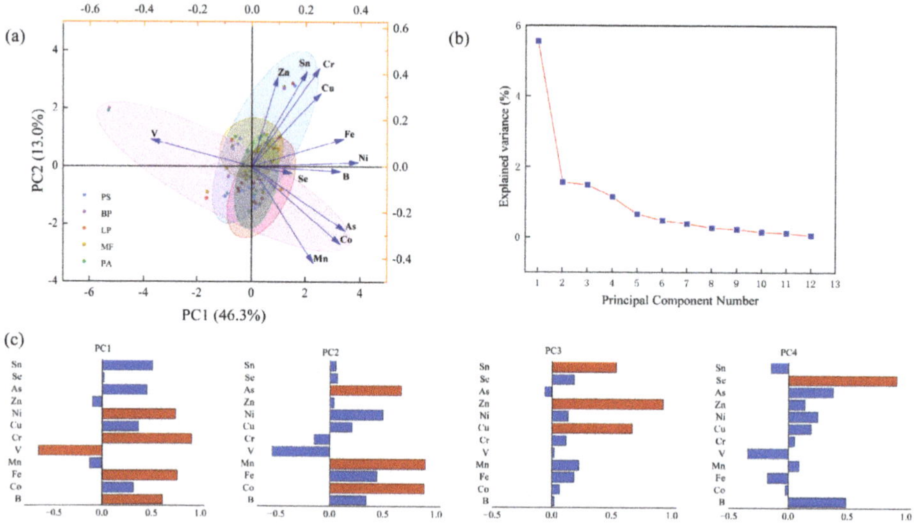

Figure 3. Principal component analysis of minor and trace elements in five forests. (**a**) Loading plot, (**b**) variance explained. (**c**) Bar plots illustrating the loadings of each minor or trace element. The orange bars denote the loadings and contributions that are deemed significant.

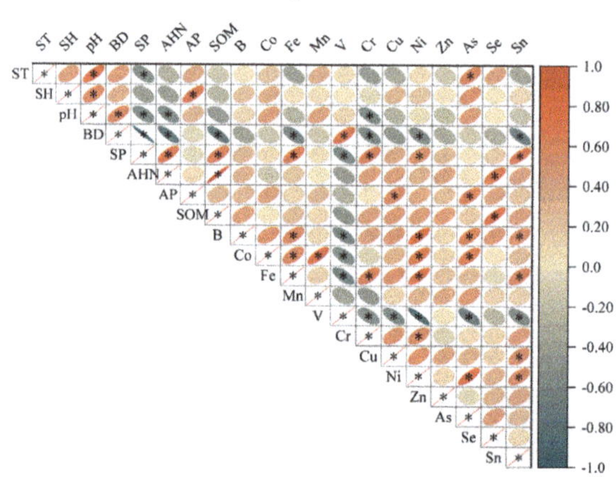

Figure 4. Pearson correlation between selected soil properties and soil minor and trace elements. Note: *, correlation is significant at the 0.05 level.

As the forest plots were close to each other and the annual mean temperature and annual precipitation conditions were close, we used soil temperature and soil humidity instead of temperature and precipitation for piecewise SEM analyses. The results obtained from SEM demonstrated the influence of climatic factors (ST and SH) and soil properties on each of the principal components related to the concentrations of trace elements. ST directly and negatively affects the forest type and PC1 (Figure 5a). PC4 was directly and positively influenced by ST and soil properties (Figure 5d). SH indirectly and negatively affects the PC2 and PC4 (Figure 5b,d) and positively and indirectly influences the PC1 and

PC3 (Figure 5a,c). We did not observe a direct pathway for soil minor and trace elements effects of forest type in the model.

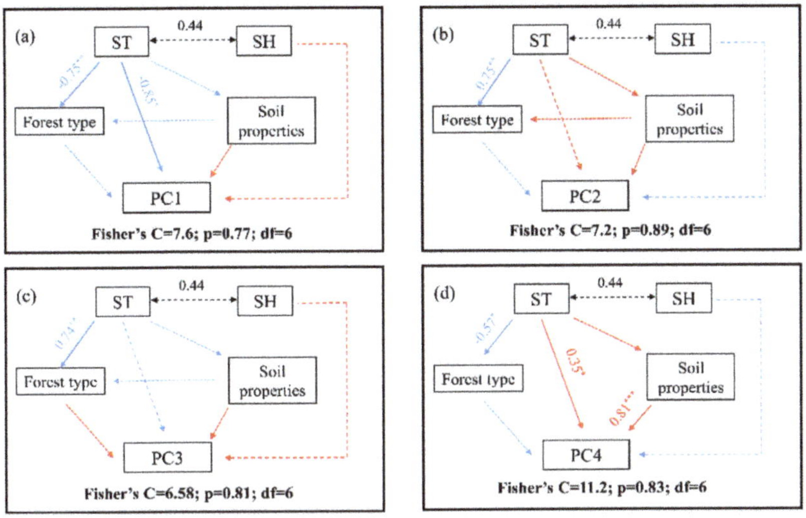

Figure 5. The influence of environmental factors on minor and trace elements in soil through structural equation modeling (SEM). (**a**) PC1 SEM plot, (**b**) PC2 SEM plot, (**c**) PC3 SEM plot, (**d**) PC4 SEM plot. Red solid arrows indicate positive effects, blue solid arrows indicate negative effects and dashed lines indicate non-significant paths. Significance levels of each predictor are * $p < 0.05$, ** $p < 0.01$, *** $p < 0.001$.

4. Discussion

4.1. Forest Type and Soil Depth Effect on Minor and Trace Elements in Soils

Soil has wide variations in minor and trace element concentrations due to soil proportions, vegetation type and bedrock weathering. Tree species can influence soil through various mechanisms, including the quantity and composition of litter, root activity, the microclimatic conditions they create and the ground vegetation that is established beneath their canopy [32]. The elements of B, Fe, Cr, Cu, Ni and Sn exhibited significant enrichment within the broadleaf forest and mixed broadleaf–conifer forest zones. Coniferous forests possess a high concentration of keratin, which serves to inhibit the adhesion and invasion of microorganisms on leaves that are abundant in keratin [33]. The rate of decomposition of leaves in coniferous forests is much smaller than that in broad-leaved forests. Broadleaf forests, such as *Larix principis-rupprechtii*, which thrive in cold climates characterized by high precipitation, exhibit the development of robust tracheids and fine root structures. This adaptation facilitates the formation of intricate and resilient root systems [34]. The resilient root systems facilitate the processes of rock fragmentation and soil weathering, thereby contributing to the release of essential nutrients, including Fe, Cu and other alkaline cations [35]. In contrast, the average concentrations of Co, Mn, V, Zn, As and Se were mainly enriched in coniferous forest zones. Coniferous forests demonstrate a greater capacity for soil acidification compared to broadleaf forests [35], resulting in enhanced leaching of cationic elements, including Zn, Mn and Co [36].

Our results show that average concentrations of Fe, Cr, Ni, Zn, As, Sn and Co increase with soil depth in BP forests. The metallic constituents present in the soil primarily originate from the weathering processes of the underlying bedrock [20]. The B concentrations present a trend of increasing then decreasing with soil depth; this phenomenon may be attributed to the presence of boron in the soil, which arises from the weathering of boron-containing minerals found in the underlying geological formations [37]. However, during the process of weathering, element B exhibits mobility and enters into solution, with its concentration being regulated by the presence of clay minerals, which can either adsorb or incorporate it [38], which results in the enrichment in B layers. We have observed that the concentrations of Se and Cu were higher in the A layer than the C layer, while the concentrations of these elements remained consistent across the study sites within the A horizon, indicating a lack of significant influence from the underlying bedrock [7]. Mn and V levels did not show differences between soil layers; thus, the stratification of soil may exhibit greater complexity, necessitating additional research to address the acceptable thresholds for this particular element.

4.2. Soil Minor and Trace Elements in Relation to Soil Properties and Climate

Temperature and precipitation represent the primary climatic variables that significantly affect plant survival and distribution [39,40]. Consequently, these factors contribute significantly to the spatial variability observed in extractable mineral elements [41,42]. Extreme temperatures, whether excessively low or high, can hinder plant growth and the functioning of soil microorganisms. This inhibition, in turn, limits the accumulation of organic matter and the absorption of minerals [43]. Precipitation also affects soil properties (e.g, SOM, AHN, AP), with increased SH due to precipitation increasing leaching along the soil profile, further affecting soil micronutrient distribution. The pH ranged from 6.34 to 7.01 (Table S2) and was generally weakly acidic, approaching neutrality. Research has indicated that the complexity and stability of microbial networks tend to decrease as the pH approaches neutrality. Moreover, research indicates that more complex microbial networks enhance both the efficiency and rate of ecosystem functioning relative to simpler networks, thereby facilitating increased leaching of cationic elements from the soil. Soil pH showed a significant positive correlation with SH and ST, and it can be hypothesized that SH and ST affect the accumulation of these elements in the soil by influencing the pH value. There was a significant negative correlation between soil BD and the concentration of Fe, Cr, Ni and Sn. This phenomenon can be ascribed to the strong correlation between biodiversity and various factors, including soil moisture, vegetation communities, soil texture and the content of organic matter [44,45]. Increases in bulk weight lead to poorer air permeability and less oxygen in the soil, limiting the activity of soil microorganisms and resulting in lower metal element concentrations. Se contents were significantly positive correlations with the concentrations of AHN and SOM. In general, the majority of Se present in soil is associated with organic matter, resulting in an increase in Se concentration corresponding to higher levels of SOM [46,47]. The concentration of As had significantly positive correlations with the contents of AP, as arsenic's soil cycling and behavior are often linked to phosphorus [48]. We observed that Cu also had a strong significant correlation with AP. The complexation of copper by organic matter has been identified as the primary and most efficient mechanism for the retention of copper in soil environments [49], and the concentration of AP in the soil promotes the formation of copper complexes with soluble organic matter.

The SEM results showed that climate (ST and SH), forest type and soil properties have an effect on each of the principal components related to trace element concentrations. ST has a direct negative effect on variation in B, Fe, V, Cr and Ni concentrations. Several studies have shown that micronutrient availability is low under high-temperature conditions in semi-arid regions [50]. These findings help to explain the negative effect of temperature on B, Fe, V, Cr and Ni content in this study. The concentration of Se was directly and positively influenced by ST and soil properties. Selenium is classified as a chalcophile element, which makes it susceptible to secondary enrichment or dilution as a result of supergene geochemical processes [51]. Research indicates that the concentration of this substance in surface sediments is typically influenced by various geological and geographical factors, including the lithology of the bedrock, the types of vegetation present, climatic conditions, and elemental distributions [52–54]. In this study, we did not observe any direct pathway for soil minor and trace elements effects of forest type. This phenomenon can be ascribed to the dual origins of soil minor and trace elements, which include the erosion of local bedrock (lithogenic fraction) and contributions from atmospheric sources (atmospheric fraction) [7]. Little or no influence by forest type was observed.

5. Conclusions

In summary, the concentrations of B, Fe, Cr, Cu, Ni and Sn were mainly enriched in the broadleaf forest and mixed broadleaf–conifer forest zones. The average concentrations of Co, Mn, V, Zn, As and Se were mainly enriched in coniferous forest zones. We have observed that the concentrations of Se and Cu were higher in the A layer than the C layer, and Fe, Cr, Ni, Zn, As, Sn and Co concentrations increased with soil depth in BP forests. The B, Fe, V, Cr and Ni concentrations were directly and negatively affected by soil temperature, while the concentrations of Se are mainly influenced by soil temperature and soil properties.

The core area of the Winter Olympics is a hotspot for national ecological environmental protection and water conservation in the Beijing–Tianjin–Hebei region. Understanding the distribution of minor and trace elements in forest soils across the core area can provide important information on soil succession patterns in the boreal forest. Investigations of the principal drivers of elemental concentrations may help elucidate the response of elemental distributions to climate change.

Supplementary Materials: The following supporting information can be downloaded at https://www.mdpi.com/article/10.3390/biology14010082/s1, Table S1: Soil temperature and soil humidity of sampling sites (n = 9); Table S2: Variance analysis of soil chemical and physical properties between different forests and soil layers; Table S3: Descriptive statistics of minor and trace element concentrations in soils of the five forests.

Author Contributions: Conceptualization, X.W. and H.Z.; Methodology, X.W., H.Z., Z.W. and W.T.; Formal analysis, Z.W.; Resources, Z.L.; Data curation, X.W.; Writing—original draft, X.W.; Writing—review and editing, Z.W., W.T. and Z.L. All authors have read and agreed to the published version of the manuscript.

Funding: This research was funded by the National Major Science and Technology Program for Water Pollution Control and Treatment (No. 2017ZX07101-002) and the Discipline Construction Program of Huayong Zhang, Distinguished Professor of Shandong University, School of Life Sciences (61200082363001).

Institutional Review Board Statement: Not applicable.

Informed Consent Statement: Not applicable.

Data Availability Statement: The data presented in this study are available on request from the corresponding author.

Acknowledgments: We would like to thank Feng, Y., Liu, Z., and Zhang, S.J. for their assistance in data collection during the experiment.

Conflicts of Interest: The authors declare that they have no known competing financial interest or personal relationships that could have appeared to influence the work reported in this paper.

References

1. Sexton, J.O.; Noojipady, P.; Song, X.-P.; Feng, M.; Song, D.-X.; Kim, D.-H.; Anand, A.; Huang, C.; Channan, S.; Pimm, S.L.; et al. Conservation policy and the measurement of forests. *Nat. Clim. Change* **2016**, *6*, 192–196. [CrossRef]
2. Pan, Y.; Birdsey, R.A.; Fang, J.; Houghton, R.; Kauppi, P.E.; Kurz, W.A.; Phillips, O.L.; Shvidenko, A.; Lewis, S.L.; Canadell, J.G.; et al. A Large and Persistent Carbon Sink in the World's Forests. *Science* **2011**, *333*, 988–993. [CrossRef] [PubMed]
3. Wei, X.; Giles-Hansen, K.; Spencer, S.A.; Ge, X.; Onuchin, A.; Li, Q.; Burenina, T.; Ilintsev, A.; Hou, Y. Forest harvesting and hydrology in boreal Forests: Under an increased and cumulative disturbance context. *For. Ecol. Manag.* **2022**, *522*, 120468. [CrossRef]
4. Burton, P.J.; Messier, C.; Adamowicz, W.L.; Kuuluvainen, T. Sustainable management of Canada's boreal forests: Progress and prospects. *Écoscience* **2006**, *13*, 234–248. [CrossRef]
5. Scheffer, M.; Hirota, M.; Holmgren, M.; Van Nes, E.H.; Chapin, F.S. Thresholds for boreal biome transitions. *Proc. Natl. Acad. Sci. USA* **2012**, *109*, 21384–21389. [CrossRef]
6. Huang, C.; Liang, Y.; He, H.S.; Wu, M.M.; Liu, B.; Ma, T. Sensitivity of aboveground biomass and species composition to climate change in boreal forests of Northeastern China. *Ecol. Model.* **2021**, *445*, 109472. [CrossRef]
7. Agnan, Y.; Courault, R.; Alexis, M.A.; Zanardo, T.; Cohen, M.; Sauvage, M.; Castrec-Rouelle, M. Distribution of trace and major elements in subarctic ecosystem soils: Sources and influence of vegetation. *Sci. Total Environ.* **2019**, *682*, 650–662. [CrossRef] [PubMed]
8. Thornton, I.; Webb, J.S. Trace Elements in Soils and Plants. In *Food Chains and Human Nutrition*; Blaxter, K., Ed.; Springer: Dordrecht, The Netherlands, 1980; pp. 273–315. [CrossRef]
9. Mehri, A. Trace Elements in Human Nutrition (II)—An Update. *Int. J. Prev. Med.* **2020**, *11*, 2. [CrossRef]
10. Chen, Q.Y.; Costa, M. Arsenic: A Global Environmental Challenge. *Annu. Rev. Pharmacol. Toxicol.* **2021**, *61*, 47–63. [CrossRef]
11. Trengove, C.L.; Judson, G.J. Trace element supplementation of sheep: Evaluation of various copper supplements and a soluble glass bullet containing copper, cobalt and selenium. *Aust. Vet. J.* **1985**, *62*, 321–324. [CrossRef] [PubMed]
12. Gifford, S.; Dunstan, R.H.; O'Connor, W.; Roberts, T.; Toia, R. Pearl aquaculture—Profitable environmental remediation? *Sci. Total Environ.* **2004**, *319*, 27–37. [CrossRef] [PubMed]
13. Sunda, W.G.; Huntsman, S.A. Processes regulating cellular metal accumulation and physiological effects: Phytoplankton as model systems. *Sci. Total Environ.* **1998**, *219*, 165–181. [CrossRef]
14. Augusto, L.; Achat, D.L.; Jonard, M.; Vidal, D.; Ringeval, B. Soil parent material—A major driver of plant nutrient limitations in terrestrial ecosystems. *Glob. Change Biol.* **2017**, *23*, 3808–3824. [CrossRef]
15. Sinsabaugh, R.L.; Lauber, C.L.; Weintraub, M.N.; Ahmed, B.; Allison, S.D.; Crenshaw, C.; Contosta, A.R.; Cusack, D.; Frey, S.; Gallo, M.E.; et al. Stoichiometry of soil enzyme activity at global scale. *Ecol. Lett.* **2008**, *11*, 1252–1264. [CrossRef]
16. Vitousek, P.M.; Menge, D.N.; Reed, S.C.; Cleveland, C.C. Biological nitrogen fixation: Rates, patterns and ecological controls in terrestrial ecosystems. *Philos. Trans. R. Soc. Lond. B Biol. Sci.* **2013**, *368*, 20130119. [CrossRef]
17. Delgado-Baquerizo, M.; Maestre, F.T.; Gallardo, A.; Bowker, M.A.; Wallenstein, M.D.; Quero, J.L.; Ochoa, V.; Gozalo, B.; García-Gómez, M.; Soliveres, S.; et al. Decoupling of soil nutrient cycles as a function of aridity in global drylands. *Nature* **2013**, *502*, 672–676. [CrossRef] [PubMed]
18. Simpson, A.H.; Richardson, S.J.; Laughlin, D.C. Soil–climate interactions explain variation in foliar, stem, root and reproductive traits across temperate forests. *Glob. Ecol. Biogeogr.* **2016**, *25*, 964–978. [CrossRef]
19. Tan, Q.; Wang, G. Decoupling of nutrient element cycles in soil and plants across an altitude gradient. *Sci. Rep.* **2016**, *6*, 34875. [CrossRef] [PubMed]
20. Sun, D.-L.; Yao, B.-M.; Yang, G.; Sun, G.-X. Climate and soil properties regulate the vertical heterogeneity of minor and trace elements in the alpine topsoil of the Hengduan Mountains. *Sci. Total Environ.* **2023**, *899*, 165653. [CrossRef] [PubMed]
21. Romero-Mujalli, G.; Hartmann, J.; Börker, J. Temperature and CO2 dependency of global carbonate weathering fluxes—Implications for future carbonate weathering research. *Chem. Geol.* **2019**, *527*, 118874. [CrossRef]
22. Waring, B.G.; Weintraub, S.R.; Sinsabaugh, R.L. Ecoenzymatic stoichiometry of microbial nutrient acquisition in tropical soils. *Biogeochemistry* **2014**, *117*, 101–113. [CrossRef]
23. Derakhshan-Babaei, F.; Mirchooli, F.; Mohammadi, M.; Nosrati, K.; Egli, M. Tracking the origin of trace metals in a watershed by identifying fingerprints of soils, landscape and river sediments. *Sci. Total Environ.* **2022**, *835*, 155583. [CrossRef]
24. Ren, H.; Zhou, Q.; He, J.; Hou, Y.; Jiang, Y.; Rodrigues, J.L.M.; Cobb, A.B.; Wilson, G.W.T.; Hu, J.; Zhang, Y. Determining landscape-level drivers of variability for over fifty soil chemical elements. *Sci. Total Environ.* **2019**, *657*, 279–286. [CrossRef] [PubMed]
25. Moreno-Jiménez, E.; Maestre, F.T.; Flagmeier, M.; Guirado, E.; Berdugo, M.; Bastida, F.; Dacal, M.; Díaz-Martínez, P.; Ochoa-Hueso, R.; Plaza, C.; et al. Soils in warmer and less developed countries have less micronutrients globally. *Glob. Change Biol.* **2023**, *29*, 522–532. [CrossRef]
26. Huggett, A.J. The concept and utility of 'ecological thresholds' in biodiversity conservation. *Biol. Conserv.* **2005**, *124*, 301–310. [CrossRef]

27. Li, W.; Xu, F.; Zheng, S.; Taube, F.; Bai, Y. Patterns and thresholds of grazing-induced changes in community structure and ecosystem functioning: Species-level responses and the critical role of species traits. *J. Appl. Ecol.* **2017**, *54*, 963–975. [CrossRef]
28. Nelson, D.W.; Sommers, L.E. Total Carbon, Organic Carbon, and Organic Matter. In *Methods of Soil Analysis. Part 2. Chem. Microbiological Properties*; Wiley: Hoboken, NJ, USA, 1982; pp. 539–579. [CrossRef]
29. Frank, K.; Beegle, D.; Denning, J. Phosphorus. In *Recommended Chemical Soil Test Procedures*; Brown, J.R., Ed.; North Central Regional Research Publication: Ames, IO, USA, 1998; Volume 221, pp. 21–30.
30. Lefcheck, J.S. PiecewiseSEM: Piecewise structural equation modelling in r for ecology, evolution, and systematics. *Methods Ecol. Evol.* **2015**, *7*, 573–579. [CrossRef]
31. Tian, P.; Liu, S.; Zhao, X.; Sun, Z.; Yao, X.; Niu, S.; Crowther, T.W.; Wang, Q. Past climate conditions predict the influence of nitrogen enrichment on the temperature sensitivity of soil respiration. *Commun. Earth Environ.* **2021**, *2*, 251. [CrossRef]
32. Augusto, L.; De Schrijver, A.; Vesterdal, L.; Smolander, A.; Prescott, C.; Ranger, J. Influences of evergreen gymnosperm and deciduous angiosperm tree species on the functioning of temperate and boreal forests. *Biol. Rev.* **2014**, *90*, 444–466. [CrossRef] [PubMed]
33. Garnier, E.; Laurent, G. Leaf anatomy, specific mass and water content in congeneric annual and perennial grass species. *New Phytol.* **1994**, *128*, 725–736. [CrossRef]
34. Ding, J.; Kong, D.; Zhang, Z.; Cai, Q.; Xiao, J.; Liu, Q.; Yin, H. Climate and soil nutrients differentially drive multidimensional fine root traits in ectomycorrhizal-dominated alpine coniferous forests. *J. Ecol.* **2020**, *108*, 2544–2556. [CrossRef]
35. Schroth, A.W.; Friedland, A.J.; Bostick, B.C. Macronutrient Depletion and Redistribution in Soils under Conifer and Northern Hardwood Forests. *Soil Sci. Soc. Am. J.* **2007**, *71*, 457–468. [CrossRef]
36. Bing, H.; Wu, Y.; Zhou, J.; Li, R.; Luo, J.; Yu, D. Vegetation and Cold Trapping Modulating Elevation-dependent Distribution of Trace Metals in Soils of a High Mountain in Eastern Tibetan Plateau. *Sci. Rep.* **2016**, *6*, 24081. [CrossRef]
37. Boyle, R.W. *Elemental Associations in Mineral Deposits and Indicator Elements of Interest in Geochemical Prospecting (Revised)*; Queen's Printer: Ottawa, ON, Canada, 1969.
38. Négrel, P.; Ladenberger, A.; Demetriades, A.; Reimann, C.; Birke, M.; Sadeghi, M. GEMAS: Boron as a geochemical proxy for weathering of European agricultural soil. *J. Geochem. Explor.* **2024**, *267*, 107618. [CrossRef]
39. Le Houérou, H.N. Biogeography of the arid steppeland north of the Sahara. *J. Arid. Environ.* **2001**, *48*, 103–128. [CrossRef]
40. Slimani, H.; Aidoud, A.; Rozé, F. 30 Years of protection and monitoring of a steppic rangeland undergoing desertification. *J. Arid. Environ.* **2010**, *74*, 685–691. [CrossRef]
41. Muñoz, M.Á.; Faz, Á. Soil and vegetation seasonal changes in the grazing Andean Mountain grasslands. *J. Mt. Sci.* **2014**, *11*, 1123–1137. [CrossRef]
42. Zhang, Y.; Li, Y.; Shi, F.; Sun, X.; Lin, G. Seasonal and spatial variation in species diversity, abundance, and element accumulation capacities of macroalgae in mangrove forests of Zhanjiang, China. *Acta Oceanol. Sin.* **2014**, *33*, 73–82. [CrossRef]
43. Ning, J.; Liu, S.; Kamran, M.; Sun, Y.; Xu, L.; Wang, H.; Zhang, M.; Chang, S.; West, C.P.; Hou, F. Trace elements apportionment in forage, soil, and livestock in rangeland ecosystems along climatic gradients. *Environ. Res.* **2022**, *215*, 114222. [CrossRef]
44. Wu, X.; Fang, H.; Zhao, Y.; Smoak, J.M.; Li, W.; Shi, W.; Sheng, Y.; Zhao, L.; Ding, Y. A conceptual model of the controlling factors of soil organic carbon and nitrogen densities in a permafrost-affected region on the eastern Qinghai-Tibetan Plateau. *J. Geophys. Res. Biogeosci.* **2017**, *122*, 1705–1717. [CrossRef]
45. Wu, X.; Zhao, L.; Fang, H.; Zhao, Y.; Smoak, J.M.; Pang, Q.; Ding, Y. Environmental controls on soil organic carbon and nitrogen stocks in the high-altitude arid western Qinghai-Tibetan Plateau permafrost region. *J. Geophys. Res. Biogeosci.* **2016**, *121*, 176–187. [CrossRef]
46. Jones, G.D.; Droz, B.; Greve, P.; Gottschalk, P.; Poffet, D.; McGrath, S.P.; Seneviratne, S.I.; Smith, P.; Winkel, L.H.E. Selenium deficiency risk predicted to increase under future climate change. *Proc. Natl. Acad. Sci. USA* **2017**, *114*, 2848–2853. [CrossRef]
47. Tolu, J.; Bouchet, S.; Helfenstein, J.; Hausheer, O.; Chékifi, S.; Frossard, E.; Tamburini, F.; Chadwick, O.A.; Winkel, L.H.E. Understanding soil selenium accumulation and bioavailability through size resolved and elemental characterization of soil extracts. *Nat. Commun.* **2022**, *13*, 6974. [CrossRef]
48. Meharg, A.A.; Meharg, C. The Pedosphere as a Sink, Source, and Record of Anthropogenic and Natural Arsenic Atmospheric Deposition. *Environ. Sci. Technol.* **2021**, *55*, 7757–7769. [CrossRef]
49. Bradl, H.B. Adsorption of heavy metal ions on soils and soils constituents. *J. Colloid. Interface Sci.* **2004**, *277*, 1–18. [CrossRef] [PubMed]
50. Khan, Z.I.; Ashraf, M.; Hussain, A.; McDowell, L.R. Seasonal Variation of Trace Elements in a Semiarid Veld Pasture. *Commun. Soil. Sci. Plant Anal.* **2006**, *37*, 1471–1483. [CrossRef]
51. Liu, W.; Wu, Y.; Zhong, Y.; Zhao, H. Concentrations, distribution and influencing factors of selenium (Se) in soil of arid and semi-arid climate: A case from Zhangye-Yongchang region, north-western China. *J. Geochem. Explor.* **2023**, *250*, 107239. [CrossRef]
52. Dinh, Q.T.; Cui, Z.; Huang, J.; Tran, T.A.T.; Wang, D.; Yang, W.; Zhou, F.; Wang, M.; Yu, D.; Liang, D. Selenium distribution in the Chinese environment and its relationship with human health: A review. *Environ. Int.* **2018**, *112*, 294–309. [CrossRef] [PubMed]

53. Mehdi, Y.; Hornick, J.L.; Istasse, L.; Dufrasne, I. Selenium in the environment, metabolism and involvement in body functions. *Molecules* **2013**, *18*, 3292–3311. [CrossRef]
54. Jie, L.; Zhiqiang, Y.; Zhigang, L.; Guodong, Z.; Yinghui, W.; Biao, C. Distribution of selenium in soils of Nanning city and its influencing factors. *Acta Pedol.* **2012**, *49*, 1012–1020. [CrossRef]

biology

MDPI

Article

Heavy Metal Contamination and Risk Assessment in Soil–Wheat/Corn Systems near Metal Mining Areas in Northwestern China

Shenghui Xu [1], Mingyang Yun [1], Yan Wang [1], Kaiwang Liu [1], Ao Wu [1], Shuning Li [1], Yanfang Su [2], Shengli Wang [3,*] and Hongmei Kang [4]

1 School of Civil and Hydraulic Engineering, Lanzhou University of Technology, Langongping Road 287, Lanzhou 730050, China; xush@lut.edu.cn (S.X.); yun15937813190@163.com (M.Y.); ywang0209@lut.edu.cn (Y.W.); gslzlkw@gmail.com (K.L.); wuao555@163.com (A.W.); lishuning0508@foxmail.com (S.L.)
2 Key Laboratory of Industrial Waste(s) Recycling and Regulation, Gansu Academy of Eco-Environmental Sciences, Lanzhou 730030, China; 13893481671@163.com
3 Gansu Key Laboratory for Environmental Pollution Prediction and Control, College of Earth and Environmental Sciences, Lanzhou University, Tianshui South Road 222, Lanzhou 730000, China
4 Institute of Biology, Gansu Academy of Sciences, Lanzhou 730000, China; lzkanghm@163.com
* Correspondence: wangshengl@lzu.edu.cn

Simple Summary

With the accelerating pace of modernization and industrialization, the rapid development of mining and metallurgical sectors has led to farmland soil contamination, consequently jeopardizing local ecosystems, food security, and residents' health. Heavy metals are known to pose risks to human health through the food chain. To comprehensively assess the potential hazards within the soil–crop system near smelting areas in Jinchang City, China, this study conducted multi-point field investigations and analyses of heavy metal distribution in farmland soils and crops around smelting facilities in this mining city. The research reveals the spatial distribution characteristics of Ni (nickel), Cu (copper), and Co (cobalt) in local soils, wheat, and corn, along with associated human health risks, while identifying key contaminated zones. These findings provide valuable reference for subsequent studies on the soil–crop system in the investigated area and establish fundamental data for remediation of heavy metal contamination in local farmland soils.

Abstract

Heavy metals in farmland soils pose severe threats to agricultural productivity and food safety. To investigate contamination in the soil–wheat/corn system, 24 sets of adjacent farmland soil, wheat, and corn plant samples were collected near metal smelting facilities in Jinchang City, a typical urban oasis in northwestern China. Concentrations of Ni (nickel), Cu (copper), and Co (cobalt) were measured. Results indicated mean soil concentrations of 143.66 mg kg^{-1} (Ni), 130.00 mg kg^{-1} (Cu), and 24.04 mg kg^{-1} (Co), all exceeding background values for Gansu Province, confirming that the sampling sites exhibit varying degrees of contamination with Ni, Cu, and Co. Correlation analyses revealed strong intermetal relationships (Ni, Cu, Co; $p < 0.01$), while spatial distribution patterns showed that Ni in wheat and corn grains closely mirrored soil Ni distribution. The bio-concentration factor (BCF) for wheat roots surpassed that of corn roots, highlighting wheat's greater susceptibility to heavy metal uptake. Heavy metal levels in crop organs exceeded limits set by the Safety Guidelines for Feed Additives. Geo-accumulation indices and potential ecological risk assessments demonstrated substantial metal accumulation and varying ecological risks, with contamination levels ranked as Cu > Ni > Co. Non-carcinogenic hazard indices indicated elevated health risks for children consuming locally grown wheat and corn. This study provides a scientific foundation for crop rotation strategies and soil remediation in the region.

Keywords: farmland soils; heavy metal contamination; spatial distribution; risk assessment; urban oasis

1. Introduction

The extraction, beneficiation, and smelting of non-ferrous metal deposits have driven industrial growth and economic development, yet simultaneously inflicted severe environmental pollution [1,2]. Rapid industrialization has led to soil heavy metal concentrations far exceeding regulatory thresholds, raising public concern over soil environmental safety [3–5]. Heavy metals from smelting slag and tailings infiltrate adjacent soils via surface runoff during rainfall, persisting indefinitely due to their non-biodegradable nature. Such pollution not only alters soil microbial community structure and function but may also contaminate groundwater through vertical permeation [6,7]. From an agricultural perspective, heavy metal pollution reduces crop yields or induces total harvest failure. Studies demonstrated that soil copper (Cu) contamination reduced wheat and corn germination rates by 10% and 16%, respectively, with wheat root length stunted to 0.03 cm and corn root growth rates declining 12-fold under Cu stress [8]. Human health risks are equally critical: excessive heavy metal intake triggers acute or chronic toxicity [9,10], causing irreversible damage to immune, reproductive, and nervous systems [11,12]. Notably, heavy metals enter humans via food chain accumulation or inhalation, with toxic element uptake in vegetables and crops linked not only to soil contamination but also to exogenous inputs such as atmospheric deposition, transportation emissions, and sewage sludge application [13]. For instance, long-term wastewater irrigation has led to the accumulation of Ni, Cd, and Cr in wheat grains in a suburban agricultural town in eastern Pakistan [14]. Wang et al. [15] identified atmospheric deposition and transportation as the dominant sources of heavy metals in the western Hexi Corridor. We employed the term "heavy metal" in this study. While we acknowledge the ongoing debate regarding its precise definition, the term has been retained to maintain consistency and ensure direct comparability with previous research.

Areas surrounding metal smelters and mines, subjected to prolonged heavy metal dust deposition and wastewater infiltration, represent critical hotspots for soil heavy metal accumulation. In Jiyuan City, Henan Province, heavy metal contamination in agricultural soils surrounding the Yuguang Smelter has been directly linked to prolonged metallurgical activities. This persistent pollution has resulted in non-carcinogenic health risks for all local children, constituting a significant health threat to residents [16]. Further investigations in Hunan Province revealed that manganese (Mn), cadmium (Cd), lead (Pb), copper (Cu), zinc (Zn), and chromium (Cr) levels in soils near manganese mines surpassed background values by factors of 16.3, 15.4, 15.0, 9.9, 6.1, and 1.1, respectively. Among these, Mn, Cd, and Pb exhibited severely contaminated levels, correlating with elevated potential ecological risks [17]. Notably, studies by Xiang et al. [3] in the Yangtze River Delta highlighted that cereals exhibited markedly stronger heavy metal bioaccumulation than vegetables or fruits, while children exhibit significantly greater health sensitivity to heavy metal exposure than adults. In response, the World Health Organization (W.H.O.) and national regulatory bodies have rigorously revised maximum permissible heavy metal concentrations in food products. To quantify non-carcinogenic health risks in mining and smelting zones, researchers widely employ hazard quotient (*HQ*) and hazard index (*HI*) models established by the US Environmental Protection Agency (USEPA), underscoring their global applicability in health risk assessment frameworks [15,18–20].

According to Wang et al. [15], metal processing and smelting activities are the primary sources of heavy metal contamination in agricultural soils across the Hexi Corridor, with the highest concentrations observed in central regions. Mining areas represent the most heavily contaminated land-use type [20]. Jinchang City, a major production base for nickel (Ni), copper (Cu), and cobalt (Co) in China and renowned as the "Nickel Capital", has experienced significant environmental degradation due to prolonged resource extraction [19,21]. Emissions from heavy metal-laden exhaust gases and dust from tailings ponds have driven the accumulation of toxic elements in farmland soils, posing potential risks to crops [17]. Wheat and corn, as staple crops in Jinchang, are critical subjects for investigating heavy metal pollution characteristics and conducting health risk assessments in peri-mining areas. Such research not only enhances public and scientific understanding of regional contamination but also provides actionable insights for mitigating heavy metal exposure and advancing environmental conservation, underscoring its practical significance.

This study aimed to (1) determine the concentrations and spatial distribution patterns of Ni, Cu, and Co in soils and staple crops (wheat and corn), and (2) evaluate and compare the degree of soil pollution and associated health risks posed by wheat and corn consumption.

2. Materials and Methods

2.1. Study Area

The study area was located in Jinchuan District, Jinchang City, Gansu Province, northwestern China (Figure 1), situated in the central Hexi Corridor and characterized by predominant northwest and southeast winds. The soil type was classified as sierozem. Jinchuan District comprises 348,700 mu (23,250 hectares) of urban, rural, and industrial/mining land, with cultivated land spanning 1.84 million mu (122,667 hectares). Crops, including wheat, corn, rapeseed, and legumes, are cultivated across 1.36 million mu (≈90,667 hectares) of sown area, of which wheat dominates, accounting for over 60% of total cultivation. The region hosts globally significant nickel–copper sulfide deposits, ranking third in reserves worldwide and first in China. Industrial activity began in the 1960s with the commissioning of China's first nickel smelting line. Today, the area contains one metallurgical plant, one copper slag concentrator facility, and two tailings ponds. Prolonged resource exploitation has reduced the region's environmental carrying capacity, compounded by agricultural practices such as fertilizer, pesticide, and livestock manure application. Consequently, soils near smelting facilities exhibit varying degrees of heavy metal contamination, with metallurgical dust and slag heaps identified as the most severe pollution sources [22].

2.2. Sample Collection and Pre-Treatment

Field sampling was conducted in 2022 prior to harvest in farmland soils adjacent to metal smelting facilities and along transportation routes in Jinchuan District (38°26′4″–38°38′16″ N, 102°11′21″–102°26′25″ E). Surface soil samples (0~15 cm depth) and corresponding wheat/corn plants were collected using a stainless-steel sampler. A total of 24 sampling sites were established, with 12 sites each for wheat (labeled w1–w12) and corn (c1–c12) (Figure 1). Soil and crop samples were transported to the laboratory in polyethylene bags. Soil samples were air-dried at room temperature (30 °C), gently ground, sieved through a 10-mesh sieve, and stored in resealable bags. Wheat and corn plants were dissected into five components: roots, stems, leaves, husks, and grains. Plant tissues were rinsed with deionized water, oven-dried at 70 °C to constant weight, pulverized using a grinder, homogenized into powder, and sieved through a 10-mesh sieve. Processed samples were stored in resealable bags for subsequent analytical testing.

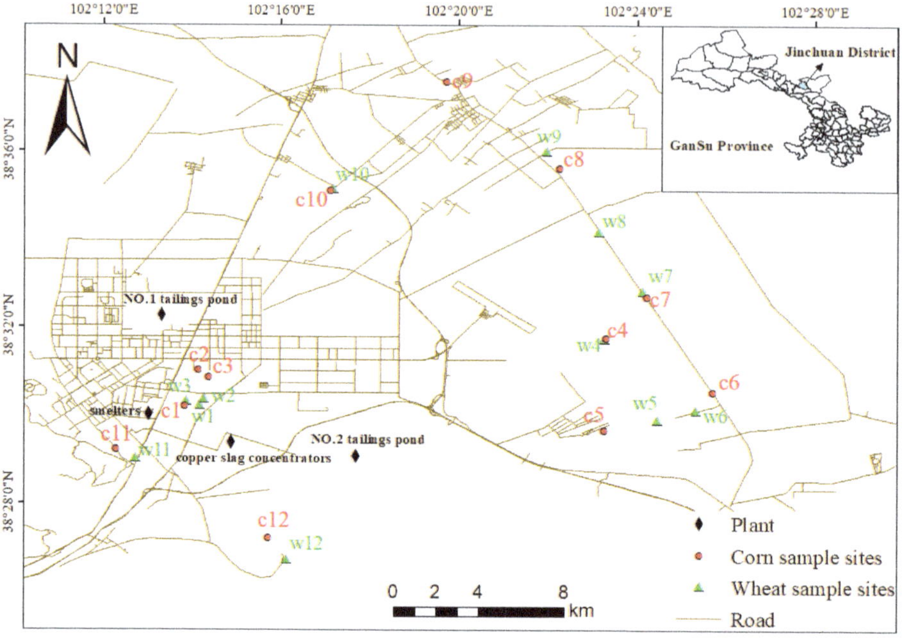

Figure 1. Distribution maps of sampling sites.

2.3. Chemical Analysis of Soils and Grains

Heavy metal concentrations in soil and grain samples were determined by atomic absorption spectroscopy (AAS, ZEEnit 700P, Jena Analytical Instruments, Jena, Germany). Soil samples were digested with a HF-HNO_3-H_2O_2 (2:3:1, $v/v/v$) mixture, while grain samples were digested with HNO_3-H_2O_2 (10:3, v/v). Following digestion and dilution, the solutions were analyzed by AAS. Soil pH and electrical conductivity (EC) were measured in a 1:2.5 soil–water suspension. Soil organic matter (OM) was quantified by the potassium dichromate volumetric method, and cation exchange capacity (CEC) was determined by the hexaamminecobalt(III) chloride method. Quality control was ensured by nationally certified reference materials GBW07386 (GSS-30) and GBW10023 (GSB-14), with recovery rates maintained at 95–105%. Triplicate measurements were conducted for all soil and crop tissue samples, and results were averaged.

2.4. Soil Heavy Metal Pollution Indices

2.4.1. Geo-Accumulation Index

The geo-accumulation index method intuitively evaluates the degree of heavy metal pollution, focusing on the impact of geological factors related to heavy metals and exogenous heavy metals generated by human activities [23,24]. The calculation formula is as follows:

$$I_{geo} = [C_i/(K \cdot B_i)] \quad (1)$$

where *Igeo* represents the geo-accumulation index for heavy metal i; Ci is the measured concentration of heavy metal i, mg kg^{-1}; Bi denotes the local geochemical background value of heavy metal i in the study area, mg kg^{-1}; K is the correction coefficient, set to 1.5. Based on *Igeo* values, soil heavy metal pollution levels are classified into six grades: no pollution, mild pollution, moderate pollution, moderately heavy pollution, severe pollution, and extreme pollution, and the corresponding ranges of *Igeo* values are: $Igeo < 0$, $0 \leq Igeo < 1$, $1 \leq Igeo < 2$, $2 \leq Igeo < 3$, $3 \leq Igeo < 4$, $Igeo \geq 4$.

2.4.2. Potential Ecological Risk Index

The potential ecological risk index (RI) evaluates the ecological risks posed by heavy metals in soils by integrating their ecological effects, toxicity, and environmental impacts [25,26]. The methodology is defined as follows:

$$RI = \sum_{i=1}^{n} E_r^i = T_i \times C_i \times B_i \quad (2)$$

where RI is the comprehensive potential ecological risk index, *Eri* is the potential ecological risk index for heavy metal i, Ti is the toxicity response coefficient of heavy metal i (assigned as Ti = 5 for Cu, Ni, and Co), and Ci is the measured concentration of heavy metal i, mg kg^{-1}. Bi denotes the local geochemical background value of heavy metal i in the study area, mg kg^{-1}. RI is judged as follows: low ecological risk (RI $\leq$ 120), moderate ecological risk (120 < RI $\leq$ 240), and considerable ecological risk (240 < RI $\leq$ 480). *Eri* is judged as: low ecological risk (*Eri* $\leq$ 40), moderate ecological risk (40 < *Eri* $\leq$ 80), and considerable ecological risk (80 < *Eri* $\leq$ 160).

2.5. Bio-Concentration Factor

The bioconcentration factor (BCF), a dimensionless metric, quantifies the transfer of heavy metals from soil to crop tissues and is calculated as follows:

$$BCF = \frac{C_w}{C_s} \quad (3)$$

where Cw (mg kg^{-1}) is the heavy metal concentration in crop tissues; Cs (mg kg^{-1}) is the heavy metal concentration in the rhizospheric soil.

2.6. Health Risk Assessment

Adopting deterministic risk assessment methods. The non-carcinogenic health risks associated with grain consumption were assessed using the health risk evaluation model established by the US Environmental Protection Agency (USEPA), quantified via the hazard index (*HI*) [27].

The long-term chronic daily intake of heavy metals was calculated as:

$$CDI_i = \frac{C_i \times IR \times EF \times ED}{BW \times AT} \quad (4)$$

where *CDI* is the chronic daily intake of heavy metals, mg $kg^{-1}d^{-1}$; Ci is the measured concentration of heavy metal i, mg kg^{-1}; *IR* is the grain ingestion rate, which was 0.412 kg day^{-1} and 0.206 kg day^{-1} for adults and children, respectively, for wheat; and 0.200 kg day^{-1} and 0.100 kg day^{-1} for adults and children, respectively, for corn, determined according to the China Statistical Yearbook 2022; *EF* is exposure frequency, 365 days $year^{-1}$; *ED* is exposure duration: 34 years for adults and 6 years for children; *BW* is the average body weight: 69.6 kg for adults and 15.0 kg for children, respectively; and *AT* stands for averaging time for exposure, with values of *ED* $\times$ 365 days $year^{-1}$.

The non-carcinogenic hazard index (*HI*) is calculated as follows:

$$HI = \sum_{i=1}^{n} HQ_i = \sum_{i=1}^{n} \frac{CDI_i}{RfD_i} \quad (5)$$

where *HI* is the non-carcinogenic hazard index, dimensionless; *HQi* is the non-carcinogenic hazard quotient for heavy metal i, dimensionless; *RfDi* is the reference dose for heavy metal I and the *RfD* values of Ni, Cu, and Co are 0.02, 0.04, and 0.0003 mg $kg^{-1}day^{-1}$, respectively. These values correspond to the soluble ionic forms. Typically, when $HI \leq 1$, it indicates negligible non-carcinogenic health risks to humans, while when $1 < HI \leq 10$, it suggests potential non-carcinogenic health risks to humans.

2.7. Statistical Analysis

Statistical analyses of soil and plant chemical data were conducted using Origin 2022, SPSS 27.0.1, and Microsoft Excel 2021. Sampling site distribution maps and spatial distribution patterns of heavy metals were generated with ArcGIS 10.8.

3. Results and Discussion

3.1. Soil Physicochemical Properties and Heavy Metal Concentrations

Descriptive statistics of soil properties and heavy metal concentrations in the study area are summarized in Table 1. For wheat-cultivated soils, Ni, Cu, and Co concentrations ranged from 31.65~367.40, 21.50~336.00, and 15.52~36.64 mg kg^{-1}, respectively, with mean values of 126.69, 118.41, and 23.61 mg kg^{-1}. Coefficients of variation (CV) for these metals followed the order Cu (96%) > Ni (91%) > Co (35%), where the high CVs for Cu and Ni indicate pronounced influence of local mining and metal smelting activities on soil Cu and Ni concentrations [28]. In corn-cultivated soils, Ni, Cu, and Co concentrations spanned 25.43~916.10, 24.10~682.00, and 13.70~50.38 mg kg^{-1}, respectively, with mean concentrations of 160.63, 141.58, and 24.47 mg kg^{-1}. The CV values exhibited a distinct order: Ni (152%) > Cu (134%) > Co (40%). Notably, Ni and Cu displayed exceptionally high variability (CV > 130%), reflecting extreme site-specific differences in their concentrations, likely attributable to anthropogenic inputs from industrial operations [29].

Table 1. Descriptive statistics of soil properties and heavy metal concentrations in agricultural soils of study area and related soil quality standards (mg kg^{-1}).

	Statistical Values	Ni	Cu	Co	pH	EC (μS cm^{-1})	OM (g kg^{-1})	CEC (cmol^{+} kg^{-1})
Wheat soil (n = 12)	Max	367.40	336.00	36.64	8.15	347.00	40.90	12.11
	Min	31.65	21.50	15.52	7.83	170.80	20.67	0.49
	Mean	126.69	118.41	23.61	8.00	253.53	29.90	5.00
	Standard deviation (SD)	115.57	113.23	8.17	0.10	56.75	6.41	3.94
	Coefficient of variation (CV)	0.91	0.96	0.35	0.01	0.22	0.21	0.79
Corn soil (n = 12)	Max	916.10	682.00	50.38	8.23	324.00	38.87	8.97
	Min	25.43	24.10	13.70	7.92	147.00	14.26	0.12
	Mean	160.63	141.58	24.47	8.13	193.72	28.74	3.92
	Standard deviation (SD)	244.83	189.64	9.77	0.08	54.10	6.65	2.62
	Coefficient of variation (CV)	1.52	1.34	0.40	0.01	0.28	0.23	0.67
Reference standard	Soil background values of Gansu Province	35.20	24.10	12.10				
	Risk screening value of agricultural land soil	190.00	100.00	——				

Compared to regional background values (Table 1), the mean concentrations of Ni, Cu, and Co in Jinchuan District significantly exceeded the soil background levels of Gansu Province. In wheat-cultivated soils, mean Ni, Cu, and Co concentrations were 3.60-, 4.91-, and 1.95-fold higher than local background values, respectively. For corn-cultivated soils, these ratios increased, respectively, to 4.56-, 5.87-, and 2.02-fold; this indicates that local soils have been contaminated to varying degrees. Furthermore, the mean heavy metal concentrations in this study surpassed those reported in industrially impacted regions such as Sargodha, eastern Pakistan [14], and Guilin City, Guangxi Province, China [30]. Notably, Ni concentrations at sampling sites w8, c6, c7, c8, and c10 were lower than the regional background value, likely due to their greater distance from smelting facilities and reduced industrial influence. According to Table 1, 29.17% of sampling sites exceeded China's risk screening values for Ni, while 37.50% surpassed the Cu threshold, suggesting potential risks to agricultural product safety. These anomalies are attributable to decades of intensive mining and metallurgical activities in the region. Previous work by Batool et al. [14] reported that in a suburban agricultural town of eastern Pakistan, prolonged wastewater irrigation has resulted in cadmium (Cd) accumulation in locally grown wheat exceeding safety thresholds, posing potential health risks to consumers. Studies by Lv et al. [17] identified metal smelting activities as the primary cause of heavy metal contamination in soils surrounding manganese mining areas in Hunan Province, China. Sierozem soils, characterized by organic matter and calcium carbonate content, can immobilize toxic heavy metals through adsorption, limiting their environmental mobility [31]. For instance, in karst soils of southwestern China, Cu predominantly binds to organic matter and carbonate phases, whereas Ni is primarily sequestered in goethite mineral matrices [32,33].

Soil analysis (Table 1) revealed weakly alkaline conditions in the study area, with pH ranging from 7.83 to 8.23. Low electrical conductivity (EC) and organic matter (OM) levels aligned with typical sierozem characteristics. The cation exchange capacity (CEC), reflecting soil nutrient retention potential, ranged from 0.12 to 12.11 cmol$^+$ kg^{-1} (mean: 4.46 cmol$^+$ kg^{-1}), indicating moderate fertility retention capacity. Heavy metal enrichment in soils is influenced by multiple factors, including geogenic background, pedogenic processes, and anthropogenic activities [34,35]. Correlation analyses (Table 2) were conducted to explore relationships among Ni, Cu, Co, pH, EC, OM, and CEC. Pearson correlation coefficients (Table 2) showed significant positive correlations ($p < 0.01$) among Ni, Cu, and Co in agricultural soils, suggesting shared anthropogenic origins and potential synergistic contamination effects. In wheat-cultivated soils, Ni, Cu, and Co exhibited positive correlations with OM ($p < 0.05$), indicating co-accumulation with organic matter. Conversely, no such trend was observed in corn soils. These divergent patterns align with findings by Lasota et al. [6], who reported that correlations between heavy metals and OM may vary across soil datasets. Humic substances in Silesia, Poland, can stabilize heavy metals via metal–organic ligand complexation, potentially explaining the OM-dependent accumulation in wheat soils [6]. Ni and Co in wheat soils showed strong positive correlations with CEC ($p < 0.01$), while Cu correlated moderately ($p < 0.05$). This phenomenon is likely attributable to the greater abundance of negatively charged sites on soil particles at higher CEC levels, which strengthens their adsorption capacity for heavy metals [36]. In corn soils, EC and pH displayed a significant negative correlation ($p < 0.01$), likely driven by reduced soluble salt content in high-pH soils, which lowers ionic conductivity [37].

Table 2. Pearson correlation coefficients between heavy metals and other parameters in soils.

Soils	Ni	Cu	Co	pH	EC	OM	CEC
Ni	1	0.986 **	0.935 **	−0.172	−0.339	0.240	−0.174
Cu	**0.987 ****	1	0.952 **	−0.158	−0.370	0.305	−0.201
Co	**0.972 ****	**0.970 ****	1	−0.233	−0.312	0.183	−0.204
pH	**0.091**	**−0.009**	**−0.048**	1	−0.781 **	0.437	−0.200
EC	**−0.191**	**−0.157**	**−0.174**	**−0.123**	1	−0.414	0.438
OM	**0.646 ***	**0.612 ***	**0.577 ***	**0.156**	**0.260**	1	0.295
CEC	**0.726 ****	**0.662 ***	**0.722 ****	**0.194**	**−0.197**	**0.484**	1

Bold represents the correlation coefficients of wheat, and the other stands for corn. ** $p < 0.01$. * $p < 0.05$.

3.2. Spatial Distribution of Heavy Metals in Soils and Crop Grains

Compared to other spatial interpolation methods, the inverse distance weighting (IDW) method more accurately predicts maximum and minimum values and provides better boundary details [38]. Figure 2 displays the spatial distribution maps of heavy metals in soils and crop grains generated using the IDW method. Co in wheat grains was not analyzed due to undetectable levels. Figure 2 shows that the spatial distribution trends of Ni, Cu, and Co in soils were similar, indicating shared sources of heavy metals, consistent with the correlation analysis results. The most severely contaminated soil sampling sites were located near the smelting plant. Based on the prevailing wind direction, the severe heavy metal contamination in soils northwest and southeast of the mining area was likely caused by dust deposition from tailings ponds under wind transport. Other sampling sites, farther from mining activities, showed lesser impacts. In agricultural soils of Anxin County, Hebei Province, the primary sources of Cd, Pb, Zn, and Cu contamination are industrial activities and mobile sources, demonstrating that heavy metals disperse from sources to surrounding areas under human influence [35]. Studies by Anaman et al. [39]. and Xu et al. [40]. highlighted transportation and atmospheric deposition as major factors affecting heavy metal distribution. The spatial distribution results confirm that local metal smelting activities dominate heavy metal contamination.

The spatial distribution of heavy metals in crop grains revealed that Ni distribution in grains closely resembled that in soils, while Cu and Co distributions differed significantly, suggesting soil Ni concentrations can predict grain Ni levels. For wheat grains, Cu hotspots were concentrated near the smelting plant and extended eastward, whereas corn grains showed Cu hotspots near the smelting plant and northeastward, with Co hotspots located south of the plant. The heavy metal content in wheat and corn grains is influenced by soil physicochemical properties, metal speciation, soil microbes and minerals, and crop root exudates [41], which likely explain the spatial discrepancies in Cu and Co distributions between grains and soils.

To further elucidate factors influencing heavy metal translocation from soils to wheat and corn grains, a correlation analysis was conducted between heavy metal concentrations in grains, soils, and soil properties (Table 3). First, Ni and Cu in corn grains showed a significant positive correlation ($p < 0.01$), suggesting shared origins. Notably, while Ni is toxic, Cu is an essential micronutrient for corn, implying potential overlap in their uptake or transport regulatory pathways [42]. This contrasts with findings by Pekel et al. [43], who reported no Ni-Cu correlation in corn grains, possibly due to varietal differences in metal absorption [10,44]. Second, Ni in grains correlated strongly with soil Ni, Cu, and Co ($p < 0.01$), indicating that elevated soil metal levels enhance Ni accumulation in grains, consistent with spatial distribution predictions. Conversely, grain Cu and Co showed no significant correlations with soil metals (Ni, Cu, Co), aligning with results from Wu et al. [45], which may reflect sub-threshold soil metal concentrations for phytotoxicity. According to Deng et al. [46], Ni and Co undergo continuous translocation between phloem and xylem, ultimately redistributing via phloem to plant tissues. Third, Ni in wheat grains correlated significantly with soil OM ($p < 0.05$) and CEC ($p < 0.01$), suggesting that organic chelators in soil enhance Ni bioavailability and plant uptake [47]. While Cu in grains and Co in corn grains showed no significant links to soil properties, the relationship between grain metals and bioavailable soil fractions warrants further exploration.

Table 3. Pearson correlation coefficients for metal concentrations in wheat and corn grains, total soil metal contents and soil properties.

Grain	Ni	Cu	Co
		Grain	
Ni	1	0.714 **	0.138
Cu	**0.429**	1	0.428
Co	**NA**	**NA**	1
		Soil	
Ni	**0.887 **** (0.831 **)	**0.540** (0.504)	**NA** (−0.243)
Cu	**0.817 **** (0.877 **)	**0.524** (0.559)	**NA** (−0.211)
Co	**0.782 **** (0.885 **)	**0.573** (0.499)	**NA** (−0.056)
		other	
pH	**0.466** (−0.243)	**0.168** (0.048)	**NA** (−0.059)
EC	**−0.240** (−0.339)	**0.260** (−0.486)	**NA** (−0.090)
OM	**0.686 *** (0.050)	**0.324** (0.190)	**NA** (−0.352)
CEC	**0.728 **** (−0.363)	**0.191** (−0.244)	**NA** (−0.078)

Bold represents the correlation coefficients of wheat, and the other stands for corn. ** $p < 0.01$. * $p < 0.05$. NA: not available.

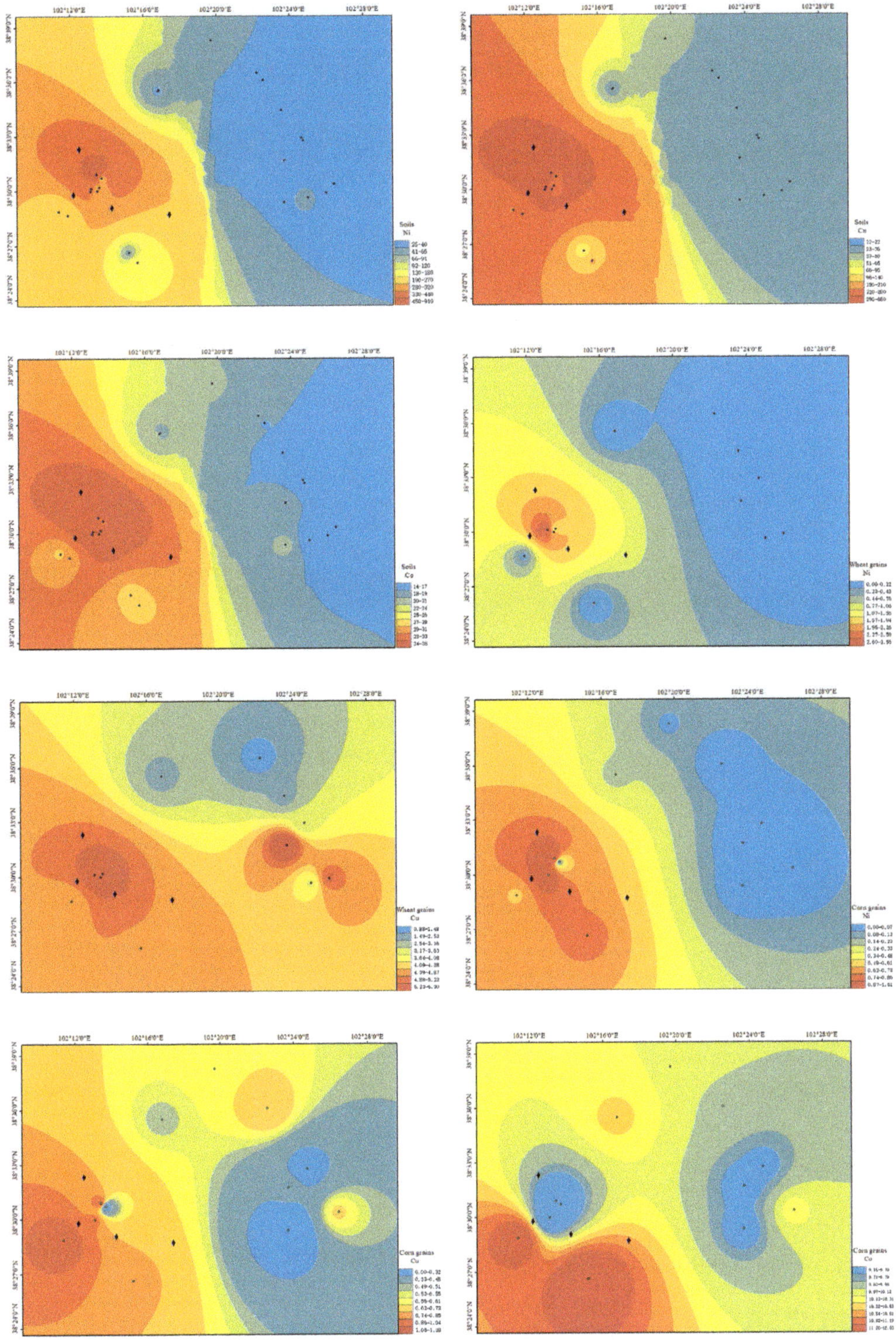

Figure 2. Spatial distribution maps of heavy metals in soils and grains.

3.3. BCFs of Ni, Cu and Co

Table 4 presents the mean bioconcentration factor (BCF) values across different tissues of wheat and maize plants. For wheat plants, the average BCFs of Ni, Cu, and Co followed the order: leaves > roots > husks > stems > grains, roots > leaves > grains > husks > stems and roots > leaves > stems > husks > grains, respectively. In corn plants, the orders were, respectively, leaves > roots > stems > husks > grains (Ni), leaves > roots > stems > husks > grains (Cu) and grains > husks > leaves > stems > roots (Co). BCF values indicated higher accumulation of Ni and Cu in roots and leaves compared to other tissues. Podar and Maathuis [48] reported that cereals inherently restrict heavy metal translocation from roots to edible aerial parts. Liu et al. [49] demonstrated that atmospheric heavy metals can infiltrate plants via foliar uptake. In this study, elevated Co levels in corn grains may stem from atmospheric dust deposition, with taller corn plants being more susceptible to airborne contamination. Studies by Ma et al. [50] and Xu et al. [19] consistently demonstrate that plant roots serve as the primary site of heavy metal accumulation. This may be attributed to the immobilization of heavy metals through chelation by root cell walls, which restricts their entry into the cytoplasm, or the presence of the Casparian strip that further impedes metal translocation [41].

Table 4. Average bio-concentration factor (BCF) of Ni, Cu and Co in different parts of crop plants.

Roots			**Stems**			**Leaves**			**Husks**			**Grains**		
Ni	Cu	Co	Ni	Cu	Co	Ni	Cu	Co	Ni	Cu	Co	Ni	Cu	Co
0.164	**0.215**	**0.381**	**0.011**	**0.061**	**0.149**	**0.251**	**0.204**	**0.258**	**0.078**	**0.071**	**0.119**	**0.002**	**0.079**	**0.000**
0.112	0.096	0.066	0.030	0.090	0.068	0.124	0.300	0.180	0.014	0.058	0.368	0.002	0.010	0.471

Bold represents the BCF of wheat, and the other stands for corn.

As shown in Table 4, the BCFs of Ni, Cu, and Co in wheat roots were 1.46-, 2.24-, and 5.77-fold higher than those in corn roots, respectively, indicating that wheat is more susceptible to heavy metal contamination. Similar conclusions were reported by Wu et al. [16] and Xue et al. [51] in agricultural regions of Henan and Hebei Provinces, China, suggesting that wheat exhibits greater sensitivity to heavy metal stress than corn, posing higher health risks to humans. Rezapour et al. [52] demonstrated that wheat exhibits stronger translocation of heavy metals from soil to roots than from roots to grains. In contrast, Romdhane et al. [53] found that corn preferentially transfers heavy metals to stems and leaves. To assess the safety of locally produced crop straw as livestock feed, mean heavy metal concentrations in wheat and corn tissues are presented in Table 5. Results indicated that, according to the Safety Guidelines for Feed Additives, Co concentrations in all wheat and corn tissues exceeded regulatory limits except for wheat grains and corn roots/stems. Similarly, Cu concentrations in wheat roots, corn leaves, and wheat leaves surpassed permissible thresholds, confirming that straw from these crops is unsuitable for use as standalone animal feed.

Table 5. Mean heavy metal concentrations in different parts of wheat and corn (mg kg^{-1}).

	Roots			**Stems**			**Leaves**			**Husks**			**Grains**		
	Ni	Cu	Co	Ni	Cu	Co	Ni	Cu	Co	Ni	Cu	Co	Ni	Cu	Co
Wheat	20.57 ± 21.11	22.93 ± 24.20	8.56 ± 2.41	1.50 ± 1.61	3.77 ± 2.30	3.00 ± 2.18	28.87 ± 25.82	22.88 ± 22.52	6.48 ± 4.16	7.25 ± 5.36	5.56 ± 4.42	2.73 ± 0.98	0.60 ± 1.16	4.11 ± 1.69	0.00 ± 0.00
Corn	13.18 ± 17.24	10.39 ± 11.35	1.75 ± 1.70	2.31 ± 1.81	5.37 ± 3.87	1.53 ± 0.64	16.19 ± 22.02	25.94 ± 26.90	4.44 ± 2.51	3.08 ± 5.53	5.38 ± 5.94	8.05 ± 0.73	0.36 ± 0.52	0.60 ± 0.43	10.20 ± 1.12

The data are expressed as the means ± standard deviations (mean ± SDs) of 12 sampling sites (n = 12). Code of practice for the safe use of feed additives: 15 mg kg^{-1} for Cu and 2 mg kg^{-1} for Co in feed, with no limit for Ni.

3.4. Soil Pollution and Health Risk Assessment

Table 6 summarizes the percentages of sampling sites classified under different geo-accumulation index (*Igeo*) and single-element potential ecological risk index (*Eri*) levels. According to *Igeo* criteria, 21% and 37% of sampling sites for Ni and Cu, respectively, exhibited contamination levels of moderately heavy pollution or higher, with 4% of sites for both metals reaching extreme pollution, indicating significant accumulation of Ni and Cu in regional soils. All Co-contaminated sites were classified as moderate pollution or lower. For *Eri*, 12% (Ni) and 29% (Cu) of sampling sites posed moderate ecological risk or higher, with 4% of sites for both metals categorized as considerable ecological risk, suggesting potential threats to human health via food chain accumulation. Co-associated risks remained low across all sites. Overall, the descending order of contamination and ecological risk levels for all metals was: Cu > Ni > Co.

Table 6. Percentage of sampling points in different *Igeo* and *Eri* risk levels in Jinchuan District.

	Class	Ni (%)	Cu (%)	Co (%)
Igeo	0	54	54	33
	1	8	8	58
	2	17	0	8
	3	17	25	0
	4	0	8	0
	5	4	4	0
E_r^i	<40	88	71	100
	40~80	8	25	0
	80~160	4	4	0

As shown in Table 6, the three elements in soils were significantly altered by production activities from mining enterprises near farmland, leading to severe spatial heterogeneity in heavy metal distribution. The potential ecological risk (RI) assessment (Figure 3A) revealed the highest ecological risk at sampling site c2, classified as considerable ecological risk, followed by c1 and w3 with moderate ecological risk, all located near local mining areas. Sites adjacent to mining zones (w1, w2, c3, w11, c11, w12, c12) exhibited elevated RI values within the low ecological risk category, further confirming mining activities as the primary contamination source. Elevated risks at other sites likely stem from persistent dust deposition linked to metal smelting and vehicular emissions [4]. Given the irreversible ecological impacts of heavy metal pollution in agricultural soils, immediate remediation measures are urgently required for this region.

To better understand contamination characteristics, *Igeo*, *Eri* and total heavy metal concentrations in soils were incorporated into principal component analysis (PCA). As shown in Figure 3B, two principal components were extracted: PC1, dominated by *Eri* and total metal concentrations, explained 93.6% of the total variance, while PC2, driven by *Igeo* values, accounted for 5.5%. Based on Figure 3B, the *Igeo* levels approximately ranked as Co > Cu > Ni, indicating multi-metal composite pollution with high similarity among the three metals, being consistent with spatial distribution predictions. The PCA results suggest that spatial heterogeneity of heavy metal concentrations is the primary factor influencing contamination patterns. The high explanatory power of PC1 underscores anthropogenic interventions, particularly mining and smelting activities, as key drivers of heavy metal distribution across the study area.

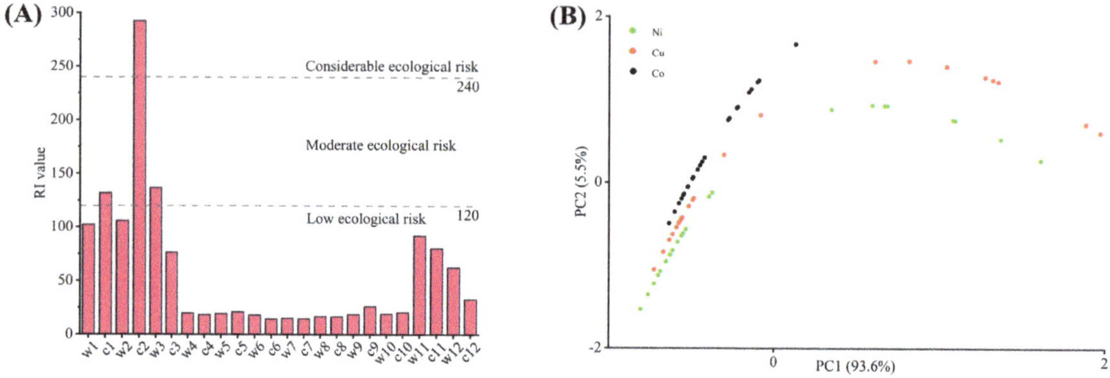

Figure 3. Results of the evaluation of potential ecological risks (**A**). PCA plot of total heavy metal concentration in soil and risks of *Igeo* and *Eri* (**B**).

Wheat and corn, as staple crops in Jinchuan District, Jinchang City, are closely linked to the health of local residents. Table 7 presents the chronic daily intake (*CDI*) of heavy metals for adults and children. Most studies indicated that children face higher non-carcinogenic risks from heavy metals in grains compared to adults [5,54]. In this study, wheat was identified as the primary source of Ni and Cu exposure, while Co exposure predominantly originated from corn. The *CDI* values for children were 2.32-fold higher than those for adults. The non-carcinogenic hazard index (*HI*) results are shown in Figure 4. *HI* values in this study revealed that children in Jinchuan District face greater non-carcinogenic health risks than adults, with the highest risk observed for wheat from site w3 (*HI* = 4.309), followed by w1 (3.279) and w2 (2.881). Notably, 16.67% of adults and 87.50% of children in the study area were exposed to non-carcinogenic health risks. Approximately 75% of children exhibited hazard quotients (*HQ*) > 1 for Cu in wheat, while all children showed *HQ* > 1 for Co in corn, indicating that Cu in wheat and Co in corn are the primary contributors to *HI* among the three heavy metals. Health risk assessments indicate that children face greater harm, likely due to their lower body weight and higher metabolic rates [55]. Furthermore, long-term heavy metal exposure may cause irreversible impairments to neurological and immune systems in children [11]—an urgent public health priority requiring immediate attention.

Table 7. *CDI* mean values of wheat and corn grains (mg kg^{-1} d^{-1}).

	CDI (Ni)		*CDI* (Cu)		*CDI* (Co)	
	Adults	**Children**	**Adults**	**Children**	**Adults**	**Children**
Wheat	3.55×10^{-3}	8.23×10^{-3}	2.43×10^{-2}	5.64×10^{-2}	0.00	0.00
Corn	1.02×10^{-3}	2.37×10^{-3}	1.73×10^{-3}	4.02×10^{-3}	2.93×10^{-2}	6.80×10^{-2}

For individuals living near metal smelting zones, heavy metals can also enter the human body through inhalation and dermal contact, potentially leading to higher non-carcinogenic health risks [56]. Based on this study, the following recommendations are proposed for local agricultural practices: In areas corresponding to sampling sites c8, c9, and c10, corn cultivation poses health risks exceeding safety thresholds, while wheat cultivation presents lower risks. Therefore, replacing corn with wheat or alternative crops in these regions is advised to reduce heavy metal exposure for residents. Other sampling sites are no longer suitable for growing wheat or corn intended for human consumption, and soil remediation measures should be implemented in these farmlands.

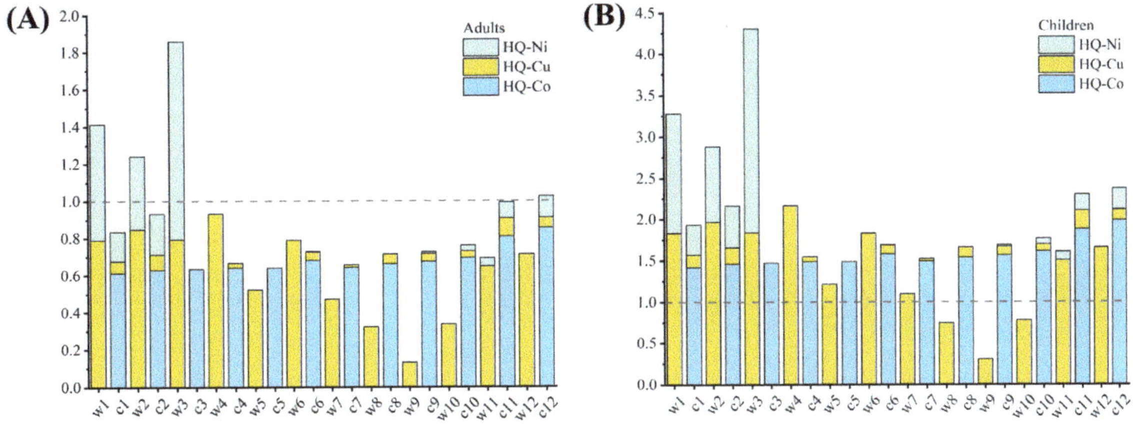

Figure 4. Non-Carcinogenic Health Risk Index (*HI*) for Adults (**A**) and Children (**B**).

4. Conclusions

The mean concentrations of Ni and Cu in the study area exceeded China's risk screening values for agricultural soils, while Co surpassed the background levels of Gansu Province. Correlation analyses revealed a high homology among Ni, Cu, and Co in soils, indicating shared anthropogenic origins. Spatial distribution patterns identified metal smelting facilities as the primary source of soil contamination. Ni in crop grains closely mirrored its spatial distribution in soils, enabling soil Ni levels to predict grain Ni contamination, whereas Cu and Co exhibited divergent grain-soil distribution trends. The correlation between grains and soils confirmed that soil is the dominant source of heavy metals in crops. Wheat demonstrated greater susceptibility to heavy metal contamination than corn, and straw from both crops was unsuitable for direct use as animal feed. Soil contamination levels followed the descending order: Cu > Ni > Co. Health risk assessments identified wheat as the principal contributor to Ni and Cu exposure, while corn dominated Co exposure. Given the elevated non-carcinogenic risks, cultivating wheat and corn for human consumption in the study area is no longer viable, necessitating immediate soil remediation.

This study not only confirms severe complex heavy metal contamination in the region, but also systematically reveals—for the first time—the shared sources and system characteristics of Ni, Cu, and Co pollution, their crop-specific uptake patterns, and the differential health risks they pose particularly to children. This study provides actionable insights for local agricultural restructuring and establishes a transferable risk management framework for mining-impacted cities globally. A key limitation lies in the use of total heavy metal concentrations for risk assessment. Future efforts should prioritize speciation analysis to determine bioavailability, alongside mechanistic investigations into crop-specific uptake and intra-grain translocation of heavy metal fractions.

Author Contributions: Conceptualization, S.X., S.W. and H.K.; methodology, M.Y.; software, M.Y. and Y.W.; validation, S.X. and S.W.; formal analysis, M.Y. and K.L.; investigation, S.X. and M.Y.; resources, S.X.; data curation, M.Y.; writing—original draft preparation, M.Y.; writing—review and editing, S.X., Y.W. and Y.S.; visualization, M.Y., A.W. and S.L.; supervision, S.X., S.W. and H.K.; project administration, S.X. and Y.W.; funding acquisition, S.X. All authors have read and agreed to the published version of the manuscript.

Funding: This research was funded by Key Research and Development Program of Gansu Academy of Sciences (2025ZDYF-02).

Institutional Review Board Statement: Not applicable.

Informed Consent Statement: Not applicable.

Data Availability Statement: The datasets generated for this study are available on request to the corresponding author.

Conflicts of Interest: The authors declare no conflicts of interest.

Abbreviations

Igeo	The geo-accumulation index
RI	The potential ecological risk index
BCF	The bioconcentration factor
HI	The hazard index
CDI	The chronic daily intake

References

1. Li, H.; Yao, J.; Min, N.; Duran, R. Comprehensive assessment of environmental and health risks of metal(loid) s pollution from non-ferrous metal mining and smelting activities. *J. Clean. Prod.* **2022**, *375*, 134049. [CrossRef]
2. Li, N.; Guo, Z.; Gao, C.; You, H.; Zhou, H.; Zhang, X.; Zheng, T. Mechanisms and pollution characteristics of heavy metal flow in China's non-ferrous metal industry. *J. Clean. Prod.* **2025**, *515*, 145791. [CrossRef]
3. Xiang, M.; Ma, J.; Cheng, J.; Lei, K.; Li, F.; Shi, Z.; Li, Y. Collaborative evaluation of heavy metal pollution of soil-crop system in the southeast of Yangtze River Delta, China. *Ecol. Indic.* **2022**, *143*, 109412. [CrossRef]
4. Liu, P.; Wu, Q.; Hu, W.; Tian, K.; Huang, B.; Zhao, Y. Effects of atmospheric deposition on heavy metals accumulation in agricultural soils: Evidence from field monitoring and Pb isotope analysis. *Environ. Pollut.* **2023**, *330*, 121740. [CrossRef]
5. Cai, L.-M.; Wang, Q.-S.; Luo, J.; Chen, L.-G.; Zhu, R.-L.; Wang, S.; Tang, C.-H. Heavy metal contamination and health risk assessment for children near a large Cu-smelter in central China. *Sci. Total Environ.* **2019**, *650*, 725–733. [CrossRef] [PubMed]
6. Lasota, J.; Błońska, E.; Łyszczarz, S.; Tibbett, M. Forest Humus Type Governs Heavy Metal Accumulation in Specific Organic Matter Fractions. *Water Air Soil Pollut.* **2020**, *231*, 80. [CrossRef]
7. Abou Fayssal, S.; Kumar, P.; Popescu, S.M.; Khanday, M.U.-D.; Sardar, H.; Ahmad, R.; Gupta, D.; Gaur, S.K.; Alharby, H.F.; Al-Ghamdi, A.G. Health risk assessment of heavy metals in saffron (*Crocus sativus* L.) cultivated in domestic wastewater and lake water irrigated soils. *Heliyon* **2024**, *10*, e27138. [CrossRef]
8. Vasilachi-Mitoseru, I.-C.; Stoleru, V.; Gavrilescu, M. Integrated Assessment of Pb(II) and Cu(II) Metal Ion Phytotoxicity on *Medicago sativa* L., *Triticum aestivum* L., and *Zea mays* L. Plants: Insights into Germination Inhibition, Seedling Development, and Ecosystem Health. *Plants* **2023**, *12*, 3754. [CrossRef]
9. Ene, A.; Moraru, S.S.; Moraru, D.I.; Pantelica, A.; Gosav, S.; Ceoromila, A.M. Major and Trace Element Accumulation in Soils and Crops (Wheat, Corn, Sunflower) around Steel Industry in the Lower Danube Basin and Associated Ecological and Health Risks. *Appl. Sci.* **2024**, *14*, 5616. [CrossRef]
10. Amjad, M.; Raza, H.; Murtaza, B.; Abbas, G.; Imran, M.; Shahid, M.; Naeem, M.A.; Zakir, A.; Iqbal, M.M. Nickel Toxicity Induced Changes in Nutrient Dynamics and Antioxidant Profiling in Two Maize (*Zea mays* L.) Hybrids. *Plants* **2020**, *9*, 5. [CrossRef]
11. Zhou, L.; Zhou, L.; Wu, H.; Li, J.; Kong, L.; Yang, H. Effects of Applying Biochar on Soil Cadmium Immobilisation and Cadmium Pollution Control in Lettuce (*Lactuca sativa* L.). *Agriculture* **2024**, *14*, 1068. [CrossRef]
12. Hussain, I.; Irshad, M.; Hussain, A.; Qadir, M.; Mehmood, A.; Rahman, M.; Alrefaei, A.F.; Almutairi, M.H.; Ali, S.; Hamayun, M. Phosphate solubilizing *Aspergillus Niger* PH1 ameliorates growth and alleviates lead stress in maize through improved photosynthetic and antioxidant response. *BMC Plant Biol.* **2024**, *24*, 642. [CrossRef]
13. Elbagory, M.; Abd El-Aziz, M.A.; Omara, A.E.-D.; Abou Fayssal, S.; Kumar, V. Assessment of Potentially Toxic Elements in Four Melon Fruit Varieties Grown in the Ganges and Yamuna River Basin. *Horticulturae* **2025**, *11*, 216. [CrossRef]
14. Batool, F.; Hussain, M.I.; Nazar, S.; Bashir, H.; Khan, Z.I.; Ahmad, K.; Alnuwaiser, M.A.; Yang, H.-H. Potential of sewage irrigation for heavy metal contamination in soil-wheat grain system: Ecological risk and environmental fate. *Agric. Water Manag.* **2023**, *278*, 108144. [CrossRef]
15. Wang, F.; Guan, Q.; Tian, J.; Lin, J.; Yang, Y.; Yang, L.; Pan, N. Contamination characteristics, source apportionment, and health risk assessment of heavy metals in agricultural soil in the Hexi Corridor. *Catena* **2020**, *191*, 104573. [CrossRef]
16. Wu, H.; Yang, F.; Li, H.; Li, Q.; Zhang, F.; Ba, Y.; Cui, L.; Sun, L.; Lv, T.; Wang, N.; et al. Heavy metal pollution and health risk assessment of agricultural soil near a smelter in an industrial city in China. *Int. J. Environ. Health Res.* **2020**, *30*, 174–186. [CrossRef] [PubMed]

17. Lv, Y.; Kabanda, G.; Chen, Y.; Wu, C.; Li, W. Spatial distribution and ecological risk assessment of heavy metals in manganese (Mn) contaminated site. *Front. Environ. Sci.* **2022**, *10*, 942544. [CrossRef]
18. Ogarekpe, N.M.; Nnaji, C.C.; Oyebode, O.J.; Ekpenyong, M.G.; Ofem, O.I.; Tenebe, I.T.; Asitok, A.D. Groundwater quality index and potential human health risk assessment of heavy metals in water: A case study of Calabar metropolis, Nigeria. *Environ. Nanotechnol. Monit. Manag.* **2023**, *19*, 100780. [CrossRef]
19. Xu, S.; Wu, A.; Liu, J.; Gao, G.; Liu, K.; Su, Y.; Jin, C.; Li, C.; Yang, Y. Spatiotemporal Distribution, and Risk Assessment of Cu and Ni Pollution in a Farmland Soil-Corn System of Arid Oasis City in Northwest China. *Soil Sediment Contam.* **2025**, *34*, 2110–2131. [CrossRef]
20. Shi, J.; Zhao, D.; Ren, F.; Huang, L. Spatiotemporal variation of soil heavy metals in China: The pollution status and risk assessment. *Sci. Total Environ.* **2023**, *871*, 161768. [CrossRef]
21. Xu, S.; Li, C.; Wang, Y.; Wu, A.; Gao, G.; Zang, F. Characteristics and evaluation of heavy metal pollution in a soil-wheat system of an arid oasis city in northwest China. *Ecotoxicol. Environ. Saf.* **2024**, *271*, 115958. [CrossRef]
22. Xu, S.; Wu, A.; Yun, M.; Liu, K.; Li, J. Tolerance and remediation potential of aquatic plants for copper, nickel, lead and zinc in heavy metal-contaminated waters. *Water Sci. Technol.* **2025**, *92*, 967–982. [CrossRef] [PubMed]
23. Jiang, W.; Meng, L.; Liu, F.; Sheng, Y.; Chen, S.; Yang, J.; Mao, H.; Zhang, J.; Zhang, Z.; Ning, H. Distribution, source investigation, and risk assessment of topsoil heavy metals in areas with intensive anthropogenic activities using the positive matrix factorization (PMF) model coupled with self-organizing map (SOM). *Environ. Geochem. Health* **2023**, *45*, 6353–6370. [CrossRef] [PubMed]
24. Zheng, F.; Guo, X.; Tang, M.; Zhu, D.; Wang, H.; Yang, X.; Chen, B. Variation in pollution status, sources, and risks of soil heavy metals in regions with different levels of urbanization. *Sci. Total Environ.* **2023**, *866*, 161355. [CrossRef]
25. Nie, G.; Tu, T.; Hu, L.; Wu, L.; Zhou, Y. Accumulation characteristics and evaluation of heavy metals in soils and vegetables of plastic-covered sheds in typical red soil areas of China. *Qual. Assur. Saf. Crops Foods* **2023**, *15*, 22–35. [CrossRef]
26. Xiang, Q.; Yu, H.; Chu, H.; Hu, M.; Xu, T.; Xu, X.; He, Z. The potential ecological risk assessment of soil heavy metals using self-organizing map. *Sci. Total Environ.* **2022**, *843*, 156978. [CrossRef]
27. Kumar, P.; Alhag, S.K.; Al-Shahari, E.A.; Al-Fakeh, M.S.; Fayssal, S.A.; Bachheti, R.K.; Siric, I.; Eid, E.M. Impact of Irrigation with Contaminated River Water on Growth, Yield, and Heavy Metals Accumulation in Planted Armenian Cucumber (*Cucumis melo var. flexuosus*(L.) Naudin.). *Water Air Soil Pollut.* **2025**, *236*, 1. [CrossRef]
28. Zhao, X.; He, B.; Wu, H.; Zheng, G.; Ma, X.; Liang, J.; Li, P.; Fan, Q. A comprehensive investigation of hazardous elements contamination in mining and smelting-impacted soils and sediments. *Ecotoxicol. Environ. Saf.* **2020**, *192*, 110320. [CrossRef]
29. Shamsaddin, H.; Jafari, A.; Jalali, V.; Schulin, R. Spatial distribution of copper and other elements in the soils around the Sarcheshmeh copper smelter in southeastern Iran. *Atmos. Pollut. Res.* **2020**, *11*, 1681–1691. [CrossRef]
30. Xiao, H.; Shahab, A.; Xi, B.; Chang, Q.; You, S.; Li, J.; Sun, X.; Huang, H.; Li, X. Heavy metal pollution, ecological risk, spatial distribution, and source identification in sediments of the Lijiang River, China. *Environ. Pollut.* **2021**, *269*, 116189. [CrossRef]
31. Zhao, X.; Takahashi, Y.; Wu, W.; Liu, C.; Fan, Q. Speciation of Zn and Cd in sierozem soil, northwest China: Bulk EXAFS and micro synchrotron X-ray fluorescence. *Environ. Sci. Process. Impacts* **2023**, *25*, 954–963. [CrossRef]
32. Wang, Y.; Jing, J.; Li, Y.; Zhang, Y.; Liu, Y. Bioavailability and Speciation of Potentially Toxic Trace Metals in Limestone-Derived Soils in a Karst Region, Southwestern China. *Water Air Soil Pollut.* **2025**, *236*, 169. [CrossRef]
33. Wei, N.; Gu, X.; Wen, Y.; Guo, C.; Ji, J. Geochemical speciation and activation risks of Cd, Ni, and Zn in soils with naturally high background in karst regions of southwestern China. *J. Hazard. Mater.* **2025**, *486*, 137100. [CrossRef]
34. Wen, Y.; Li, W.; Yang, Z.; Zhang, Q.; Ji, J. Enrichment and source identification of Cd and other heavy metals in soils with high geochemical background in the karst region, Southwestern China. *Chemosphere* **2020**, *245*, 125620. [CrossRef]
35. Zhao, G.; Ma, Y.; Liu, Y.; Cheng, J.; Wang, X. Source analysis and ecological risk assessment of heavy metals in farmland soils around heavy metal industry in Anxin County. *Sci. Rep.* **2022**, *12*, 10562. [CrossRef]
36. Wang, Y.; Zhang, Z.; Li, Y.; Liang, C.; Huang, H.; Wang, S. Available heavy metals concentrations in agricultural soils: Relationship with soil properties and total heavy metals concentrations in different industries. *J. Hazard. Mater.* **2024**, *471*, 134410. [CrossRef] [PubMed]
37. Ismayilov, A.I.; Mamedov, A.I.; Fujimaki, H.; Tsunekawa, A.; Levy, G.J. Soil Salinity Type Effects on the Relationship between the Electrical Conductivity and Salt Content for 1:5 Soil-to-Water Extract. *Sustainability* **2021**, *13*, 3395. [CrossRef]
38. Xue, W.; Pangara, C.; Aung, A.M.; Yu, S.; Tabucanon, A.S.; Hong, B.; Kurniawan, T.A. Spatial changes of nutrients and metallic contaminants in topsoil with multi-geostatistical approaches in a large-size watershed. *Sci. Total Environ.* **2022**, *824*, 153888. [CrossRef]
39. Anaman, R.; Peng, C.; Jiang, Z.; Liu, X.; Zhou, Z.; Guo, Z.; Xiao, X. Identifying sources and transport routes of heavy metals in soil with different land uses around a smelting site by GIS based PCA and PMF. *Sci. Total Environ.* **2022**, *823*, 153759. [CrossRef]
40. Xu, Z.; Mi, W.; Mi, N.; Fan, X.; Zhou, Y.; Tian, Y. Characteristics and sources of heavy metal pollution in desert steppe soil related to transportation and industrial activities. *Environ. Sci. Pollut. Res.* **2020**, *27*, 38835–38848. [CrossRef] [PubMed]

41. Vasilachi, I.C.; Stoleru, V.; Gavrilescu, M. Analysis of Heavy Metal Impacts on Cereal Crop Growth and Development in Contaminated Soils. *Agriculture* **2023**, *13*, 1983. [CrossRef]
42. Chen, W.; Li, X.; Zhang, X.; Chachar, Z.; Lu, C.; Qi, Y.; Chang, H.; Wang, Q. Genome-wide association study of trace elements in maize kernels. *BMC Plant Biol.* **2024**, *24*, 724. [CrossRef] [PubMed]
43. Pekel, A.Y.; Calik, A.; Alatas, M.S.; Kuter, E.; Cengiz, O.; Omurtag, G.Z.; Inan, G. Evaluation of Correlations Between Nutrients, Fatty Acids, Heavy Metals, and Deoxynivalenol in Corn (*Zea mays* L.). *J. Appl. Poult. Res.* **2019**, *28*, 94–107. [CrossRef]
44. Zha, Y.; Zhao, L.; Niu, T.; Yue, E.; Wang, X.; Shi, J. Multi-Target Element-Based Screening of Maize Varieties with Low Accumulation of Heavy Metals (HMs) and Metalloids: Uptake, Transport, and Health Risks. *Agriculture* **2023**, *13*, 1123. [CrossRef]
45. Wu, P.; Guo, Z.; Hua, K.; Wang, D. Long-term application of organic amendments changes heavy metals accumulation in wheat grains by affecting soil chemical properties and wheat yields. *J. Soils Sediments* **2023**, *23*, 2136–2147. [CrossRef]
46. Deng, T.-H.-B.; Chen, J.-Q.; Geng, K.-R.; van der Ent, A.; Tang, Y.-T.; Wen, D.; Wang, X.; Li, L.; Du, R.-Y.; Morel, J.-L.; et al. Quantification of nickel and cobalt mobility and accumulation via the phloem in the hyperaccumulator *Noccaea caerulescens*(Brassicaceae). *Metallomics* **2021**, *13*, mfab012. [CrossRef]
47. Wang, Q.; Huang, Q.; Guo, G.; Qin, J.; Luo, J.; Zhu, Z.; Hong, Y.; Xu, Y.; Hu, S.; Hu, W.; et al. Reducing bioavailability of heavy metals in contaminated soil and uptake by maize using organic-inorganic mixed fertilizer. *Chemosphere* **2020**, *261*, 128122. [CrossRef]
48. Podar, D.; Maathuis, F.J.M. The role of roots and rhizosphere in providing tolerance to toxic metals and metalloids. *Plant Cell Environ.* **2022**, *45*, 719–736. [CrossRef] [PubMed]
49. Liu, P.; Li, L.; Ippolito, J.A.; Xing, W.; Wang, Y.; Wang, Y.; Cheng, Y.; Qiu, K. Heavy metal distribution in wheat plant components following foliar Cd application. *Chemosphere* **2023**, *322*, 138177. [CrossRef] [PubMed]
50. Ma, J.F.; Shen, R.F.; Shao, J.F. Transport of cadmium from soil to grain in cereal crops: A review. *Pedosphere* **2021**, *31*, 3–10. [CrossRef]
51. Xue, P.; Zhao, Q.; Sun, H.; Geng, L.; Yang, Z.; Liu, W. Characteristics of heavy metals in soils and grains of wheat and maize from farmland irrigated with sewage. *Environ. Sci. Pollut. Res.* **2019**, *26*, 5554–5563. [CrossRef] [PubMed]
52. Rezapour, S.; Atashpaz, B.; Moghaddam, S.S.; Damalas, C.A. Heavy metal bioavailability and accumulation in winter wheat (*Triticum aestivum* L.) irrigated with treated wastewater in calcareous soils. *Sci. Total Environ.* **2019**, *656*, 261–269. [CrossRef] [PubMed]
53. Romdhane, L.; Panozzo, A.; Radhouane, L.; Dal Cortivo, C.; Barion, G.; Vamerali, T. Root Characteristics and Metal Uptake of Maize (*Zea mays* L.) under Extreme Soil Contamination. *Agronomy* **2021**, *11*, 178. [CrossRef]
54. Ahmad, W.; Alharthy, R.D.; Zubair, M.; Ahmed, M.; Hameed, A.; Rafique, S. Toxic and heavy metals contamination assessment in soil and water to evaluate human health risk. *Sci. Rep.* **2021**, *11*, 17006. [CrossRef]
55. Gao, S.; Yang, L.; Li, Y.; Liu, S.; Zhang, H.; Arens, E.; Zhai, Y. Metabolic rate in children and adolescents: Tabulate values for common activities and comparisons with standards and adult values. *Build. Environ.* **2023**, *244*, 110804. [CrossRef]
56. Khoshakhlagh, A.H.; Ghobakhloo, S.; Peijnenburg, W.J.; Gruszecka-Kosowska, A.; Cicchella, D. To breathe or not to breathe: Inhalational exposure to heavy metals and related health risk. *Sci. Total Environ.* **2024**, *932*, 172556. [CrossRef] [PubMed]

biology

MDPI

Article

Microbial Community Imbalance Drives Nitrous Oxide Emissions from Strongly Acidic Soil—Insights from a Laboratory Experiment with Microbial Inhibitors

Waqar Ahmed [1,2,†], Hongyang Gong [1,2,†], Xiaoxiao Xiang [2], Runze Chen [2], Yumeng Xu [2], Wenxuan Shi [3], Binzhe Li [4], Junhui Yin [1,*] and Qing Chen [2]

1 State Key Lab of Biocontrol, Guangdong Provincial Key Laboratory of Plant Stress Biology, School of Agriculture and Biotechnology, Shenzhen Campus of Sun Yat-sen University, Shenzhen 518107, China; waqarahmedupr@gmail.com (W.A.); hygong@cau.edu.cn (H.G.)
2 College of Resources and Environmental Sciences, China Agricultural University, Beijing 100193, China; xiangxx@cau.edu.cn (X.X.); crz15751136686@163.com (R.C.); 13470700223@163.com (Y.X.); qchen@cau.edu.cn (Q.C.)
3 Teagasc, Environmental Research Centre, Johnstown Castle, Y35 TC97 Co. Wexford, Ireland; wenxuan.shi@teagasc.ie
4 College of Engineering (Key Laboratory for Clean Renewable Energy Utilization Technology, Ministry of Agriculture), China Agricultural University, Beijing 100083, China; libinzhe@cau.edu.cn
* Correspondence: yinjh9@mail.sysu.edu.cn
† These authors contributed equally to this work.

Simple Summary

This study examined how adding different amounts of two microbial inhibitors (streptomycin and cycloheximide) affects nitrous oxide (N_2O) emissions in strongly acidic soil. High concentrations of streptomycin (6 and 10 mg g^{-1}) reduced N_2O emissions, whereas lower concentrations (2 and 4.5 mg g^{-1}) increased emissions. Cycloheximide initially elevated N_2O emissions compared to the control, but emissions decreased as the concentration of cycloheximide increased. The inhibitors also altered soil nitrogen and carbon levels. This study highlights that microbial inhibitors can disrupt soil microbial communities, leading to changes in N_2O emissions. High concentrations of inhibitors can reduce emissions by suppressing certain microbes, but lower concentrations may cause surviving microbes to produce more N_2O.

Abstract

Nitrous oxide (N_2O) is a potent greenhouse gas with intensive emissions from acidic soil. This study explored the impact of the disruption of the microbial balance from microbial inhibitors (streptomycin and cycloheximide) on soil's N_2O emission and nitrogen (N) dynamics. Under all the conditions examined, biotic processes accounted for 96–98% of total N_2O emissions. High concentrations of streptomycin (6 and 10 mg g^{-1}) reduced N_2O emissions from 2.24 μg kg^{-1} h^{-1} to 1.93 μg kg^{-1} h^{-1} and 2.12 μg kg^{-1} h^{-1}, respectively, whereas lower concentrations (2 and 4.5 mg g^{-1}) increased emissions from 2.24 μg kg^{-1} h^{-1} to 2.95 μg kg^{-1} h^{-1} and 3.27 μg kg^{-1} h^{-1}, respectively. Lower cycloheximide (2 and 4.5 mg g^{-1}) significantly enhanced N_2O emissions, reaching 9.15 μg kg^{-1} h^{-1} and 5.68 μg kg^{-1} h^{-1}, respectively, whereas higher dosages (6 mg g^{-1} and 10 mg g^{-1}) inhibited N_2O emissions, reducing them to 5.55 μg kg^{-1} h^{-1} and 4.84 μg kg^{-1} h^{-1}, respectively. Carbon dioxide (CO_2) emissions generally decreased with increasing inhibitor dosages but significantly increased at 2 mg g^{-1} and 4.5 mg g^{-1} streptomycin. The inhibitors also altered soil N and carbon (C) dynamics, increasing ammonium (NH_4^+-N), dissolved organic nitrogen (DON), and dissolved organic carbon (DOC) levels. Pearson correlation analysis indicated that

N_2O emission was negatively correlated with cycloheximide dosage (R = −0.68, $p < 0.001$), NH_4^+-N (R = −0.31, $p < 0.001$) and DOC content (R = −0.57, $p < 0.05$). These findings highlight the consequences of microbial disruption on N_2O emission and the complex microbial interactions in acidic soils. High concentrations of microbial inhibitors effectively reduce N_2O emissions by suppressing key microbial groups in nitrification and denitrification. Conversely, lower concentrations may prompt compensatory responses from surviving microorganisms, resulting in increased N_2O production. Future research should focus on sustainable management strategies to mitigate N_2O emissions while preserving the soil's microbial community.

Keywords: nitrous oxide; microbial inhibition; nitrogen dynamics; dissolved organic carbon; acidic soil

1. Introduction

The mitigation of greenhouse gas (GHG) emissions is one of the most pressing global challenges [1]. Nitrous oxide (N_2O) is the third most significant contributor to global warming and a major ozone-depleting substance [2,3]. Agricultural soils are the primary source of anthropogenic N_2O emissions, accounting for approximately 60% of global emissions [4]. Soil pH is the major factor affecting soil N_2O emission, with some studies showing that acidic soil is an important emission source of N_2O [5]. It has been reported that acidic soil (pH 5.0) emitted three times more N_2O than alkaline soil (pH 8.0) with a similar application of nitrogen (N) fertilizer [6]. Soil acidification is an increasing issue as more than half of worldwide arable land is becoming acidic [7]. Understanding the mechanisms driving N_2O emission in acidic soils is of paramount importance for developing effective N_2O emission mitigation strategies.

N_2O emission in acidic soils is driven by a combination of biological and abiotic pathways [8]. Chemical processes, such as chemodenitrification, where reactive N intermediates (e.g., HNO_2) decompose abiotically under low pH conditions, have been identified as significant contributors to N_2O emission in acidic environments [9]. For example, research in tropical soils has attributed over 50% of N_2O fluxes to abiotic reactions involving Fe^{2+}-mediated nitrite reduction [10,11]. However, emerging evidence highlights the resilience and adaptability of acid-tolerant microbial communities in driving N_2O production [12]. Prokaryotic processes, including bacterial nitrification, denitrification, and nitrifier denitrification, remain active even at pH < 4.5, facilitated by acidophilic taxa such as *Acidithiobacillus* and *Acidobacteriota* [13]. Eukaryotic contributions, particularly fungal denitrification, are increasingly recognized as critical, as fungi exhibit greater tolerance to acidic conditions than many bacteria [14]. However, the interplay between prokaryotic and eukaryotic communities in regulating N_2O flux, and how their functional redundancy or competition modulates overall emissions, remains unresolved. This knowledge gap hinders the development of microbial-focused mitigation strategies.

Soil microorganisms coexist in a dynamic equilibrium, each contributing to nutrient cycling and soil structure maintenance [15]. Agricultural management practices, such as liming, the application of organic amendments, and pesticide use, strongly disrupt the balance of microbial populations, thus influencing N_2O emissions [16–18]. For instance, an increased pH decreases the contribution of fungi to N_2O but increases bacterial N_2O production in acidic soil [19]. The addition of manure enhances N availability and organic C, resulting in stimulating nitrification and denitrification processes dominated by prokaryotes (e.g., *Nitrosomonas* and *Pseudomonas*) [20]. Meanwhile, the application of bactericide and fungicide influences N cycling; the effects depend on soil type and conditions [21]. Disease control measures, especially the application of pesticides, indirectly influence the interaction between prokaryotes and eukaryotes, highlighting the need to disentangle these microbial networks to refine N_2O mitigation strategies [22]. However, inhibitor effects are dose-dependent and may inadvertently stimulate non-target microbial groups or modify substrate availability, complicating their interpretation. Disentangling the roles of prokaryotes and eukaryotes in acidic soils is essential to clarify their individual and interactive effects on N_2O dynamics, particularly under scenarios of ecological imbalance induced by agricultural interventions.

The objectives of this study were (i) to distinguish the contribution of biological and abiotic processes to N_2O emissions in acidic soils; (ii) to examine the N_2O emissions response to the inhibition of prokaryotic and eukaryotic microorganisms; iii) and to explore the relationship of N_2O emissions to the dosage of prokaryotic and eukaryotic inhibitors. A microcosm experiment was conducted using acidic vegetable soil treated with gradient concentrations of cycloheximide (a eukaryotic inhibitor) and streptomycin (a prokaryotic inhibitor), both individually and in combination. Gas (N_2O and CO_2) fluxes, soil physicochemical properties, and substrate dynamics were monitored to evaluate dose–response relationships. By elucidating the roles of prokaryotic–eukaryotic interactions and inhibitor dosage in regulating N_2O emission, this work advances a mechanistic understanding of N cycling in acidic agroecosystems and informs sustainable management practices to mitigate greenhouse gas emission.

2. Materials and Methods

2.1. Soil Sampling

The soil used was collected from Qunfeng Town, Zhuzhou City, Hunan Province 27°43′57″ N, 113°4′0″ E. This region's mean annual temperature is 18 °C, and the mean annual precipitation is 1200–1600 mm. Soil samples of the 0–20 cm soil layer were collected in April 2019. The sampled soil was sieved through a 2 mm mesh and air-dried. This study's reported measurements and calculations were based on oven-dry soil mass. The soil had a pH of 3.9, with a texture composed of 10.8% clay, 57.3% silt, and 31.9% sand, and it contained 1.94 g kg^{-1} of total nitrogen (TN), 18.7 g kg^{-1} of soil organic C (SOM), a C/N ratio of 9.66, 86.5 mg kg^{-1} of available phosphorus (Olsen-P), and 92 mg kg^{-1} of exchangeable potassium (AK).

2.2. Experimental Design

To quantitatively assess the relative contributions of biotic and abiotic processes to soil N_2O emissions, the soil samples were divided into two distinct groups: sterilized and non-sterilized treatments. Both groups were subjected to three N amendment conditions: (i) control (CK); (ii) ammonium sulfate $(NH_4)_2SO_4$ application at 100 mg N kg^{-1} soil (NH_4^+-N); and (iii) potassium nitrate (KNO_3) application at 100 mg N kg^{-1} soil (NO_3^--N). $(NH_4)_2SO_4$ and KNO_3 were procured from Sinopharm Chemical Reagent Co., Ltd. (Shanghai, China). Each treatment had three replicates. For sterilized incubation, soil samples were sterilized by ^{60}Co-γ irradiation with a dose of 25 kGy for 72 h, which is a widely used and effective method to sterilize soil samples [23]. Sterilization of media and other solutions was achieved by autoclaving them for 30 min at 121 °C. Twenty grams of air-dried soil was pre-incubated in 120 mL serum flasks at 25 °C and 40% water holding capacity (WHC) for 7 d. The soil water holding capacity was 28% w/w water, A pipette was used to dropwise add deionized water with dissolved $(NH_4)_2SO_4$ or KNO_3 to the bottles, and a syringe was used to spray deionized water evenly onto the soil surface so that the soil moisture reached 70% of WHC. The bottles were then sealed with a rubber stopper and cultured in a shading incubator at 25 °C.

The microbial inhibition experiment was designed to investigate the effects of prokaryotic and eukaryotic inhibitors on soil N_2O emission using streptomycin and cycloheximide (Sigma-Aldrich Corporation, St. Louis, MO, USA) individually. The experimental design consisted of 25 treatments, each with three replicates, including (1) a control group (CK) without inhibitors (0 mg g^{-1} streptomycin, 0 mg g^{-1} cycloheximide); (2) 4 streptomycin-only treatments (ST) at concentrations of 2, 4.5, 6, and 10 mg g^{-1} without cycloheximide; (3) 4 cycloheximide-only treatments (CY) at concentrations of 2, 4.5, 6, and 10 mg g^{-1} without streptomycin; and (4) 16 combined treatments (ST + CY) representing all possible concentration combinations of both inhibitors (2, 4.5, 6, and 10 mg g^{-1} for each inhibitor). This comprehensive design enabled a systematic evaluation of individual and interactive effects across the complete concentration gradient of both microbial inhibitors. The gas in the headspace of the flask was sampled, after 4 and 8 h of incubation, using a syringe to transfer the sample into a 12 mL headspace bottle (which had been evacuated). Gas samples were stored until analysis. After gas sampling, destructive sampling was performed on all the treated soils, and the leaching of soil inorganic N and soluble organic C was immediately conducted on the collected soil samples.

2.3. *Measurements and Analysis*

2.3.1. Gas Sampling and Analysis for Sterilization Experiment

Following the treatment's application in the sterilization experiment, all the flasks were sealed with bromobutyl rubber septa and aluminum caps (Macherey-Nagel, GmbH & Co. KG, Düren, Germany) and flushed with an O_2-He mixture (21% O_2, *v*/*v*) using five vacuum–replenishment cycles to minimize background gases. Headspace pressure was adjusted to atmospheric levels using a water-filled syringe. The flasks were incubated at 20 ± 0.5 °C in an automated incubation-monitoring system (Robot system) [24], which continuously measured N_2O concentrations at 8 h intervals [25]. Gas sampling was performed using a peristaltic pump (Gilson Model 222, Gilson S.A.S., Villiers-le-Bel, France) connected to an Agilent 7890A gas chromatograph (Agilent Technologies Inc., Santa Clara, CA, USA). N_2O fluxes were calculated from concentration changes between sampling intervals, following the NMBU methodology (Norwegian University of Life Sciences).

2.3.2. Gas Sampling and Analysis for Biology Inhibitor Experiment

The N_2O concentrations in the gas samples from the biology inhibitor experiment were analyzed using an Agilent 6890 gas chromatograph equipped with an electron capture detector (ECD) (Agilent Technologies Inc., Santa Clara, CA, USA). Soil gas fluxes were calculated using the following equation:

$$F = \frac{\rho \times V \times \Delta c \times 273}{\Delta t \times m \times (273 + T)}, \quad (1)$$

where F is the gas production rate (mg N $g^{-1}h^{-1}$); ρ is the gas density in the standard state—N_2O is 1.964 kg^{-1} m^3, CO_2 is 1.997 kg^{-1} m^3; V is the volume occupied by the gas in the culture bottle, which was 100 mL in this experiment; Δc is the gas concentration difference (ppmv); Δt is the time interval between two adjacent measurements (h); and T is the temperature at which the gas is measured, which in this experiment was 25 °C.

2.3.3. Chemical and Physical Analysis

The soil was extracted using a 1 M KCl (Sinopharm Chemical Reagent Co., Ltd., Shanghai, China) solution at a soil/KCl solution ratio of 1:5 (*w*/*v*) immediately after incubation, and concentrations of NH_4^+-N and NO_3^--N were measured using a continuous flow analyzer (AA3, Seal Analytical, Norderstedt, Germany).

DOC and DON were extracted with a 0.5 M K_2SO_4 solution and then measured by a TOC/TN instrument (TOC-VCSH, Shimadzu, Kyoto, Japan).

The calculation method of N conversion intensity (mg N kg^{-1} h^{-1}) is as follows:

$$\text{Net Nitrification rate} = \frac{NO_3\text{-}\text{-N (AI)} - NO_3\text{-}\text{-N (BI)}}{\text{incubation time}}, \quad (2)$$

$$\text{Net mineralization rate} = \frac{[NH_4^+\text{-N (AI)} + NO_3\text{-}\text{-N (AI)}] - [NH_4^+\text{-N (BI)} + NO_3\text{-}\text{-N (BI)}]}{\text{incubation time}}, \quad (3)$$

where (AI) = after incubation, (BI) = before incubation.

2.4. Statistical Analysis

Statistical analyses were performed using one-way ANOVA in SPSS 20.0 (IBM Corporation, Armonk, NY, USA), with post hoc LSD tests for multiple comparisons ($p < 0.05$). Data are presented as means ± standard error (SE) in all figures and tables, which were generated using SigmaPlot 15.0 (Systat Software Inc., San Jose, CA, USA) and Origin 2024 (OriginLab Corporation, Northampton, MA, USA) software.

3. Results

3.1. Effects of Biotic and Abiotic Interactions on Soil N_2O Emission

The experimental results revealed significant differences in N_2O emissions between sterilized and unsterilized treatments, as illustrated in Figure 1. Specifically, unsterilized soils exhibited significantly higher N_2O emissions, ranging from 486 to 713 μg N kg^{-1}, compared to sterilized soils, which only produced 12.6 to 18.9 μg N kg^{-1}. Within the unsterilized treatments, the NO_3^--N amendment resulted in the highest cumulative N_2O emissions (713 μg N kg^{-1}), followed by NH_4^+-N (532 μg N kg^{-1}) and the control (486 μg N kg^{-1}). In sterilized treatments, N_2O emissions remained minimal, with NH_4^+-N yielding 18.9 μg N kg^{-1}, the control yielding 17.6 μg N kg^{-1}, and NO_3^--N yielding 12.6 μg N kg^{-1}. Quantitative analysis indicated that biotic processes accounted for 96–98% of total N_2O emissions, while abiotic contributions were negligible (1–3%). The NO_3^--N treatment had the highest relative biotic contribution (98%), followed by the NH_4^+-N and control treatments (96% each).

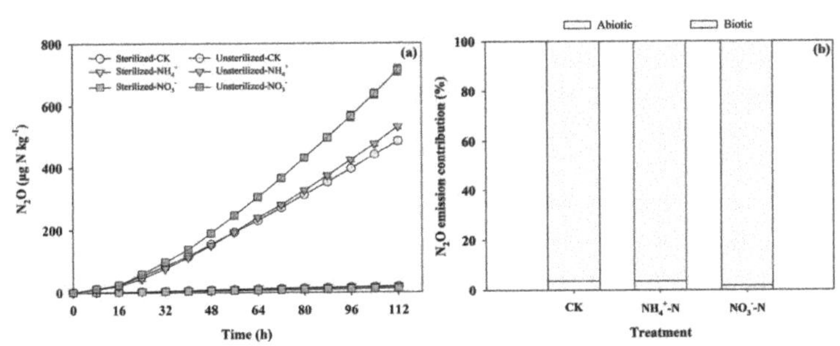

Figure 1. (**a**) Cumulative N_2O emissions after 112 h of incubation in highly acidic soils under the application of different treatments, including treatments with the application of microbial inhibitors (1: Sterilized-CK, 2: Sterilized-NH_4^+-N, 3: Sterilized-NO_3^--N) and treatments without applying microbial inhibitors (1: Sterilized-CK, 2: Sterilized-NH_4^+-N, 3: Sterilized-NO_3^--N). (**b**) Effect of biotic and abiotic interaction on soil N_2O emission in different treatments including: CK (control), NH_4^+-N (ammonium nitrogen), NO_3^--N (nitrate nitrogen).

3.2. Effects of Microbial Inhibitors on Soil N_2O Flux

The application of microbial inhibitors, streptomycin and cycloheximide, significantly influenced N_2O emissions across treatments, as illustrated in Figure 2. N_2O fluxes ranged from 0.541 to 2.45 μg kg^{-1} h^{-1} at 4 h and from 1.93 to 9.38 μg kg^{-1} h^{-1} at 8 h. In the absence of inhibitors, N_2O emissions were 0.701 μg kg^{-1} h^{-1} at 4 h and 2.24 μg kg^{-1} h^{-1} at 8 h. High concentrations of streptomycin (6 and 10 mg g^{-1}) reduced N_2O emissions from 2.24 μg kg^{-1} h^{-1} to 1.93 μg kg^{-1} h^{-1} and 2.12 μg kg^{-1} h^{-1}, respectively, whereas lower concentrations (2 and 4.5 mg g^{-1}) increased emissions from 2.24 μg kg^{-1} h^{-1} to 2.95 μg kg^{-1} h^{-1} and 3.27 μg kg^{-1} h^{-1}, respectively. The application rate of 6 mg g^{-1} showed the lowest emission of 1.93 μg kg^{-1} h^{-1} at 8 h among all the microbial inhibitor treatments, as the CK treatment showed emissions of 2.24 μg kg^{-1} h^{-1} at 8 h of incubation. Cycloheximide at 2 mg g^{-1} and 4.5 mg g^{-1} significantly enhanced N_2O emissions at 8 h, reaching 9.15 μg kg^{-1} h^{-1} and 5.68 μg kg^{-1} h^{-1}, respectively. However, higher dosages (6 mg g^{-1} and 10 mg g^{-1}) inhibited N_2O emissions compared to lower-dosage treatments, reducing them to 5.55 μg kg^{-1} h^{-1} and 4.84 μg kg^{-1} h^{-1}, respectively. Furthermore, applying both inhibitors in combination showed significantly different results. The treatment ST_2CY_2 exhibited the highest N_2O emissions of 9.38 μg kg^{-1} h^{-1}. In contrast, treatments ST_2CY_{10} and $ST_{4.5}CY_6$ showed inhibitory effects, with the lowest N_2O emissions recorded at 5.30 μg kg^{-1} h^{-1} and 5.43 μg kg^{-1} h^{-1} at 8 h. The combined application of the two inhibitors significantly increased N_2O emissions, with flux ranging from 1.39 to 9.38 μg kg^{-1} h^{-1}. However, the combined application of the two inhibitors resulted in lower N_2O emissions than the sum of their individual effects. For example, the ST_2CY_2 treatment produced 9.38 μg kg^{-1} h^{-1} N_2O at 8 h, lower than the sum of individual ST_2CY_0 (2.95 μg kg^{-1} h^{-1}) and ST_0CY_2 (9.15 μg kg^{-1} h^{-1}) effects. The addition of cycloheximide to streptomycin treatment significantly increased N_2O emissions, whereas the reverse combination had no such effect. For instance, the ST_2CY_2 treatment exhibited significantly higher N_2O emissions than ST_2CY_0 at 8 h, while showing no significant difference from ST_0CY_2.

3.3. Effects of Microbial Inhibitors on Soil CO_2 Flux

The application of microbial inhibitors, streptomycin and cycloheximide, significantly influenced CO_2 emissions across different treatments, as shown in Figure 3. The CO_2 fluxes ranged from 5.62 to 11.4 mg kg^{-1} h^{-1} at 4 h and from 21.1 to 39.3 mg kg^{-1} h^{-1} at 8 h. For soils without the addition of an inhibitor, CO_2 emission was 9.42 mg kg^{-1} h^{-1} and 33.9 mg kg^{-1} h^{-1} at 4 and 8 h, respectively. In general, CO_2 emission decreased with inhibitor dosages. However, when treated with 2 mg g^{-1} and 4.5 mg g^{-1} streptomycin, the CO_2 emissions increased significantly, reaching 39.3 mg kg^{-1} h^{-1} and 37.5 mg kg^{-1} h^{-1} at 8 h, respectively. The combined application of the two inhibitors significantly decreased CO_2 emissions (ranging from 5.62 to 31.7 mg kg^{-1} h^{-1}), except ST_6CY_2 at 4 h. The combined application of the two inhibitors resulted in lower CO_2 emissions than the sum of their individual effects. For example, the ST_6CY_2 treatment produced 31.7 mg kg^{-1} h^{-1} at 8 h, lower than the sum of individual ST_6CY_0 (23.1 mg kg^{-1} h^{-1}) and ST_0CY_2 (30.9 mg kg^{-1} h^{-1}) effects.

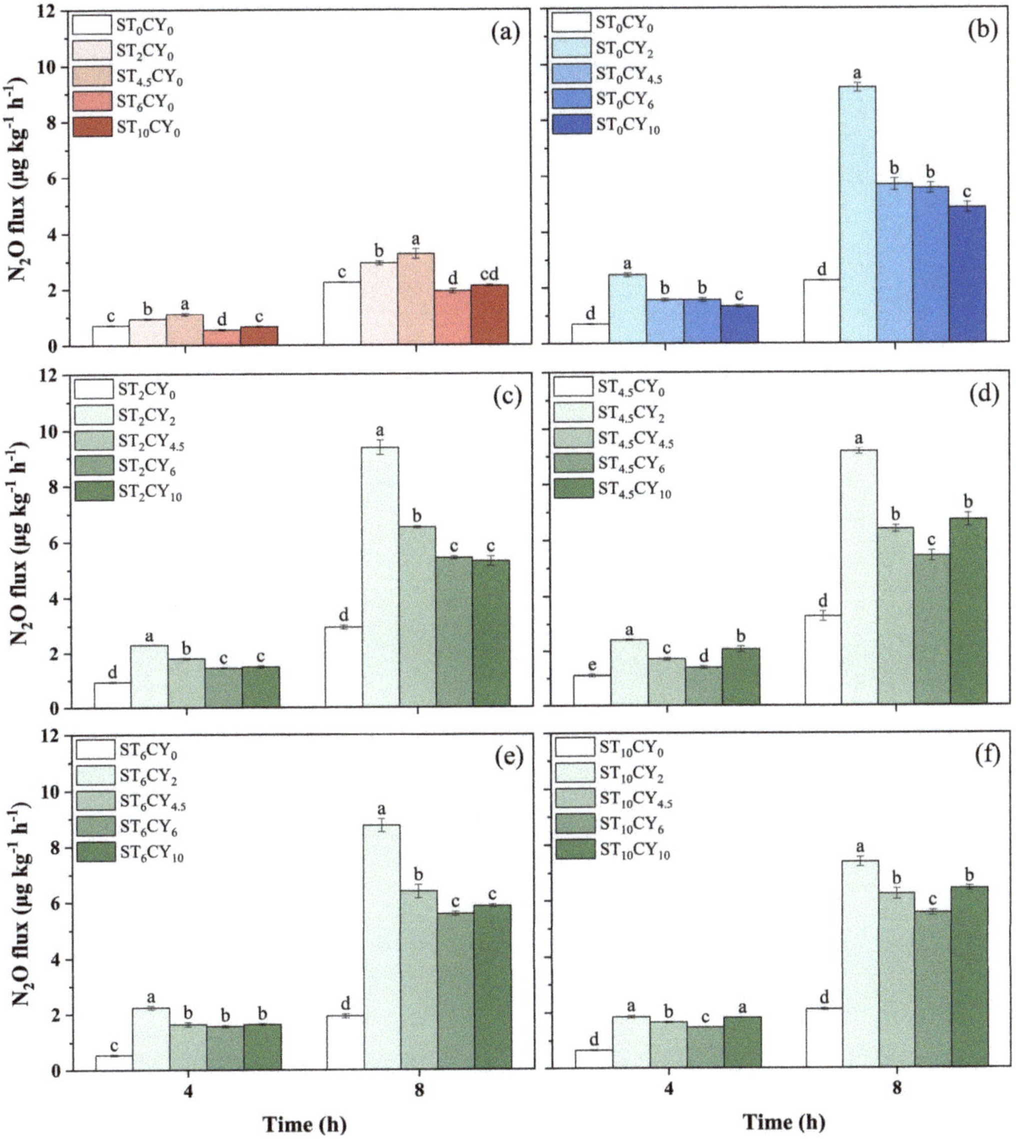

Figure 2. The N_2O flux after 4 and 8 h of incubation in highly acidic soils under the application of microbial inhibitors (streptomycin and cycloheximide) with different concentrations, including (0, 2, 4.5, 6, 10 mg g^{-1}) applied alone and with different combinations of both microbial inhibitors. Cycloheximide concentration was maintained at 0 mg g^{-1} (**a**) with varying streptomycin concentrations, while streptomycin concentrations were maintained at 0 (**b**), 2 (**c**), 4.5 (**d**), 6 (**e**), and 10 (**f**) mg g^{-1} with varying cycloheximide concentrations. Error bars represent standard errors, with different lowercase letters indicating statistically significant differences among treatments ($p < 0.05$).

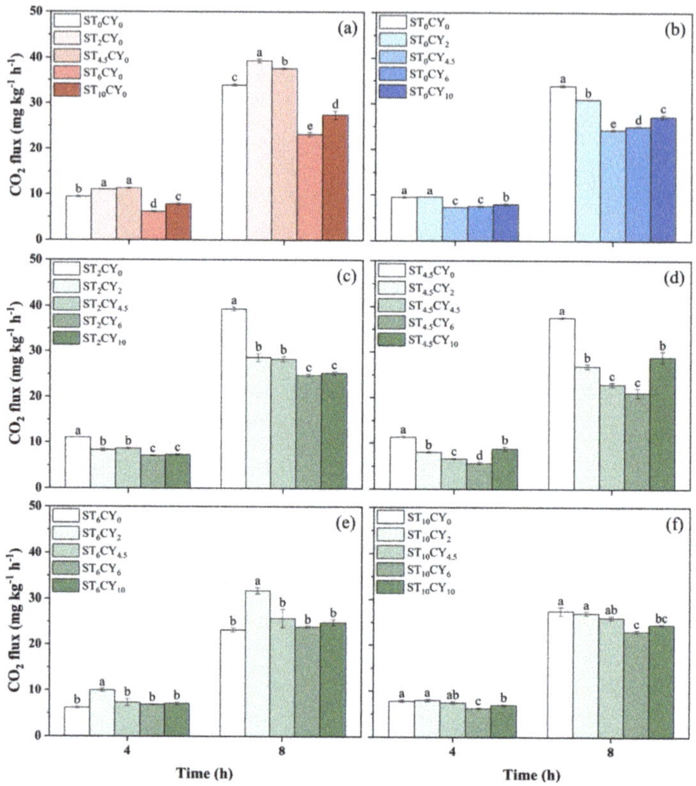

Figure 3. The CO_2 flux after 4 and 8 h of incubation in highly acidic soils under the application of microbial inhibitors (streptomycin and cycloheximide) with different concentrations, including (0, 2, 4.5, 6, 10 mg g^{-1}) applied alone and with different combinations of both microbial inhibitors. Cycloheximide concentration was maintained at 0 mg g^{-1} (**a**) with varying streptomycin concentrations, while streptomycin concentrations were maintained at 0 (**b**), 2 (**c**), 4.5 (**d**), 6 (**e**), and 10 (**f**) mg g^{-1} with varying cycloheximide concentrations. Error bars represent standard errors, with different lowercase letters indicating statistically significant differences among treatments ($p < 0.05$).

3.4. *Effects of Microbial Inhibitors on Soil Available Nitrogen and Carbon Contents*

The NH_4^+-N and NO_3^--N content in soils was affected by the application of microbial inhibitors, as shown in Figures S1 and S2. Without inhibitors, NH_4^+-N content was 11.0 mg kg^{-1}. Streptomycin increased the NH_4^+-N to 15.2–21.7 mg kg^{-1}, while cycloheximide elevated it to 24.6–25.3 mg kg^{-1}. Among different streptomycin dosages without cycloheximide, the NH_4^+-N content was significantly higher with 10 mg g^{-1} than that with lower dosages. NO_3^--N content without inhibitors was 181 mg kg^{-1}, varying between 170 and 186 mg kg^{-1} across treatments. The treatments involving the application of 10 mg g^{-1} of streptomycin combined with cycloheximide, as well as the ST_6CY_6 and ST_6CY_{10} treatments, significantly reduced the NO_3^--N content.

The soil DON and DOC content was significantly influenced by the application of microbial inhibitors, as shown in Figures 4 and 5. Soil DON content in the soils without inhibitors was 3.59 mg kg^{-1}, increasing to 278–717 mg kg^{-1} with streptomycin and 202–457 mg kg^{-1} with cycloheximide treatment, showing an overall upward trend with an increasing inhibitor dosage. The same was true when the two inhibitors were applied in combination, with DON content ranging from 298 to 1049 mg kg^{-1}. The combined application of the two inhibitors resulted in a lower DON content than the sum of their individual effects. For example, the DON concentration in treatment ST_2CY_2 was 298 mg kg^{-1}, lower than the sum of individual ST_2CY_0 (278 mg kg^{-1}) and ST_0CY_2 (202 mg kg^{-1}) effects. Under the same addition levels, the DON content in samples treated with streptomycin was higher than that in samples treated with cycloheximide. The highest DON content was observed in the combined treatment of 10 mg g^{-1} streptomycin and 10 mg g^{-1} cycloheximide ($ST_{10}CY_{10}$). The DOC content in the soils without inhibitors' addition was 444 mg kg^{-1}. The DOC content significantly increased with inhibitors' addition. The DOC content of different treatments varied from 444 mg kg^{-1} to 7018 mg kg^{-1}. Under the same addition levels, the DOC content in samples treated with streptomycin was lower than that in samples treated with cycloheximide. Both streptomycin and cycloheximide individually enhanced DOC accumulation in a dose-dependent manner, with higher concentrations resulting in greater increases. And the combination of the two inhibitors further amplified this effect. The highest DOC content was shown in soil treated with 10 mg g^{-1} streptomycin combined with 10 mg g^{-1} cycloheximide, with values of 7018 mg kg^{-1}. The combined application of the two inhibitors resulted in lower DOC content than the sum of their individual effects, except $ST_{4.5}CY_2$. For example, the DOC concentration in treatment ST_2CY_2 was 3258 mg kg^{-1}, lower than the sum of individual ST_2CY_0 (1024 mg kg^{-1}) and ST_0CY_2 (2511 mg kg^{-1}) effects.

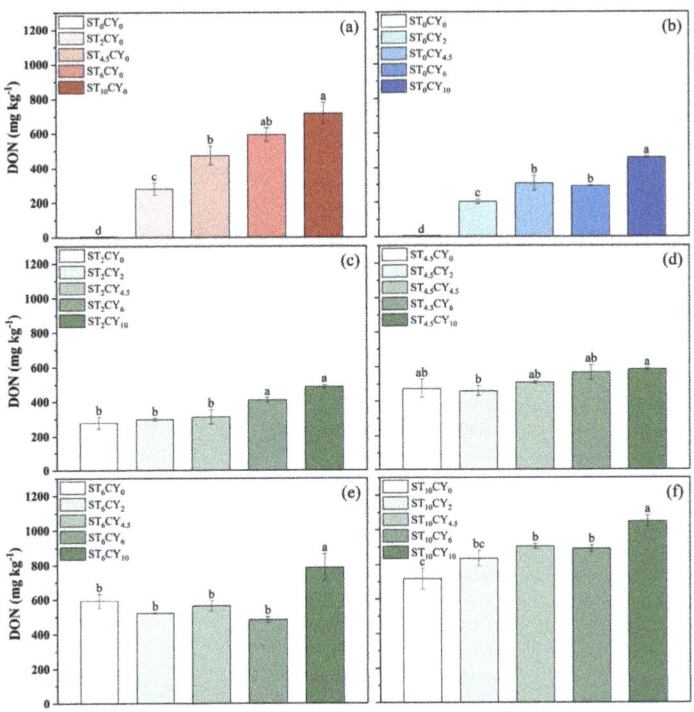

Figure 4. Influence on the concentration of dissolved organic nitrogen (DON) after incubation in highly acidic soils of the application of microbial inhibitors (streptomycin and cycloheximide) with different concentrations, including (0, 2, 4.5, 6, 10 mg g^{-1}) applied alone and with different combinations of both microbial inhibitors. Cycloheximide concentration was maintained at 0 mg g^{-1} (**a**) with varying streptomycin concentrations, while streptomycin concentrations were maintained at 0 (**b**), 2 (**c**), 4.5 (**d**), 6 (**e**), and 10 (**f**) mg g^{-1} with varying cycloheximide concentrations. Error bars represent standard errors, with different lowercase letters indicating statistically significant differences among treatments ($p < 0.05$).

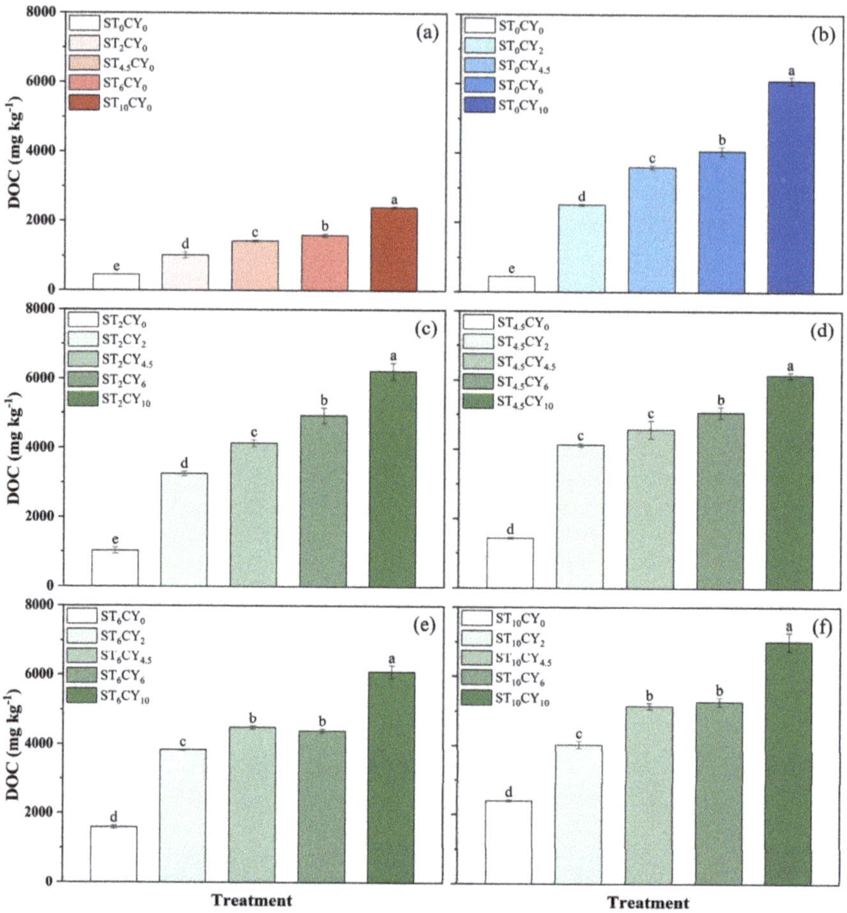

Figure 5. Influence on the concentration of dissolved organic carbon (DOC) after incubation in highly acidic soils of the application of microbial inhibitors (streptomycin and cycloheximide) with different concentrations, including (0, 2, 4.5, 6, 10 mg g^{-1}) applied alone and with different combinations of both microbial inhibitors. Cycloheximide concentration was maintained at 0 mg g^{-1} (**a**) with varying streptomycin concentrations, while streptomycin concentrations were maintained at 0 (**b**), 2 (**c**), 4.5 (**d**), 6 (**e**), and 10 (**f**) mg g^{-1} with varying cycloheximide concentrations. Error bars represent standard errors, with different lowercase letters indicating statistically significant differences among treatments ($p < 0.05$).

3.5. Correlation of N_2O and Soil Properties with Microbial Inhibitor Dosages

N_2O emissions and soil properties were significantly influenced by microbial inhibitors, as shown in Figure 6. The Pearson correlation analysis indicated that N_2O emission was negatively correlated with cycloheximide dosage (R = −0.68, $p < 0.001$), and it was also negatively associated with NH_4^+-N (R = −0.31, $p < 0.001$) and DOC (R = −0.57, $p < 0.05$). In terms of soil properties, soil's NH_4^+-N content was significantly positively correlated with streptomycin (R = 0.57, $p < 0.05$) and cycloheximide (R = 0.47, $p < 0.01$) dosage. Soil's DON content was significantly positively correlated with streptomycin (R = 0.87, $p < 0.001$) and cycloheximide (R = 0.28, $p < 0.05$) dosage. Soil's DOC content was significantly positively correlated with cycloheximide (R = 0.88, $p < 0.001$) dosage.

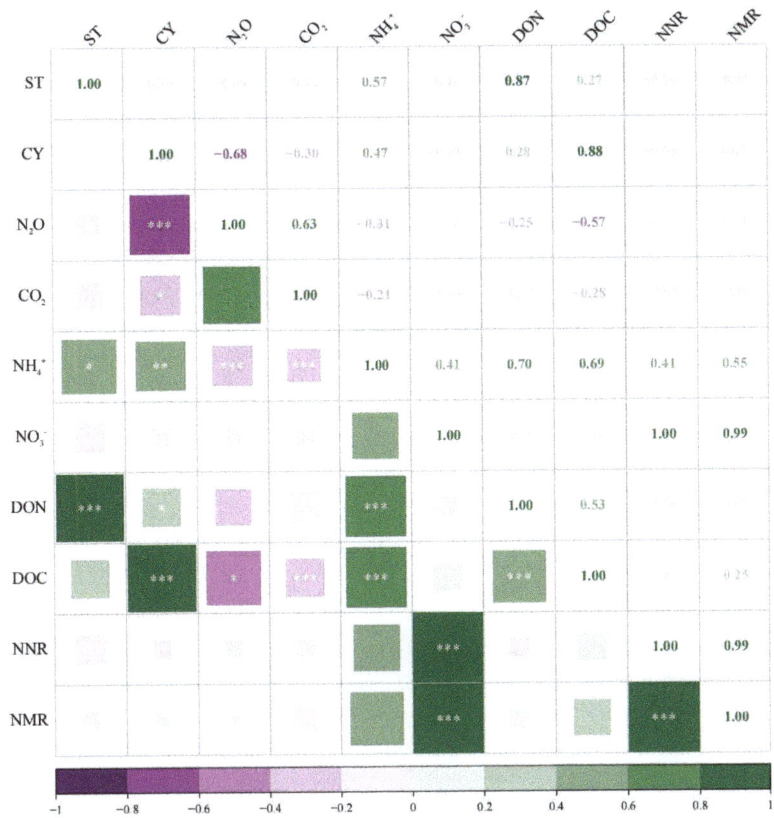

Figure 6. Correlation heatmap of gaseous and soil properties with microbial inhibitor dosages. The size of each square is proportional to the absolute value of the correlation coefficient. ST: streptomycin dosage; CY: cycloheximide dosage; NNR: net nitrification rate; NMR: net mineralization rate. *, $p < 0.05$; **, $p < 0.01$; ***, $p < 0.001$.

4. Discussion

4.1. Effects of Microbial Inhibitors on N_2O Production Pathways

N_2O is primarily produced through microbially driven N cycling processes [26]. These processes can be significantly influenced by microbial inhibitors, which selectively target specific microbial groups and alter their activity [27]. In strongly acidic vegetable soils, fungal contributions to N_2O production are particularly pronounced due to their greater tolerance to acidic conditions compared to many bacteria [14]. The inhibition of microbial activity is typically expected to reduce N_2O emission by suppressing key microbial groups involved in nitrification and denitrification [28,29]. For instance, streptomycin, a prokaryotic inhibitor, is expected to suppress ammonia-oxidizing bacteria (AOB) and nitrite-oxidizing bacteria (NOB), thereby reducing N_2O production through nitrification [8]. Similarly, cycloheximide, a eukaryotic inhibitor, is expected to inhibit fungal activity, thereby reducing heterotrophic nitrification [30] and fungal denitrification [31]. However, the findings of this study revealed a more complex pattern of responses to microbial inhibitors.

High concentrations of streptomycin (6 mg g^{-1} and 10 mg g^{-1}) resulted in a decrease in N_2O emissions (Figure 2). This finding aligns with previous studies demonstrating that streptomycin can effectively suppress bacterial activity, leading to decreased N_2O

production [18]. However, the N_2O generation rate of most soil samples treated with this inhibitor did not decrease but increased. This pattern was consistent across individual applications of streptomycin and cycloheximide, as well as their combined treatments. Previous research indicated that microbial inhibition could trigger compensatory responses from surviving microorganisms [32,33]. The decomposition of microbial necromass can provide additional substrates for N_2O production pathways such as nitrification and denitrification [34]. Additionally, microbial inhibitors themselves can serve as sources of C and N, further fueling microbial activity and the generation of N_2O [20,35]. The increase in N_2O production can thus be attributed to a balance between microbial death-induced substrate release and compensatory microbial responses [36,37]. The results of this study indicate that the effects of microbial inhibitors on N_2O emission are not straightforward and can vary depending on the concentration and type of inhibitor used. This complexity arises from the intricate interplay between microbial community structure, functional redundancy, and the availability of substrates for the production of N_2O.

The observed effects of microbial inhibitors on N_2O emission in strongly acidic vegetable soils highlight the importance of considering the unique characteristics of these environments. Acidic soils are known to harbor distinct microbial communities with specialized adaptations to low pH conditions, including acidophilic taxa [13]. These acidophilic microorganisms can maintain active nitrification and denitrification processes even at pH levels below 4.5, contributing significantly to N_2O emission [12]. Therefore, microbial imbalance in acidic soils can have unintended consequences, such as shifts in the microbial community structure and enhanced N_2O production through compensatory mechanisms. It also highlights the importance of considering the broader ecological context in which these inhibitors are applied. Agricultural practices, such as the use of fertilizers and organic amendments, can significantly influence soil's microbial communities and their associated biogeochemical processes. Understanding these complex interactions is crucial for developing sustainable strategies to reduce N_2O emissions while maintaining soil fertility and ecosystems' functionality.

4.2. *Effects of Microbial Inhibitors on the Dynamics of Dissolved Nitrogen and Carbon Fractions*

NH_4^+-N and NO_3^--N serve as substrates for N_2O production [38], and microbial inhibitors can influence N_2O emissions by affecting the microbial utilization of these nitrogen substrates. The significant increase in NH_4^+-N content following the application of microbial inhibitors suggests that the inhibition of microbial N transformation (as shown in Figures S3 and S4) processes, such as nitrification, leads to the accumulation of NH_4^+-N in soil. This is likely due to the reduced activity of ammonia-oxidizing bacteria and archaea, which are key players in the nitrification pathway [39]. The inhibitors impede protein synthesis in certain fungi and bacteria, thereby reducing their biological activities, slowing nitrification, and decreasing NH_4^+-N consumption. Additionally, archaea contribute significantly to nitrification in acidic soil [40,41]. Cycloheximide may inhibit ammonia-oxidizing archaea [42], reducing the nitrification rate and further contributing to the accumulation of NH_4^+-N.

The concurrent increase in the soil's DON and DOC indicates that microbial inhibitors not only affect N cycling but also alter the availability of organic substrates (Figures 4 and 5). The decomposition of microbial biomass and the release of intracellular compounds likely contribute to these elevated levels, influencing microbial respiration and C mineralization rates [21]. Furthermore, the death of some microorganisms due to the application of an inhibitor creates a biological pulse, providing a substrate for surviving microorganisms [43]. Lower concentrations of cycloheximide may cause fungal cell lysis [44], supplying C and N substrates to the remaining microbial community, which enhances NH_4^+-N availability, microbial respiration, and mineralization rates [33], consistent with the enhanced CO_2 production and mineralization rates observed in our experiments for soil samples treated with inhibitors.

The observed increase in soil's DON and DOC content following the application of microbial inhibitors suggests that the decomposition of microbial biomass and release of intracellular compounds contribute to these elevated levels. This has significant implications for microbial respiration, C mineralization rates, and the dynamics of soil organic matter [45]. Increased DOC content can enhance microbial activity and respiration, leading to elevated CO_2 emissions and potential changes in soil C storage [46]. This highlights the importance of considering the broader impacts of microbial inhibitors on soil C cycling and the potential feedbacks to climate change.

In this study, we conducted experiments in strongly acidic vegetable soils. While our findings offer valuable insights into N_2O emissions under microbial suppression conditions in such highly acidic soils, the results may not be directly applicable to other soil types or management practices. A meta-analysis has demonstrated that the impact of inhibitors on N_2O emissions, NH_4^+-N, ammonification, nitrate, and nitrification is contingent upon soil type [21]. Specifically, fungal inhibitors have been shown to mitigate N_2O emissions in agricultural soils, to significantly diminish N_2O emissions in grassland soils, and to augment N_2O emissions in forest soils [21]. Fungal communities display a higher degree of sensitivity to tillage and cover cropping practices compared to bacterial communities [47]. Consequently, when inhibitors are applied, distinct agricultural management practices can elicit divergent effects on soil N_2O emissions.

5. Conclusions

This study concludes that microbial communities' imbalance, induced by selective inhibition in strongly acidic vegetable soils, has significant implications for N_2O emissions and N dynamics. High concentrations of microbial inhibitors can effectively reduce N_2O emission, likely by suppressing key microbial groups involved in nitrification and denitrification. However, lower concentrations may trigger compensatory responses from surviving microorganisms, leading to increased N_2O production. The observed increases in DOC and DON levels suggest that microbial inhibitors can stimulate microbial decomposition, providing additional substrates for N_2O production and potentially exacerbating community imbalances. These findings underscore the complexity of microbial interactions in acidic soils and highlight the importance of considering the broader ecological context when applying microbial inhibitors. Future research should focus on long-term ecological impacts and explore sustainable management strategies to mitigate N_2O emissions while preserving soil's health and agricultural productivity.

Supplementary Materials: The following supporting information can be downloaded at: https://www.mdpi.com/article/10.3390/biology14060621/s1, Figure S1: Influence on the concentration of ammonium nitrogen (NH_4^+-N) after incubation in highly acidic soils of the application of microbial inhibitors (streptomycin and cycloheximide) with different concentrations, including (0, 2, 4.5, 6, 10 mg g^{-1}) applied alone and with different combinations of both microbial inhibitors. Figure S2: Influence on the concentration of nitrate nitrogen (NO_3^--N) after incubation in highly acidic soils of the application of microbial inhibitors (cycloheximide and streptomycin) with different concentrations, including (0, 2, 4.5, 6, 10 mg g^{-1}) applied alone and with different combinations of both microbial inhibitors. Figure S3: Influence on the concentration of net nitrification rate (NNR) after incubation in highly acidic soils of the application of microbial inhibitors (streptomycin and cycloheximide) with different concentrations, including (0, 2, 4.5, 6, 10 mg g^{-1}) applied alone and with different combinations of both microbial inhibitors. Figure S4: Influence on the concentration of net mineralization rate (NMR) after incubation in highly acidic soils of the application of microbial inhibitors (streptomycin and cycloheximide) with different concentrations, including (0, 2, 4.5, 6, 10 mg g^{-1}) applied alone and with different combinations of both microbial inhibitors.

Author Contributions: Conceptualization, W.A. and J.Y.; methodology, W.A. and H.G.; validation, H.G., X.X. and R.C.; formal analysis, W.A. and Y.X.; investigation, H.G., X.X. and R.C.; data curation, Y.X., W.S. and B.L.; writing—original draft preparation, W.A.; writing—review and editing, W.A., J.Y. and Q.C.; visualization, W.A. and Q.C.; supervision, J.Y. and Q.C.; project administration, J.Y.; funding acquisition, J.Y. All authors have read and agreed to the published version of the manuscript.

Funding: This work was supported by the National Natural Science Foundation of China (Nos. 42407312) and the Postdoctoral Fellowship Program of CPSF (GZC20233290).

Institutional Review Board Statement: Not applicable.

Informed Consent Statement: Not applicable.

Data Availability Statement: Data will be made available on request.

Conflicts of Interest: The authors declare no conflicts of interest.

References

1. Rockström, J.; Gupta, J.; Qin, D.; Lade, S.J.; Abrams, J.F.; Andersen, L.S.; Armstrong McKay, D.I.; Bai, X.; Bala, G.; Bunn, S.E.; et al. Safe and just Earth system boundaries. *Nature* **2023**, *619*, 102–111. [CrossRef] [PubMed]
2. Palmer, K.; Biasi, C.; Horn, M.A. Contrasting denitrifier communities relate to contrasting N_2O emission patterns from acidic peat soils in arctic tundra. *ISME J.* **2012**, *6*, 1058–1077. [CrossRef] [PubMed]
3. Ravishankara, A.R.; Daniel, J.S.; Portmann, R.W. Nitrous oxide (N_2O): The dominant ozone-depleting substance emitted in the 21st century. *Science* **2009**, *326*, 123–125. [CrossRef] [PubMed]
4. Tian, H.; Xu, R.; Canadell, J.G.; Thompson, R.L.; Winiwarter, W.; Suntharalingam, P.; Davidson, E.A.; Ciais, P.; Jackson, R.B.; Janssens-Maenhout, G.; et al. A comprehensive quantification of global nitrous oxide sources and sinks. *Nature* **2020**, *586*, 248–256. [CrossRef]
5. Tokuda, S.; Hayatsu, M. Nitrous oxide production from strongly acid tea field soils. *Soil Sci. Plant Nutr.* **2000**, *46*, 835–844. [CrossRef]
6. Wang, Y.; Guo, J.; Vogt, R.D.; Mulder, J.; Wang, J.; Zhang, X. Soil pH as the chief modifier for regional nitrous oxide emissions: New evidence and implications for global estimates and mitigation. *Glob. Change Biol.* **2018**, *24*, 617–626. [CrossRef]
7. Guo, J.; Liu, X.; Zhang, Y.; Shen, J.; Han, W.; Zhang, W.; Christie, P.; Goulding, K.; Vitousek, P.M.; Zhang, F. Significant acidification in major Chinese croplands. *Science* **2010**, *327*, 1008–1010. [CrossRef]
8. Butterbach-Bahl, K.; Baggs, E.M.; Dannenmann, M.; Kiese, R.; Zechmeister-Boltenstern, S. Nitrous oxide emissions from soils: How well do we understand the processes and their controls? *Philos. Trans. R. Soc. Biol. Sci.* **2013**, *368*, 20130122. [CrossRef]
9. Homyak, P.M.; Kamiyama, M.; Sickman, J.O.; Schimel, J.P. Acidity and organic matter promote abiotic nitric oxide production in drying soils. *Glob. Change Biol.* **2017**, *23*, 1735–1747. [CrossRef]
10. Chen, D.; Yuan, X.; Zhao, W.; Luo, X.; Li, F.; Liu, T. Chemodenitrification by Fe(II) and nitrite: pH effect, mineralization and kinetic modeling. *Chem. Geol.* **2020**, *541*, 119586. [CrossRef]
11. Robinson, T.C.; Latta, D.E.; Notini, L.; Schilling, K.E.; Scherer, M.M. Abiotic reduction of nitrite by Fe(ii): A comparison of rates and N_2O production. *Environ. Sci. Process. Impacts* **2021**, *23*, 1531–1541. [CrossRef] [PubMed]
12. Hashidoko, Y.; Takakai, F.; Toma, Y.; Darung, U.; Melling, L.; Tahara, S.; Hatano, R. Emergence and behaviors of acid-tolerant Janthinobacterium sp. that evolves N_2O from deforested tropical peatland. *Soil Biol. Biochem.* **2008**, *40*, 116–125. [CrossRef]
13. Xu, S.; Yu, Y.; Fan, H.; Bilyera, N.; Meng, X.; Xue, J.; Lu, Z.; Yang, Z.; Chapman, S.J.; Gao, F.; et al. Microbial communities overwhelm environmental controls in explaining nitrous oxide emission in acidic soils. *Soil Biol. Biochem.* **2024**, *194*, 109453. [CrossRef]
14. Bösch, Y.; Pold, G.; Saghaï, A.; Karlsson, M.; Jones, C.M.; Hallin, S. Distribution and environmental drivers of fungal denitrifiers in global soils. *Microbiol. Spectr.* **2023**, *11*, 23–61. [CrossRef]
15. Wu, H.; Cui, H.; Fu, C.; Li, R.; Qi, F.; Liu, Z.; Yang, G.; Xiao, K.; Qiao, M. Unveiling the crucial role of soil microorganisms in carbon cycling: A review. *Sci. Total Environ.* **2024**, *909*, 168627. [CrossRef]
16. Chen, S.; Edwards, C.A.; Subler, S. Effects of the fungicides benomyl, captan and chlorothalonil on soil microbial activity and nitrogen dynamics in laboratory incubations. *Soil Biol. Biochem.* **2001**, *33*, 1971–1980. [CrossRef]
17. Chen, X.; Xu, Y.; Sun, R.; Ye, X.; Ma, C.; Mao, J.; Zhang, C.; Gao, H.; Zhang, W. Soil microbial communities under wheat and maize straw incorporation are closely associated with soil organic carbon fractions and chemical structure. *Appl. Soil Ecol.* **2023**, *182*, 104724. [CrossRef]

18. Yin, J.; Chen, H.; Duan, P.; Zhu, K.; Li, N.; Ma, Y.; Xu, Y.; Guo, J.; Liu, R.; Chen, Q. Soil microbial communities as potential regulators of N_2O sources in highly acidic soils. *Soil Ecol. Lett.* **2023**, *5*, 230178. [CrossRef]
19. Yin, J.; Cui, W.; Xu, Y.; Ma, Y.; Chen, H.; Guo, J.; Liu, R.; Chen, Q. Understanding the relative contributions of fungi and bacteria led nitrous oxide emissions in an acidic soil amended with industrial waste. *Ecotoxicol. Environ. Saf.* **2023**, *255*, 114727. [CrossRef]
20. Lazcano, C.; Zhu-Barker, X.; Decock, C. Effects of organic fertilizers on the soil microorganisms responsible for N_2O emissions: A review. *Microorganisms* **2021**, *9*, 983. [CrossRef]
21. Ullah, M.R.; Dijkstra, F.A. Fungicide and bactericide effects on carbon and nitrogen cycling in soils: A meta-analysis. *Soil Syst.* **2019**, *3*, 23. [CrossRef]
22. Wang, Q.; Chai, Q.; Dou, X.; Zhao, C.; Yin, W.; Li, H.; Wei, J. Soil Microorganisms in Agricultural Fields and Agronomic Regulation Pathways. *Agronomy* **2024**, *14*, 669. [CrossRef]
23. Wang, Y.; Cao, W.; Guo, J.; Zhang, M. Effects of Increasing pH on Nitrous Oxide and Dinitrogen Emissions from Denitrification in Sterilized and Unsterilized Forest Soils. *Forests* **2022**, *13*, 1589. [CrossRef]
24. Molstad, L.; Dörsch, P.; Bakken, L.R. Robotized incubation system for monitoring gases (O_2, NO, N_2O N_2) in denitrifying cultures. *J. Microbiol. Methods* **2007**, *71*, 202–211. [CrossRef]
25. Yin, J.; Liu, R.; Cao, W.; Zhu, K.; Fenton, O.; Guo, J.; Chen, Q. Nitrogen and carbon addition changed nitrous oxide emissions from soil aggregates in straw-incorporated soil. *J. Soil Sediments* **2022**, *22*, 617–629. [CrossRef]
26. Qian, H.; Yuan, Z.; Chen, N.; Zhu, X.; Huang, S.; Lu, C.; Liu, K.; Zhou, F.; Smith, P.; Tian, H.; et al. Legacy effects cause systematic underestimation of N_2O emission factors. *Nat. Commun.* **2025**, *16*, 2775. [CrossRef] [PubMed]
27. Yasemi, M.; Jalali, A.; Asadzadeh, M.; Komijani, M. Organophosphate pesticides and their potential in the change of microbial population and frequency of antibiotic resistance genes in aquatic environments. *Chemosphere* **2025**, *376*, 144296. [CrossRef]
28. Castellano-Hinojosa, A.; Le Cocq, K.; Charteris, A.F.; Abadie, M.; Chadwick, D.R.; Clark, I.M.; González-López, J.; Bedmar, E.J.; Cardenas, L.M. Relative contributions of bacteria and fungi to nitrous oxide emissions following nitrate application in soils representing different land uses. *Int. Biodeterior. Biodegrad.* **2021**, *159*, 105199. [CrossRef]
29. Wang, L.; Wang, Z.; Wang, F.; Guan, Y.; Meng, D.; Li, X.; Zhou, H.; Li, X.; Chen, Y.; Tan, Z. Response of performance, antibiotic resistance genes and bacterial community exposure to compound antibiotics stress: Full nitrification to shortcut nitrification and denitrification. *Chem. Eng. J.* **2023**, *451*, 138750. [CrossRef]
30. Martikainen, P.J. Heterotrophic nitrification—An eternal mystery in the nitrogen cycle. *Soil Biol. Biochem.* **2022**, *168*, 108611. [CrossRef]
31. Aldossari, N.; Ishii, S. Fungal denitrification revisited—Recent advancements and future opportunities. *Soil Biol. Biochem.* **2021**, *157*, 108250. [CrossRef]
32. Laughlin, R.J.; Rütting, T.; Müller, C.; Watson, C.J.; Stevens, R.J. Effect of acetate on soil respiration, N_2O emissions and gross N transformations related to fungi and bacteria in a grassland soil. *Appl. Soil Ecol.* **2009**, *42*, 25–30. [CrossRef]
33. Rex, D.; Clough, T.J.; Richards, K.G.; de Klein, C.; Morales, S.E.; Samad, M.S.; Grant, J.; Lanigan, G.J. Fungal and bacterial contributions to codenitrification emissions of N_2O and N_2 following urea deposition to soil. *Nutr. Cycl. Agroecosyst.* **2018**, *110*, 135–149. [CrossRef]
34. Thakur, M.P.; van Groenigen, J.W.; Kuiper, I.; De Deyn, G.B. Interactions between microbial-feeding and predatory soil fauna trigger N_2O emissions. *Soil Biol. Biochem.* **2014**, *70*, 256–262. [CrossRef]
35. Rousk, J.; Demoling, L.A.; Bahr, A.; Bååth, E. Examining the fungal and bacterial niche overlap using selective inhibitors in soil. *FEMS Microbiol. Ecol.* **2008**, *63*, 350–358. [CrossRef]
36. McLain, J.E.T.; Martens, D.A. N_2O production by heterotrophic N transformations in a semiarid soil. *Appl. Soil Ecol.* **2006**, *32*, 253–263. [CrossRef]
37. Muñoz-Leoz, B.; Garbisu, C.; Charcosset, J.; Sánchez-Pérez, J.M.; Antigüedad, I.; Ruiz-Romera, E. Non-target effects of three formulated pesticides on microbially-mediated processes in a clay-loam soil. *Sci. Total Environ.* **2013**, *449*, 345–354. [CrossRef]
38. Liu, S.; Wu, D.; Ju, X.; Shen, J.; Cheng, Y.; Deng, N.; Song, X.; Di, H.; Li, P.; Han, L.; et al. Nitrification inhibitor induced microbial NH_4^+-N immobilization improves maize nitrogen use efficiency in strong ammonia oxidation soil. *Soil Biol. Biochem.* **2025**, *202*, 109687. [CrossRef]
39. Laughlin, R.J.; Stevens, R.J. Evidence for Fungal Dominance of Denitrification and Codenitrification in a Grassland Soil. *Soil Sci. Soc. Am. J.* **2002**, *66*, 1540–1548. [CrossRef]
40. Huang, L.; Chakrabarti, S.; Cooper, J.; Perez, A.; John, S.M.; Daroub, S.H.; Martens-Habbena, W. Ammonia-oxidizing archaea are integral to nitrogen cycling in a highly fertile agricultural soil. *ISME Commun.* **2021**, *1*, 19. [CrossRef]
41. Taylor, A.E.; Zeglin, L.H.; Wanzek, T.A.; Myrold, D.D.; Bottomley, P.J. Dynamics of ammonia-oxidizing archaea and bacteria populations and contributions to soil nitrification potentials. *ISME J.* **2012**, *6*, 2024–2032. [CrossRef] [PubMed]
42. Xiong, W.; Li, J.; Liu, J.; Li, C.; He, Z.; Li, X. Quantification of microbially induced soil N_2O emissions by an inhibitory cocktail in mountain forest ecosystems. *Geoderma* **2024**, *443*, 116835. [CrossRef]

43. Ullah, M.R.; Carrillo, Y.; Dijkstra, F.A. Biocides provide a source of carbon and nitrogen directly to surviving microbes and indirectly through a pulse in microbial necromass. *Appl. Soil Ecol.* **2021**, *160*, 103862. [CrossRef]
44. Badalucco, L.; Pomare, F.; Grego, S.; Landi, L.; Nannipieri, P. Activity and degradation of streptomycin and cycloheximide in soil. *Biol. Fert. Soils* **1994**, *18*, 334–340. [CrossRef]
45. Trivedi, P.; Anderson, I.C.; Singh, B.K. Microbial modulators of soil carbon storage: Integrating genomic and metabolic knowledge for global prediction. *Trends Microbiol.* **2013**, *21*, 641–651. [CrossRef]
46. Kiikkilä, O.; Kanerva, S.; Kitunen, V.; Smolander, A. Soil microbial activity in relation to dissolved organic matter properties under different tree species. *Plant Soil* **2014**, *377*, 169–177. [CrossRef]
47. Wipf, H.M.; Xu, L.; Gao, C.; Spinner, H.B.; Taylor, J.; Lemaux, P.; Mitchell, J.; Coleman-Derr, D. Agricultural Soil Management Practices Differentially Shape the Bacterial and Fungal Microbiome of Sorghum bicolor. *Appl. Environ. Microbiol.* **2021**, *87*, e02345-20. [CrossRef]

biology

MDPI

Article

Liming-Induced Nitrous Oxide Emissions from Acidic Soils Dominated by Stimulative Nitrification

Xiaoxiao Xiang [1], Hongyang Gong [1], Waqar Ahmed [1], Rodney B. Thompson [2], Wenxuan Shi [3], Junhui Yin [1,4,*] and Qing Chen [1]

1 College of Resources and Environmental Sciences, China Agricultural University, Beijing 100193, China
2 Department of Agronomy, University of Almeria, Carretera de Sacramento s/n, La Cañada de San Urbano, 04120 Almería, Spain
3 Environmental Research Centre, Teagasc, Johnstown Castle, Y35 TC97 Wexford, Ireland
4 State Key Lab of Biocontrol, Guangdong Provincial Key Laboratory of Plant Stress Biology, School of Agriculture and Biotechnology, Shenzhen Campus of Sun Yat-sen University, Shenzhen 518107, China
* Correspondence: yjhui@cau.edu.cn

Simple Summary

This study examined how liming influences nitrous oxide emissions by increasing pH in acidic soils. The results showed that soil nitrous oxide emissions increased following alkaline amendment, especially when combined with urea addition, by stimulating nitrification. These findings indicated that while liming alleviates soil acidity, it can also increase greenhouse gas emissions. Future work is suggested to develop optimal management of alkaline amendments for soil pH management while restricting nitrous oxide emissions.

Abstract

Nitrous oxide (N_2O) is a potent greenhouse gas, with emissions occurring mostly from agricultural soils, especially acidic soils. This research aimed to elucidate the response of soils dominated by nitrification-driven N_2O production to alkaline amendments, given that nitrification is a key process in N_2O emission. This study investigated the impact of an alkaline mineral amendment (CSMP) on N_2O emission, nitrification rate, and functional gene abundance. Using a robotic automated incubation system, CSMP both alone and in combination with urea was applied to two acidic soils (CL: pH 5.81; WS: pH 4.91). The results demonstrated that, relative to the CK, the CSMP-only treatment significantly increased N_2O emissions by 18.4-fold in these acidic soils, with a 61.6-fold increase in the U + CSMP treatment. This very large increase was driven by a rise in AOB-*amoA* abundance and a concurrent decline in AOA-*amoA*, which was confirmed by structural equation modeling, which showed that the increase in pH strongly influenced N_2O emission primarily through AOB-*amoA*. Although CSMP is effective for reversing soil acidification, its use must be carefully managed to prevent stimulation of N_2O emissions. Future strategies should explore combining CSMP with approaches that can mitigate nitrification while maintaining its soil improvement benefits. This study provides critical insights for developing balanced management practices that address both soil health and climate change mitigation in acidic agricultural systems.

Keywords: soil amendments; acidic soil; nitrous oxide; ammonia-oxidizing bacteria

1. Introduction

N_2O is a potent greenhouse gas that is also projected to remain as the dominant anthropogenic ozone-depleting agent this century [1–3]. Modern crop production systems, relying on intensive N fertilization to sustain high yields, represent the largest anthropogenic source of global N_2O emissions [4–7], with atmospheric concentrations increasing at an annual rate of 0.75–1.0 ppb [8]. N_2O emissions are generally appreciably larger from acidic compared with alkaline soils [9]. Emission factors (EFs) of N_2O from acidic soils are more than three times those from alkaline soils [10,11]. Consequently, mitigating N_2O emissions from acidic agricultural soils is an urgent priority.

Alkaline amendments have long been used to deal with low-pH soils. While effectively addressing soil acidification, these amendments can concurrently affect N_2O emissions [12–14]. These emissions are primarily regulated by soil pH [15–17]. A meta-analysis of 121 field studies reported an average reduction of 21.3% in N_2O emission following liming [18]. This reduction was primarily attributed to pH-enhanced abundance of the bacterial N_2O reductase enzyme (*nosZ*), the sole known biological sink for N_2O, which is impaired under low pH [19–21] and which reduces the N_2O/N_2 ratio during denitrification [22–24]. Additionally, the suppression of acid-tolerant fungal denitrification further enhances the mitigation effect at higher pH [25,26].

However, there are reports that indicate that alkalizing agents can have the opposite effect and can stimulate N_2O emission [23,27]. This stimulatory effect can result from enhanced nitrification [23,28,29] and elevated nitrate (NO_3^-) availability fueling nitrification-derived N_2O formation [30,31] under alkaline conditions. This apparent contradiction occurs because of the opposing influence of pH on distinct N_2O production pathways: while elevated pH enhances the capacity for bacterial N_2O consumption [24,32], it can also promote the abundance of nitrifiers, particularly ammonia-oxidizing bacteria (AOB) [33,34]. Although nitrification was traditionally regarded as restricted in acidic soils due to limited ammonium (NH_4^+) availability [35], recent findings demonstrate significant nitrification-derived N_2O fluxes under acidic conditions [23,27,36]. A comprehensive global analysis of 5438 observations identified a distinct hump-shaped relationship between soil pH and N_2O EFs, peaking in acidic soils [10]. This demonstrates nitrification as a major source of N_2O across a wider pH range than previously assumed, highlighting the complex interplay between pH and both nitrification and denitrification processes governing net N_2O flux.

Consequently, the net impact of pH amendment on N_2O emissions is contingent upon which N_2O production pathway is dominant. This study hypothesized that (i) in acidic soils dominated by nitrification, application of the CSMP alkaline amendment will significantly increase N_2O emission; (ii) co-application of CSMP with urea will further stimulate the increased N_2O emission; and (iii) the stimulation of N_2O emission is facilitated by an increased population of ammonium oxidizers (AOA/AOB). To test these hypotheses, this study employed an alkaline amendment (CSMP) in two different acidic soils [37].

2. Materials and Methods

2.1. Soil Sampling and Physicochemical Characteristics

In the context of this research, topsoil samples (0–20 cm) were collected from two different sites (CL and WS) in Zhuzhou City, Hunan Province, China, with coordinates of 27°48′46″ N, 113°28′8″ E (CL) and 27°48′49″ N, 113°25′56″ E (WS). The WS soil (26.6 g kg^{-1}) appears to be sandier and richer in organic carbon compared with the CL soil (18.6 g kg^{-1}), which has a higher proportion of clay. The region falls within a subtropical humid monsoon climate zone, characterized by annual rainfall between 1200 and 1600 mm and a yearly average temperature of approximately 18 °C. Soil samples collected from the area were acidic, with pH values of 5.81 for CL and 4.91 for WS. Detailed information about the properties of the soils can be found in Table S1. Prior to incubation, the soil samples were passed through a 2 mm sieve and air-dried. Throughout the study, all measurements and data calculations were based on the oven-dried soil weight.

2.2. *The Alkaline Mineral Amendment*

The alkaline mineral additive employed in this study was obtained from Kingenta Ecological Engineering Group Co., Ltd., based in Linyi, Shandong Province, China. This product is composed of a blend of partially decomposed phosphate rock and insoluble potassium-containing silicate minerals, marketed commercially available as a Calcium–Silicon–Magnesium–Potassium (CSMP) product. It has been approved for agricultural applications and meets the national standards set forth in GB/T 36207–2018 [38]. The pH of the amendment was 10.3. The detailed composition is provided in Table S2.

2.3. *Robotized Batch Cultivations for Respiratory Phenotype Experiment*

Thirty grams of oven-dried equivalent soil was allocated to four different treatments, each performed in triplicate. The treatments were as follows: (i) without urea and CSMP (CK), (ii) urea added at 100 mg N per kilogram of soil (U), (iii) CSMP applied at 10 g per kilogram of soil (CSMP), and (iv) a combined treatment of 10 g kg^{-1} CSMP with 100 mg N kg^{-1} urea (U + CSMP). The urea dose (60 mg N kg^{-1} soil) was equivalent to a field application rate of 92 kg N ha^{-1}. Similarly, the 10 g kg^{-1} soil CSMP dose was equivalent to a field application of 20 t ha^{-1} [39,40].

The urea, containing 46% nitrogen and obtained from Sinopharm Chemical Reagent Co., Ltd., Shanghai, China, was dissolved in deionized water before being applied [39,40]. The CSMP amendment was thoroughly mixed into the soil to ensure even distribution. Following addition of the amendments, the treated soil samples were transferred into 120 mL serum bottles, incubated at 60% water-holding capacity (WHC), and maintained at a steady 20 ± 0.5 °C for 30 d. To create airtight conditions, each bottle was gastight-sealed with bromobutyl rubber septa and aluminum caps sourced from Macherey-Nagel, GmbH & Co. KG, Düren, Germany. Bottle headspaces were purged with an oxygen–helium mixture (21% oxygen by volume) through five vacuum and refill cycles, which effectively eliminated any residual gases. Afterward, a syringe without a plunger filled with 5 mL of distilled water was inserted to balance the gas pressure to one atm [19].

Automated headspace gas sampling was performed at 0, 1, 3, 7, 15, and 30 days using a robotic incubation system described in a previous work [37]. Isostatic pressure maintenance was achieved by replenishing each sampled volume with ultra-pure helium (99.999%). The gases were drawn out with a peristaltic pump (Gilson Model 222, Gilson S.A.S., Villiers-le-Bel, France) and subsequently analyzed with an Agilent 7890A gas chromatograph (Agilent Technologies Inc., Santa Clara, CA, USA), which had both electron capture and thermal conductivity detectors. The entire sampling and analytical process was automatically controlled using a tailored Python 2.6.3 script. The gas emission was calculated using the method of Lars R. Bakken [41].

Six batches of parallel incubation were concurrently conducted for 0, 1, 3, 7, 15, and 30 days for a suite of analyses. The analyses included the determination of soil pH, ammonium nitrogen (NH_4^+-N), and nitrate nitrogen (NO_3^--N); determination of the net nitrification rate; and identification of changes in functional gene populations.

2.4. *Measurements and Analysis*

2.4.1. Chemical Analysis

Immediately following incubation, soil was extracted with a 1 mol L^{-1} KCl (Sinopharm Chemical Reagent Co., Ltd., Shanghai, China) solution at a 1:5 (w/v) soil-to-solution ratio.

The concentrations of ammonium N (NH_4^+-N) and nitrate N (NO_3^--N) in the extracts were determined using a continuous flow analyzer (model AA3, Seal Analytical, Norderstedt, Germany). The net nitrification rate (mg N kg^{-1} h^{-1}) was calculated as follows:

$$\text{Net nitrification rate} = \frac{NO_3^-\text{-N (AI)} - NO_3^-\text{-N (BI)}}{\text{incubation time}}, \quad (1)$$

where (AI) = after incubation, (BI) = before incubation.

2.4.2. DNA Extraction and qPCR of Targeted Genes

Total DNA from the soil was isolated using the FastDNA® Spin Kit (MP Biomedicals, Irvine, CA, USA) for Soil. DNA quality was verified through gel electrophoresis on a 1% agarose gel, while concentration of DNA and purity were measured with a NanoDrop 2000 spectrophotometer (Thermo Fisher Scientific Inc., Wilmington, NC, USA). To determine the abundance of specific functional genes, quantitative real-time PCR (qPCR) was employed.

The quantitative PCR (qPCR) reaction setup, totaling 20 μL, included the following ingredients: 4 μL of 5 × TransStart FastPfu (TransGen Biotech Co., Ltd., Beijing, China) Buffer, 2 μL of 2.5 mM dNTP mix, 0.8 μL each of forward and reverse primers at 5 μm concentration, 0.4 μL of TransStart FastPfu DNA polymerase, 10 ng of extracted template DNA, and enough sterile distilled water to bring the total volume to 20 μL. Each sample was analyzed in triplicate to ensure accuracy. The sequences of the primers employed are provided in Table S3. The thermal cycling protocol comprised an initial denaturation at 95 °C for 5 min, then 45 cycles of denaturation at 95 °C for 15 s, annealing at 55 °C for 30 s, and extension at 72 °C for 30 s.

2.5. Statistical Analysis

For statistical evaluation, differences across treatment groups were analyzed using one-way ANOVA with SPSS version 20.0 (IBM, Corporation, Armonk, NY, USA). Significant differences ($p < 0.05$) were identified by least significant difference (LSD) test. All visualizations were generated using Origin 2024 (OriginLab Corporation, Northampton, MA, USA). The lavaan package by R 4.3.1 (R Foundation for Statistical Computing, Vienna, Austria) was utilized for structural equation modeling (SEM) to assess the relationships between N_2O emissions, soil properties, and gene abundances. The modification index (MI) was used to assess whether the inclusion of omitted pathways could optimize the initial conceptual model. The chi-square test (χ^2, $p > 0.05$), comparative fit index (CFI > 0.90), and standardized root-mean-square residual (SRMR < 0.08) were employed to evaluate model fit.

3. Results

3.1. Soil N_2O Emissions in Response to CSMP and Urea Application

The experimental results showed consistent temporal patterns in N_2O emissions from the two acidic soils of an initial surge followed by a decline (Figure 1a,b). In soil CL, N_2O peak values occurred on day 1 for CK (0.51 μg N kg^{-1} h^{-1}) and U (0.84 μg N kg^{-1} h^{-1}), while CSMP and U + CSMP peaked on day 7 (0.84 and 2.92 μg N kg^{-1} h^{-1}, respectively). Conversely, soil WS displayed a peak in N_2O emissions on day 15 across all treatments, with the U + CSMP treatment achieving the highest emission rate of 17.4 μg N kg^{-1} h^{-1}.

Compared with the CK, the U treatment increased cumulative N_2O emissions by 1.07-fold in CL soil and 1.21-fold in WS soil ($p < 0.05$; Figure 1c). The CSMP-only treatment raised emissions by 1.78-fold (CL) and 18.4-fold (WS) relative to CK ($p < 0.05$). The combined U + CSMP treatment further enhanced emissions, reaching 7.13-fold (CL) and 61.6-fold

(WS) compared with CK, equivalent to 2.92-fold (CL) and 27.3-fold (WS) increases over the U treatment ($p < 0.05$).

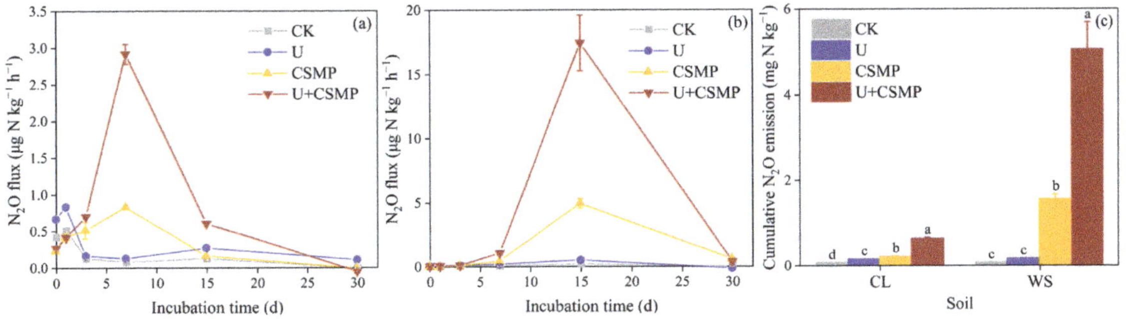

Figure 1. (**a**) N_2O flux in soil CL, (**b**) N_2O flux in soil WS, and (**c**) cumulative N_2O production in both acidic soils CL and WS, all as influenced by CSMP and urea application. CK, control; U, urea; CSMP, Calcium–Silicon–Magnesium–Potassium fertilizer; U + CSMP, combined application of Calcium–Silicon–Magnesium–Potassium fertilizer and urea. Values are means ± SE (n = 3). Different lowercase letters indicate significant differences among treatments ($p < 0.05$).

3.2. Dynamics of NH_4^+-N and NO_3^--N Concentrations and Net Nitrification Rate in Response to CSMP and Urea Application

During the incubation, NH_4^+-N concentrations in CL and WS soils followed a general pattern of initial increase and subsequent decline, which was substantially modulated by urea application (Figure 2a,b). In soil CL, NH_4^+-N peaked across all treatments on day 3, and then declined steadily. The CSMP treatment did not significantly alter NH_4^+-N concentrations in the initial 7 days; however, it significantly accelerated its reduction on day 15 relative to CK ($p < 0.05$). In soil WS, NH_4^+-N showed the same initial rise followed by decline. Relative to the CK treatment, the U treatment significantly increased NH_4^+-N ($p < 0.05$). The CSMP treatment further increased NH_4^+-N during the first 15 days ($p < 0.05$) but decreased it by day 30 ($p < 0.05$).

Throughout the incubation, NO_3^--N concentrations in both CL and WS soils progressively increased in all treatments (Figure 2c,d). The U treatment, compared with CK, significantly enhanced cumulative NO_3^--N concentration after 30 days in both soils, attaining 243 mg N kg^{-1} in CL and 18.6 mg N kg^{-1} in WS ($p < 0.05$). CSMP application substantially increased the 30-day cumulative NO_3^--N concentrations in both soils compared with the CK treatment. Notably, the U + CSMP treatment, compared with CK, increased NO_3^--N concentrations in WS soil by 9.94 times ($p < 0.05$).

The net nitrification rate in soil CL remained consistent throughout the incubation under the CK treatment, fluctuating between 2.65 and 5.23 mg N kg^{-1} d^{-1} (Figure 2e). Compared with CK, both U and CSMP treatments significantly increased the net nitrification rate, with the U + CSMP treatment showing the most substantial effect. In soil WS, the impact of all treatments on the net nitrification rate was most evident on day 30 (Figure 2f), where the CSMP treatment enhanced the rate by 19.1 times relative to CK, and the U + CSMP treatment resulted in a 10.3-fold increase relative to the U treatments.

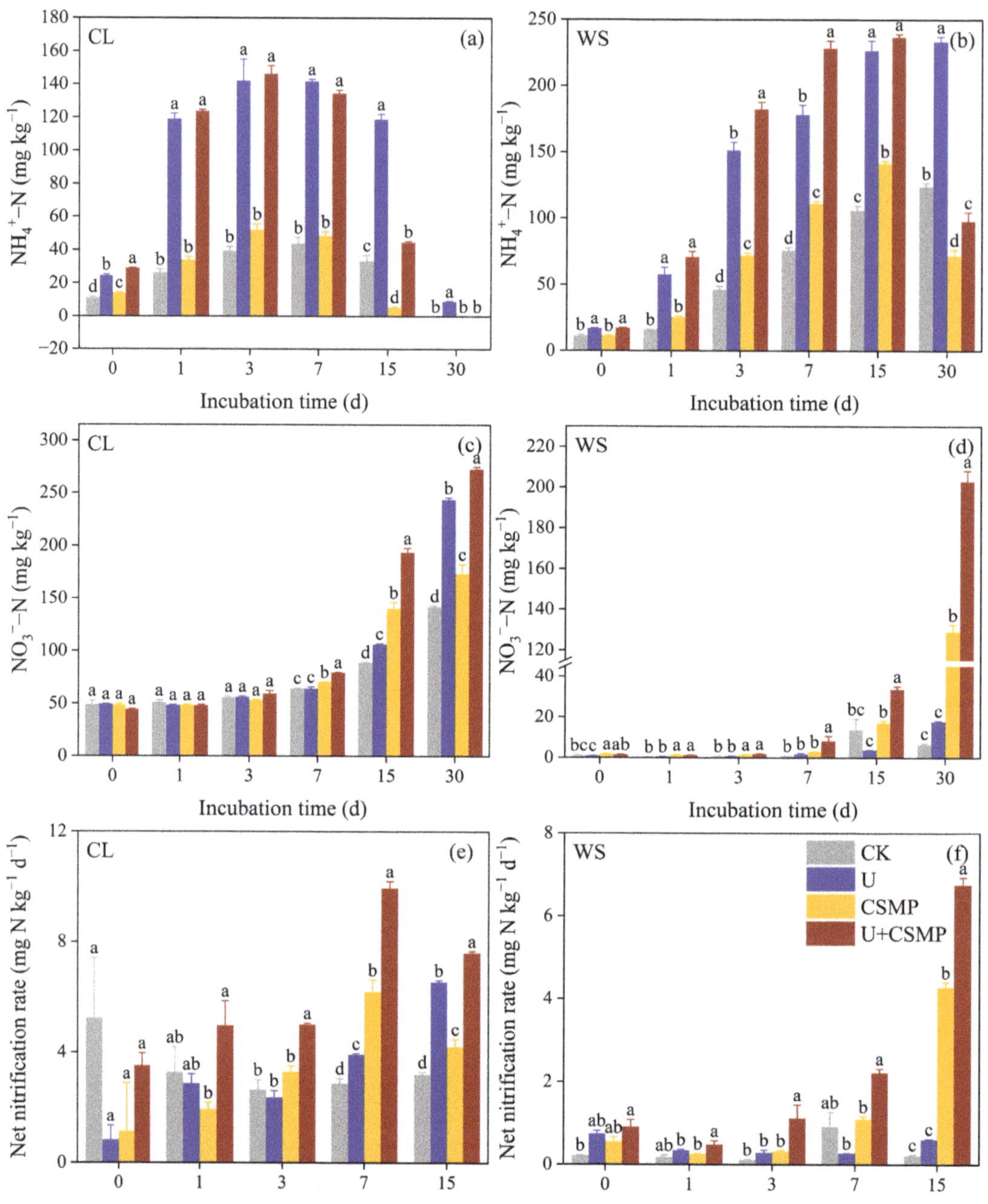

Figure 2. (**a**,**b**) Dynamics of NH_4^+-N concentrations; (**c**,**d**) NO_3^--N concentrations; and (**e**,**f**) net nitrification rate in response to CSMP and urea application in two acidic soils (CL and WS). (**a**,**c**,**e**) represent soil CL, while (**b**,**d**,**f**) represent soil WS. CK, control; U, urea; CSMP, CSMP fertilizer; U + CSMP, combined application of CSMP fertilizer and urea. Data are presented as means ± SE (n = 3). Different lowercase letters indicate significant differences among treatments ($p < 0.05$).

3.3. Abundance of Functional Genes Related to N_2O Production and Reduction

On day 3, relative to CK, CSMP had already elevated 16S rRNA gene copies, whereas U caused a decline in both acidic soils (Figure 3a). By day 30, compared with CK, gene abundance in CL had decreased by 5.30%, 3.02%, and 3.61%, while in WS it had increased by 1.12%, 1.53%, and 1.72%, respectively. The CSMP treatment significantly reduced ITS in both acidic soils on day 3 versus CK (Figure 3b). By day 30, the CSMP treatment increased fungal abundance in CL yet decreased it in WS relative to CK. In WS, the CSMP treatment also significantly lowered the ITS/16S ratio at both 3 and 30 days compared with CK (Figure 3c).

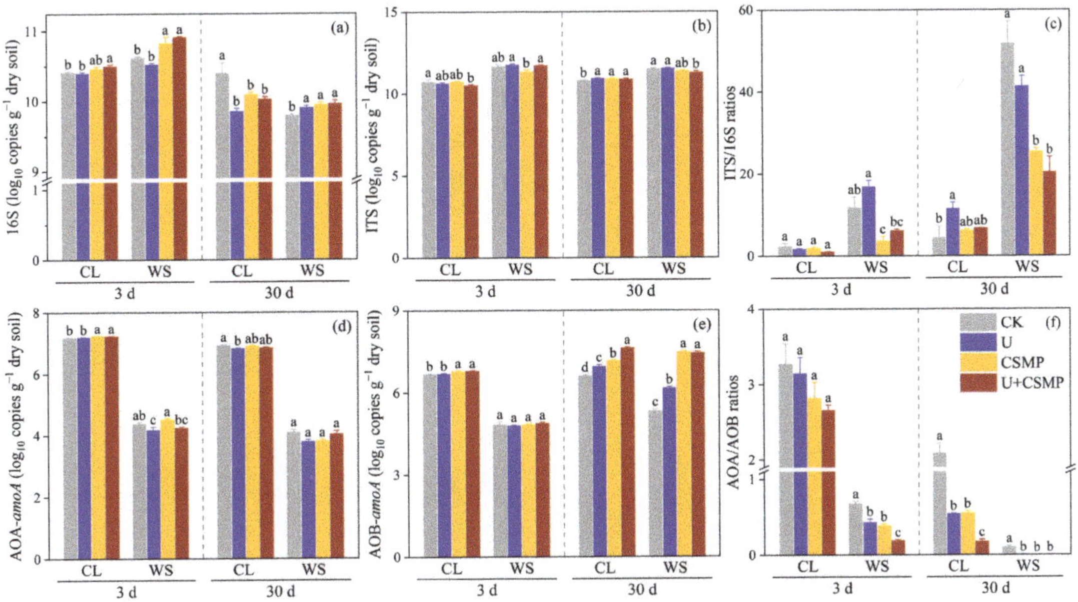

Figure 3. Effects of CSMP and urea application on microbial community structure in two acidic soils (CL and WS) over 3 d and 30 d periods. (**a**) Bacterial abundance, (**b**) fungal abundance, (**c**) fungal-to-bacterial ratio, (**d**) ammonia-oxidizing archaea abundance, (**e**) AOB abundance, and (**f**) their ratio. CK, control; U, urea; CSMP, CSMP fertilizer; U + CSMP, combined application of CSMP fertilizer and urea. Data are presented as means ± SE (n = 3). Different lowercase letters indicate significant differences among treatments ($p < 0.05$).

The AOB-*amoA* gene abundance responded more strongly to urea and CSMP applications compared with AOA-*amoA* (Figure 3d). Specifically, a significant increase of 0.44% in AOA-*amoA* gene abundance was observed in soil CL on day 3 after CSMP application compared with CK ($p < 0.05$). By day 30, urea significantly increased AOB-*amoA* gene abundance across all soil types relative to CK ($p < 0.05$), as depicted in Figure 3e. The CSMP treatment compared with CK induced significant increases in AOB-*amoA* gene abundance by 8% in CL and 40% in WS soils on day 30 ($p < 0.05$). Following the application of urea and/or CSMP, the AOA/AOB ratio significantly decreased in both soils on day 30 versus CK ($p < 0.05$), as illustrated in Figure 3f.

The initial incubation phase (day 3) revealed divergent responses of *nirK* and *nirS* genes in soil CL to all treatments. CSMP application, either alone or combined with urea, elevated *nirK* levels and reduced *nirS* levels compared with CK. By day 30, relative to CK, both *nirS* and *nirK* genes in soil CL showed a significant increase following treatments with CSMP and/or urea ($p < 0.05$), with CSMP alone causing the most substantial rise

(Figure 4a,b). In contrast, compared with CK, soil WS showed a significant decrease in *nirK* gene abundance under the CSMP treatment at day 3 ($p < 0.05$), while the U + CSMP treatment significantly enhanced *nirS* gene levels ($p < 0.05$) (Figure 4b). At day 30, relative to CK, the application of urea and/or CSMP significantly increased the abundance of the *nosZ* II gene in soil CL ($p < 0.05$), and the *nosZ* I gene in soil WS also demonstrated a significant increase upon the CSMP treatment ($p < 0.05$) (Figure 4c,d). The fungal nitrite reductase gene *nirK* displayed significant variation only at day 3, with the U + CSMP treatment in soil CL and the U and U + CSMP treatments in soil WS both showing enhanced gene expression compared with CK ($p < 0.05$) (Figure 4e). In addition, the (*nirK* + *nirS*)/(*nosZ* I + *nosZ* II) in soil WS significantly enhanced at day 3 versus CK ($p < 0.05$) (Figure 4f).

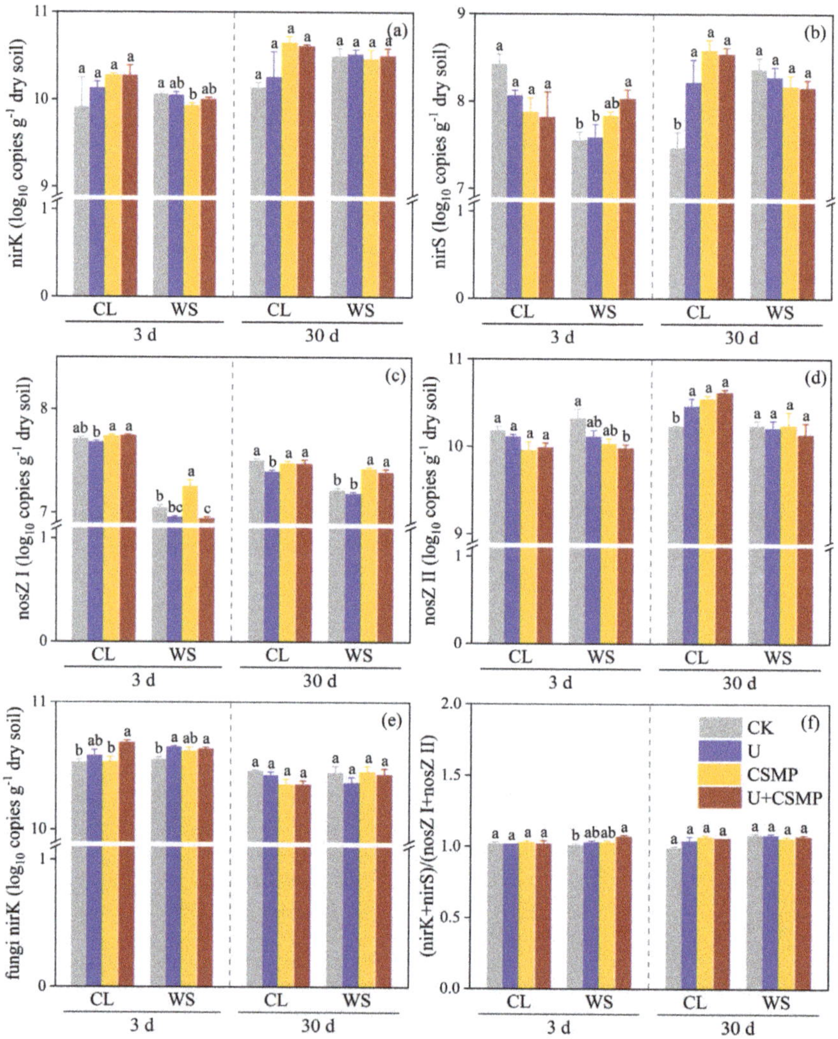

Figure 4. Effects of CSMP and urea application on microbial community structure in two acidic soils (CL and WS): (**a**) *nirK*, (**b**) *nirS*, (**c**) *nosZ* I, (**d**) *nosZ* II, (**e**) fungi *nirK*, and (**f**) (*nirK* + *nirS*)/(*nosZ* I + *nosZ* II). CK, control; U, urea; CSMP, CSMP fertilizer; U + CSMP, combined application of CSMP fertilizer and urea. Data are presented as means ± SE (n = 3). Different lowercase letters indicate significant differences among treatments ($p < 0.05$).

3.4. Structural Equation Modeling

An SEM was constructed to clarify how soil N_2O emissions are linked to pH shifts and to changes in the abundance of key functional genes in acidic soils (Figure 5). The model indicates that CSMP application significantly enhanced soil pH compared with initial levels, which subsequently stimulated the growth of both AOA ($\lambda = 0.27$) and AOB ($\lambda = 0.26$). However, their impacts on N_2O emissions exhibited fundamentally distinct patterns: AOA abundance showed a negative correlation with N_2O emissions ($\lambda = -0.72$), whereas AOB demonstrated a strong positive association ($\lambda = 0.78$). Notably, the model identified a synergistic interaction between AOA and AOB ($\lambda = 0.47$) that further increased this process.

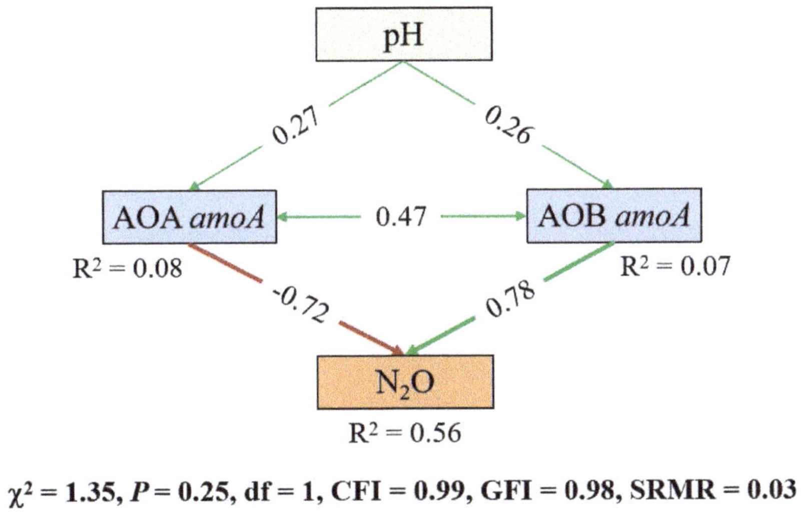

Figure 5. SEM illustrating the effects of amendments on N_2O emissions. Green arrows indicate positive relationships, while red arrows indicate negative relationships ($p < 0.05$). The R^2 values associated with response variables represent the proportion of variance explained by the model. The values on the arrows denote standardized path coefficients.

4. Discussion

4.1. N_2O Emission and N Transformations Following CSMP and Urea Application

The regulatory effects of alkaline amendments on N_2O emissions from acidic soils have acquired increasing research attention [13,42,43]. Conventional theory proposes that lime-based amendments primarily reduce greenhouse gas emissions by elevating soil pH [9,43]. In the present study, CSMP application significantly increased soil pH in both acidic soils used (Figure S1). In these two moderately acidic soils, N_2O emissions were dominated by bacterial nitrification. CSMP application, and especially the combined U + CSMP application, appreciably stimulated N_2O emissions relative to the CK (Figure 1). Although earlier studies suggested that nitrification can be less intense in acidic soils [44], a recent meta-analysis showed that N_2O emissions related to soil nitrification are higher in acidic soils compared with neutral and alkaline soils [45].

In the two acidic soils in the present study, CSMP markedly accelerated NH_4^+-N consumption versus CK (Figure 2a,b). This is because relatively high pH values stimulate the consumption of NH_4^+ in soils [23]. In the present study, accelerated NH_4^+-N consumption was accompanied by pronounced increases in net nitrification rates (up to 19.1-fold versus CK in WS soil, Figure 2f). These findings strongly suggest that CSMP influenced N_2O emissions primarily through the stimulation of nitrification rather than by conventional denitrification pathways [27]. Nitrification generally acts as the primary process of N_2O production in aerobic soils [46,47]. Therefore, an increase in the soil nitrification rate can directly promote higher N_2O emissions in aerobic soil [48,49]. This explains why CSMP promoted both soil nitrification and N_2O emission in the present study in soils in which N_2O emissions were nitrification-driven.

The U + CSMP treatment, in particular, had pronounced effects on N_2O production compared with other treatments (Figure 1). This synergistic effect appears to result from three key mechanisms: (1) Urea hydrolysis provided abundant NH_4^+ substrate for nitrification (Figure 2b). Plausible mechanisms for enhanced soil N_2O emissions following urea addition may be enhanced NH_4^+-N/NO_3^--N concentrations and stimulated N-cycling enzyme activities, thereby increasing N_2O emissions [50]. (2) Transient pH spikes from urea hydrolysis complemented sustained alkalinity from CSMP addition [51]. (3) Substantial increases occurred in AOB-*amoA* gene abundance (40% versus CK in WS soil; Figure 3e) accompanied by decreased AOA/AOB ratios (Figure 3f), reflecting the predominance of bacterial ammonia oxidation [52,53].

4.2. CSMP Reduced the Acidic Soils' N_2O Emissions Dominated by Nitrification

Soil N_2O production arises from the intricate interplay between nitrification and denitrification [54,55]. Increasing evidence indicates that nitrification-derived N_2O emissions may actually exceed those from denitrification on a global scale, accounting for an average of 51.6% of total N_2O emissions [56]. A study using ^{15}N showed that, even under acidic conditions, nitrification can contribute more to N_2O emission than denitrification [57]. pH is an important factor influencing the soil nitrification process [16]. This study observed an increase in the net nitrification rate of two acidic soils, compared with the CK, after alkaline amendment (Figure 2e,f). Following the addition of CSMP, soil H increased, which stimulated nitrification, which in turn, stimulated soil N_2O emission.

The SEM identified AOB-*amoA* abundance as the key driver of N_2O emissions in the acidic soils (Figure 5). While AOA typically mediate ammonia oxidation in unamended acidic soils and showed a stronger correlation with background N_2O fluxes in this study [58], AOB exhibited greater responsiveness to pH elevation [53]. This aligns with reports that liming preferentially stimulates AOB-mediated N_2O production due to their higher per-mole NH_3 oxidation yield [34,42]. Consistent patterns emerge across ecosystems: Norwegian pH gradient experiments documented pH-dependent AOB enrichment ($p < 0.001$) concurrent with AOA decline ($p = 0.02$) [59], while meta-analyses indicate non-significantly greater liming effects on AOB versus AOA-*amoA* genes [60]. Biochar amendments in subtropical red soils similarly increased AOB abundance versus CK without affecting AOA, despite the numerical dominance of the latter [61]. These collective findings reveal a critical dichotomy: AOA govern nitrifier-derived N_2O in native acidic soils, whereas pH-modifying amendments shift dominance to AOB. This mechanistic framework explains the contradictory biochar effects on N_2O fluxes [62], where concurrent stimulation of AOB (increasing N_2O) and *nosZ*-harboring denitrifiers (promoting N_2O reduction) yields net effects dependent on their relative activation. However, this study acknowledges the inherent limitations of DNA-based approaches in quantifying microbial activity [63]. Future studies that pair transcriptional assays with isotopic tracing are required to rigorously test the proposed mechanism [64].

Such synergistic approaches offer the potential to maintain soil quality improvements while mitigating AOB-driven N_2O emissions [65]. Despite limited existing research [66], this study proposes that coupling CSMP with approaches that mitigate nitrification could provide a dual-benefit strategy: sustaining soil-quality gains while suppressing GHG emissions [66]. This proposition is supported by previous studies in which lime was combined with DMPP (3,4-dimethylpyrazole phosphate), such as the 61.2% reduction in N_2O emissions achieved relative to lime alone [67]. Future research could therefore consider the combined application of alkaline amendments with nitrification inhibitors to reduce soil N_2O emissions and N losses while amending acidic soils.

5. Conclusions

This study demonstrated that CSMP exerted a strong yet context-dependent influence on soil N_2O emissions in acidic soils. Contrary to the prevailing assumption that alkaline amendments universally mitigate N_2O fluxes via enhanced bacterial N_2O reductase abundance activity compared with CK, these results showed that CSMP application—particularly when co-applied with urea—substantially increased nitrification-driven N_2O emissions by selectively stimulating AOB over AOA. This microbial pathway dominance compared with CK, as evidenced by elevated AOB-*amoA* gene abundance and suppressed AOA/AOB ratios, overrides the potential mitigating effects of increased *nosZ* II gene expression, thereby establishing pH-mediated AOB proliferation as the primary driver of emission intensification. Building on current field practices where nitrification inhibitors effectively reduce N_2O emissions relative to fertilizers without inhibitors, the results suggest that combining CSMP with nitrification inhibitors could offer a promising dual-benefit solution for acidic soil management, simultaneously improving soil conditions while controlling N_2O emissions. Future field studies should evaluate the feasibility of this integrated approach under realistic agricultural conditions.

Supplementary Materials: The following supporting information can be downloaded at: https://www.mdpi.com/article/10.3390/biology14091110/s1, Figure S1: (**a**) pH dynamics in soil CL, (**b**) pH dynamics in soil WS, all as influenced by CSMP and urea application. CK, control; U, urea; CSMP, CSMP fertilizer; U + CSMP, combined application of CSMP fertilizer and urea. Values are means ± SE (n = 3). Different lowercase letters indicate significant differences among treatments ($p < 0.05$); Table S1: physical and chemical properties of the test soil; Table S2: Elemental composition of silicon–calcium–potassium–magnesium fertilizer; Table S3: Primer sequences for PCR amplification. Refs. [68–76] are cited in Supplementary file.

Author Contributions: Conceptualization, X.X. and J.Y.; methodology, X.X.; software, H.G.; validation, X.X. and J.Y.; formal analysis, X.X.; investigation, W.A. and J.Y.; resources, Q.C.; data curation, X.X.; writing—original draft preparation, X.X.; writing—review and editing, H.G., J.Y., W.S., and R.B.T.; visualization, H.G.; supervision, J.Y.; project administration, Q.C.; funding acquisition, Q.C. All authors have read and agreed to the published version of the manuscript.

Funding: This work was financially supported by the National Key Research and Development Program "Intergovernmental Cooperation in International Science and Technology Innovation" (Grant Agreement Number 2023YFE0104700) and the Postdoctoral Fellowship Program of CPSF (Grant Agreement Number GZC20233290).

Data Availability Statement: Data will be made available on request.

Conflicts of Interest: The authors declare that they have no conflicts of interest.

References

1. Filonchyk, M.; Peterson, M.P.; Zhang, L.; Hurynovich, V.; He, Y. Greenhouse gases emissions and global climate change: Examining the influence of CO_2, CH_4, and N_2O. *Sci. Total Environ.* **2024**, *935*, 173359. [CrossRef]
2. Kanter, D.; Mauzerall, D.L.; Ravishankara, A.R.; Daniel, J.S.; Portmann, R.W.; Grabiel, P.M.; Moomaw, W.R.; Galloway, J.N. A post-Kyoto partner: Considering the stratospheric ozone regime as a tool to manage nitrous oxide. *Proc. Natl. Acad. Sci. USA* **2013**, *110*, 4451–4457. [CrossRef]
3. Ravishankara, A.R.; Daniel, J.S.; Portmann, R.W. Nitrous Oxide (N_2O): The Dominant Ozone-Depleting Substance Emitted in the 21st Century. *Science* **2009**, *326*, 123–125. [CrossRef]

4. Du, Y.; Gu, X.; Wang, J.; Niu, W. Yield and gas exchange of greenhouse tomato at different nitrogen levels under aerated irrigation. *Sci. Total Environ.* **2019**, *668*, 1156–1164. [CrossRef]
5. Li, L.; Hong, M.; Zhang, Y.; Paustian, K. Soil N_2O emissions from specialty crop systems: A global estimation and meta-analysis. *Glob. Chang. Biol.* **2024**, *30*, e17233. [CrossRef] [PubMed]
6. Tian, H.; Xu, R.; Canadell, J.G.; Thompson, R.L.; Winiwarter, W.; Suntharalingam, P.; Davidson, E.A.; Ciais, P.; Jackson, R.B.; Janssens-Maenhout, G.; et al. A comprehensive quantification of global nitrous oxide sources and sinks. *Nature* **2020**, *586*, 248–256. [CrossRef]
7. Zhao, Y.; Li, P.; Liu, J.; Xiao, H.; Zhang, A.; Chen, S.; Chen, J.; Liu, H.; Zhu, X.; Hussain, Q.; et al. Microbial effects of prolonged nitrogen fertilization and straw mulching on soil N_2O emissions using metagenomic sequencing. *Agric. Ecosyst. Environ.* **2025**, *382*, 109476. [CrossRef]
8. Thompson, R.L.; Lassaletta, L.; Patra, P.K.; Wilson, C.; Wells, K.C.; Gressent, A.; Koffi, E.N.; Chipperfield, M.P.; Winiwarter, W.; Davidson, E.A.; et al. Acceleration of global N_2O emissions seen from two decades of atmospheric inversion. *Nat. Clim. Chang.* **2019**, *9*, 993–998. [CrossRef]
9. Wang, Y.; Guo, J.; Vogt, R.D.; Mulder, J.; Wang, J.; Zhang, X. Soil pH as the chief modifier for regional nitrous oxide emissions: New evidence and implications for global estimates and mitigation. *Glob. Change Biol.* **2018**, *24*, e617–e626. [CrossRef]
10. Qiu, Y.; Zhang, Y.; Zhang, K.; Xu, X.; Zhao, Y.; Bai, T.; Zhao, Y.; Wang, H.; Sheng, X.; Bloszies, S.; et al. Intermediate soil acidification induces highest nitrous oxide emissions. *Nat. Commun.* **2024**, *15*, 2695. [CrossRef] [PubMed]
11. Guo, C.; Liu, X.; He, X. A global meta-analysis of crop yield and agricultural greenhouse gas emissions under nitrogen fertilizer application. *Sci. Total Environ.* **2022**, *831*, 154982. [CrossRef]
12. Baggs, E.M.; Smales, C.L.; Bateman, E.J. Changing pH shifts the microbial sourceas well as the magnitude of N_2O emission from soil. *Biol. Fert. Soils* **2010**, *46*, 793–805. [CrossRef]
13. Holland, J.E.; Bennett, A.E.; Newton, A.C.; White, P.; McKenzie, B.; George, T.; Pakeman, R.; Bailey, J.; Fornara, D.; Hayes, R. Liming impacts on soils, crops and biodiversity in the UK: A review. *Sci. Total Environ.* **2018**, *610*, 316–332. [CrossRef]
14. Zheng, Y.; Abbott, L.K.; Bolan, N.; Hu, H.; Jenkins, S.N.; Mickan, B.S. Biochar interacted with organic compounds from digestate in controlling N_2O emissions. *J. Environ. Manag.* **2025**, *385*, 125591. [CrossRef]
15. Russenes, A.L.; Korsaeth, A.; Bakken, L.R.; Dörsch, P. Spatial variation in soil pH controls off-season N_2O emission in an agricultural soil. *Soil. Biol. Biochem.* **2016**, *99*, 36–46. [CrossRef]
16. Wu, G.; Liang, F.; Wu, Q.; Feng, X.; Shang, W.; Li, H.; Li, X.; Che, Z.; Dong, Z.; Song, H. Soil pH differently affects N_2O emissions from soils amended with chemical fertilizer and manure by modifying nitrification and denitrification in wheat-maize rotation system. *Biol. Fert. Soils* **2024**, *60*, 101–113. [CrossRef]
17. Zhang, C.; Ju, X.; Zhang, J.; Rees, R.M.; Müller, C. Soil pH and long-term fertilization affect gross N transformation and N_2O production pathways in Chinese and UK croplands. *Biol. Fert. Soils* **2023**, *59*, 527–539. [CrossRef]
18. Wang, Y.; Yao, Z.; Zhan, Y.; Zheng, X.; Zhou, M.; Yan, G.; Wang, L.; Werner, C.; Butterbach-Bahl, K. Potential benefits of liming to acid soils on climate change mitigation and food security. *Glob. Chang. Biol.* **2021**, *27*, 2807–2821. [CrossRef] [PubMed]
19. Hiis, E.G.; Vick, S.H.W.; Molstad, L.; Røsdal, K.; Jonassen, K.R.; Winiwarter, W.; Bakken, L.R. Unlocking bacterial potential to reduce farmland N_2O emissions. *Nature* **2024**, *630*, 421–428. [CrossRef] [PubMed]
20. Shan, J.; Sanford, R.A.; Chee-Sanford, J.; Ooi, S.K.; Löffler, F.E.; Konstantinidis, K.T.; Yang, W.H. Beyond denitrification: The role of microbial diversity in controlling nitrous oxide reduction and soil nitrous oxide emissions. *Glob. Chang. Biol.* **2021**, *27*, 2669–2683. [CrossRef]
21. Bakken, L.R.; Bergaust, L.; Liu, B.; Frostegård, Å. Regulation of denitrification at the cellular level: A clue to the understanding of N_2O emissions from soils. *Philos. Trans. R. Soc. B Biol. Sci.* **2012**, *367*, 1226–1234. [CrossRef]
22. Shaaban, M.; Peng, Q.; Hu, R.; Wu, Y.; Lin, S.; Zhao, J. Dolomite application to acidic soils: A promising option for mitigating N_2O emissions. *Environ. Sci. Pollut. Res.* **2015**, *22*, 19961–19970. [CrossRef]
23. Wan, S.; Lin, Y.; Fan, J.; Hu, H.; Zhang, J.; Jin, S.; Deng, M.; Müller, C.; He, J. Different responses of nitrous oxide emissions to liming and manure amendment of an acidic ultisol are controlled by autotrophic and heterotrophic nitrification. *Soil Biol. Biochem.* **2023**, *178*, 108960. [CrossRef]
24. Xie, L.; Li, L.; Xie, J.; Wang, J.; Mumtaz, M.Z.; Effah, Z.; Fudjoe, S.K.; Khaskheli, M.A.; Luo, Z.; Li, L. Optimal substitution of inorganic fertilizer with organic amendment sustains rainfed maize production and decreases soil N_2O emissions by modifying denitrifying bacterial communities in Northern China. *Eur. J. Agron.* **2024**, *160*, 127287. [CrossRef]
25. Lourenço, K.S.; Costa, O.Y.D.A.; Cantarella, H.; Kuramae, E.E. Ammonia-oxidizing bacteria and fungal denitrifier diversity are associated with N_2O production in tropical soils. *Soil Biol. Biochem.* **2022**, *166*, 108563. [CrossRef]
26. Rousk, J.; Brookes, P.C.; Bååth, E. Investigating the mechanisms for the opposing pH relationships of fungal and bacterial growth in soil. *Soil Biol. Biochem.* **2010**, *42*, 926–934. [CrossRef]
27. Liu, Q.; Wu, Y.; Ma, J.; Jiang, J.; You, X.; Lv, R.; Zhou, S.; Pan, C.; Liu, B.; Xu, Q.; et al. How does biochar influence soil nitrification and nitrification-induced N_2O emissions? *Sci. Total Environ.* **2024**, *908*, 168530. [CrossRef]

28. Clough, T.J.; Bertram, J.E.; Ray, J.L.; Condron, L.; O'CAllaghan, M.; Sherlock, R.; Wells, N. Unweathered Wood Biochar Impact on Nitrous Oxide Emissions from a Bovine-Urine-Amended Pasture Soil. *Soil Sci. Soc. Am. J.* **2010**, *74*, 852–860. [CrossRef]
29. Shen, J.; Tang, H.; Liu, J.; Wang, C.; Li, Y.; Ge, T.; Jones, D.L.; Wu, J. Contrasting effects of straw and straw-derived biochar amendments on greenhouse gas emissions within double rice cropping systems. *Agric. Ecosyst. Environ.* **2014**, *188*, 264–274. [CrossRef]
30. Zhang, B.; Zhou, M.; Zhu, B.; Kemmann, B.; Pfülb, L.; Burkart, S.; Liu, H.; Butterbach-Bahl, K.; Well, R. Threshold-like effect of soil NO_3- concentrations on denitrification product $N_2O/(N_2O+N_2)$ ratio is mediated by soil pH. *Soil Biol. Biochem.* **2023**, *187*, 109213. [CrossRef]
31. Clough, T.J.; Kelliher, F.M.; Sherlock, R.R.; Ford, C.D. Lime and Soil Moisture Effects on Nitrous Oxide Emissions from a Urine Patch. *Soil Sci. Soc. Am. J.* **2004**, *68*, 1600–1609. [CrossRef]
32. Shaaban, M.; Hu, R.; Wu, Y.; Song, L.; Xu, P. Soil pH management for mitigating N_2O emissions through *nosZ* (Clade I and II) gene abundance in rice paddy system. *Environ. Res.* **2023**, *225*, 115542. [CrossRef]
33. Liang, Z.; Elsgaard, L. Nitrous oxide fluxes from long-term limed soils following P and glucose addition: Nonlinear response to liming rates and interaction from added P. *Sci. Total Environ.* **2021**, *797*, 148933. [CrossRef]
34. Hink, L.; Nicol, G.W.; Prosser, J.I. Archaea produce lower yields of N_2O than bacteria during aerobic ammonia oxidation in soil. *Environ. Microbiol.* **2017**, *19*, 4829–4837. [CrossRef]
35. Abalos, D.; Liang, Z.; Dörsch, P.; Elsgaard, L. Trade-offs in greenhouse gas emissions across a liming-induced gradient of soil pH: Role of microbial structure and functioning. *Soil Biol. Biochem.* **2020**, *150*, 108006. [CrossRef]
36. Cheng, Y.; Zhang, H.; Chen, Z.; Wang, J.; Cai, Z.; Sun, N.; Wang, S.; Zhang, J.; Chang, S.X.; Xu, M.; et al. Contrasting effects of different pH-raising materials on N_2O emissions in acidic upland soils. *Eur. J. Soil Sci.* **2021**, *72*, 432–445. [CrossRef]
37. Yin, J.; Cui, W.; Xu, Y.; Ma, Y.; Chen, H.; Guo, J.; Liu, R.; Chen, Q. Understanding the relative contributions of fungi and bacteria led nitrous oxide emissions in an acidic soil amended with industrial waste. *Ecotoxicol. Environ. Saf.* **2023**, *255*, 114727. [CrossRef]
38. *GB/T 36207-2018*; Fertilizer of Calcium Silicon Magnesium Potassium. China National Standards: Shanghai, China, 2018.
39. Zhou, K.; Zhou, Q.; Gong, K.; Zhang, J. Waste-to-resource strategy to fabricate environmentally benign flame retardants from waste phosphorus tailings. *Compos. Commun.* **2020**, *19*, 173–176. [CrossRef]
40. Shen, Y.; Xu, Y.; Liu, Z.; Lin, H.; Zhang, Y.; Song, X.; Liu, P.; Li, Y.; Chen, J.; Chen, H.; et al. Effects of byproduct amendment on enzyme activities and physicochemical properties of acidic orchard soil from Jiaodong Peninsula of China. *Commun. Soil Sci. Plant Anal.* **2018**, *49*, 913–922. [CrossRef]
41. Bakken, L. *Spreadsheet for Gas Kinetics in Batch Cultures: KINCALC*; Researchgate: Berlin, Germany, 2024.
42. Li, S.; Cai, D.; Shaaban, M.; Ma, J.; Liu, S. Liming promotes soil nitrite accumulation but reduces subsequent abiotic nitrous oxide emissions. *Res. Sq.* **2023**, preprint. [CrossRef]
43. Wu, L.; Xiao, Q.; Wang, J.; Huang, Y.; Wu, D.; Liu, J.; Wang, B.; Zhang, H.; Xu, M.; Zhang, W. Liming decreases the emission and temperature sensitivity of N_2O following labile carbon addition. *Geoderma* **2022**, *425*, 116032. [CrossRef]
44. Li, Y.; Chapman, S.J.; Nicol, G.W.; Yao, H. Nitrification and nitrifiers in acidic soils. *Soil Biol. Biochem.* **2018**, *116*, 290–301. [CrossRef]
45. Inatomi, M.; Hajima, T.; Ito, A. Fraction of nitrous oxide production in nitrification and its effect on total soil emission: A meta-analysis and global-scale sensitivity analysis using a process-based model. *PLoS ONE* **2019**, *14*, e219159. [CrossRef]
46. Lin, Y.; Hu, H.; Ye, G.; Fan, J.; Ding, W.; He, Z.; Zheng, Y.; He, J. Ammonia-oxidizing bacteria play an important role in nitrification of acidic soils: A meta-analysis. *Geoderma* **2021**, *404*, 115395. [CrossRef]
47. Pan, B.; Lam, S.K.; Wang, E.; Mosier, A.R.; Chen, D. New approach for predicting nitrification and its fraction of N_2O emissions in global terrestrial ecosystems. *Environ. Res. Lett.* **2021**, *16*, 34053. [CrossRef]
48. Yu, C.; Wang, G.; Zhang, H.; Chen, H.; Ma, Q. Biochar and nitrification inhibitor (dicyandiamide) combination had a double-win effect on saline-alkali soil improvement and soybean production in the Yellow River Delta, China. *Agronomy* **2022**, *12*, 3154. [CrossRef]
49. Farquharson, R. Nitrification rates and associated nitrous oxide emissions from agricultural soils—A synopsis. *Soil Res.* **2016**, *54*, 469–480. [CrossRef]
50. Cheng, Y.; Wang, J.; Wang, J.; Wang, S.; Chang, S.X.; Cai, Z.; Zhang, J.; Niu, S.; Hu, S. Nitrogen deposition differentially affects soil gross nitrogen transformations in organic and mineral horizons. *Earth-Sci. Rev.* **2020**, *201*, 103033. [CrossRef]
51. Samad, M.S.; Ganasamurthy, S.; Highton, M.P.; Bakken, L.R.; Clough, T.J.; de Klein, C.A.; Richards, K.G.; Lanigan, G.J.; Morales, S.E. Urea treatment decouples intrinsic pH control over N_2O emissions in soils. *Soil Biol. Biochem.* **2021**, *163*, 108461. [CrossRef]
52. Norton, J.; Ouyang, Y. Controls and Adaptive Management of Nitrification in Agricultural Soils. *Front. Microbiol.* **2019**, *10*, 1931. [CrossRef]
53. Deng, N.; Gubry-Rangin, C.; Song, X.; Ju, X.; Liu, S.; Shen, J.; Di, H.; Han, L.; Zhang, L. AOB Nitrosospira cluster 3a.2 (D11) dominates N_2O emissions in fertilised agricultural soils. *J. Environ. Manag.* **2024**, *355*, 120504. [CrossRef]
54. Hui, D.; Ray, A.; Kasrija, L.; Christian, J. Impacts of Climate Change and Agricultural Practices on Nitrogen Processes, Genes, and Soil Nitrous Oxide Emissions: A Quantitative Review of Meta-Analyses. *Agriculture* **2024**, *14*, 240. [CrossRef]

55. Dai, Z.; Zhang, Y.; Li, Y.; Xiang, T.; Wang, F.; Peng, J.; Yang, S.; Cao, W. Nitrification regulates the responses of soil nitrous oxide emissions to nitrogen addition in China: A meta-analysis from a gene perspective. *Environ. Sci. Process. Impacts* **2025**, *27*, 2367–2378. [CrossRef]
56. Han, B.; Yao, Y.; Liu, B.; Wang, Y.; Su, X.; Ma, L.; Liu, D.; Niu, S.; Chen, X.; Li, Z. Relative importance between nitrification and denitrification to N_2O from a global perspective. *Glob. Change Biol.* **2024**, *30*, e17082. [CrossRef] [PubMed]
57. Duan, P.; Song, Y.; Li, S.; Xiong, Z. Responses of N_2O production pathways and related functional microbes to temperature across greenhouse vegetable field soils. *Geoderma* **2019**, *355*, 113904. [CrossRef]
58. Prosser, J.I.; Hink, L.; Gubry Rangin, C.; Nicol, G.W. Nitrous oxide production by ammonia oxidizers: Physiological diversity, niche differentiation and potential mitigation strategies. *Glob. Chang. Biol.* **2020**, *26*, 103–118. [CrossRef]
59. Nadeem, S.; Bakken, L.R.; Frostegård, Å.; Gaby, J.C.; Dörsch, P. Contingent effects of liming on N_2O-emissions driven by autotrophic nitrification. *Front. Environ. Sci.* **2020**, *8*, 598513. [CrossRef]
60. Zhang, H.; Liang, Z.; Li, Y.; Chen, Z.; Zhang, J.; Cai, Z.; Elsgaard, L.; Cheng, Y.; van Groenigen, K.J.; Abalos, D. Liming modifies greenhouse gas fluxes from soils: A meta-analysis of biological drivers. *Agric. Ecosyst. Environ.* **2022**, *340*, 108182. [CrossRef]
61. Zhang, K.; Chen, L.; Li, Y.; Brookes, P.C.; Xu, J.; Luo, Y. The effects of combinations of biochar, lime, and organic fertilizer on nitrification and nitrifiers. *Biol. Fert. Soils* **2017**, *53*, 77–87. [CrossRef]
62. Ji, C.; Li, S.; Geng, Y.; Miao, Y.; Ding, Y.; Liu, S.; Zou, J. Differential responses of soil N_2O to biochar depend on the predominant microbial pathway. *Appl. Soil Ecol.* **2020**, *145*, 103348. [CrossRef]
63. Smith, C.J.; Osborn, A.M. Advantages and limitations of quantitative PCR (Q-PCR)-based approaches in microbial ecology. *FEMS Microbiol. Ecol.* **2009**, *67*, 6–20. [CrossRef]
64. Yu, L.; Harris, E.; Lewicka-Szczebak, D.; Barthel, M.; Blomberg, M.R.; Harris, S.J.; Johnson, M.S.; Lehmann, M.F.; Liisberg, J.; Müller, C.; et al. What can we learn from N_2O isotope data?—Analytics, processes and modelling. *Rapid Commun. Mass Spectrom.* **2020**, *34*, e8858. [CrossRef] [PubMed]
65. Lei, J.; Fan, Q.; Yu, J.; Liu, R. A meta-analysis to examine whether nitrification inhibitors work through selectively inhibiting ammonia-oxidizing bacteria. *Front. Microbiol.* **2022**, *13*, 962146. [CrossRef] [PubMed]
66. Liu, R.; Hayden, H.L.; Hu, H.; He, J.; Suter, H.; Chen, D. Effects of the nitrification inhibitor acetylene on nitrous oxide emissions and ammonia-oxidizing microorganisms of different agricultural soils under laboratory incubation conditions. *Appl. Soil Ecol.* **2017**, *119*, 80–90. [CrossRef]
67. Liu, C.; Mo, T.; Zhong, J.; Chen, H.; Xu, H.; Yang, X.; Li, Y. Synergistic benefits of lime and 3,4-dimethylpyrazole phosphate application to mitigate the nitrous oxide emissions from acidic soils. *Ecotoxicol. Environ. Saf.* **2023**, *263*, 115387. [CrossRef]
68. Mori, H.; Maruyama, F.; Kato, H.; Toyoda, A.; Dozono, A.; Ohtsubo, Y.; Nagata, Y.; Fujiyama, A.; Tsuda, M.; Kurokawa, K. Design and experimental application of a novel non-degenerate universal primer set that amplifies prokaryotic 16S rRNA genes with a low possibility to amplify eukaryotic rRNA genes. *DNA Res.* **2014**, *21*, 217–227. [CrossRef]
69. Yao, Q.; Liu, J.; Yu, Z.; Li, Y.; Jin, J.; Liu, X.; Wang, G. Three years of biochar amendment alters soil physiochemical properties and fungal community composition in a black soil of northeast China. *Soil Biol. Biochem.* **2017**, *110*, 56–67. [CrossRef]
70. Leininger, S.; Urich, T.; Schloter, M.; Schwark, L.; Qi, J.; Nicol, G.W.; Prosser, J.I.; Schuster, S.C.; Schleper, C. Archaea predominate among ammonia-oxidizing prokaryotes in soils. *Nature* **2006**, *442*, 806–809. [CrossRef]
71. Schauss, K.; Focks, A.; Leininger, S.; Kotzerke, A.; Heuer, H.; Thiele Bruhn, S.; Sharma, S.; Wilke, B.M.; Matthies, M.; Smalla, K. Dynamics and functional relevance of ammonia–oxidizing archaea in two agricultural soils. *Environ. Microbiol.* **2009**, *11*, 446–456. [CrossRef]
72. Rotthauwe, J.; Witzel, K.; Liesack, W. The ammonia monooxygenase structural gene *amoA* as a functional marker: Molecular fine-scale analysis of natural ammonia-oxidizing populations. *Appl. Environ. Microb.* **1997**, *63*, 4704–4712. [CrossRef]
73. Harter, J.; Krause, H.-M.; Schuettler, S.; Ruser, R.; Fromme, M.; Scholten, T.; Kappler, A.; Behrens, S. Linking N_2O emissions from biochar-amended soil to the structure and function of the N-cycling microbial community. *ISME J.* **2014**, *8*, 660–674. [CrossRef] [PubMed]
74. Henry, S.; Bru, D.; Stres, B.; Hallet, S.; Philippot, L. Quantitative detection of the *nosZ* gene, encoding nitrous oxide reductase, and comparison of the abundances of 16S rRNA, *narG*, *nirK*, and *nosZ* genes in soils. *Appl. Environ. Microb.* **2006**, *72*, 5181–5189. [CrossRef] [PubMed]
75. Jones, C.M.; Graf, D.R.H.; Bru, D.; Philippot, L.; Hallin, S. The unaccounted yet abundant nitrous oxide-reducing microbial community: A potential nitrous oxide sink. *ISME J.* **2013**, *7*, 417–426. [CrossRef]
76. Long, A.; Song, B.K.; Fridey, K.; Silva, A. Detection and diversity of copper containing nitrite reductase genes (*nirK*) in prokaryotic and fungal communities of agricultural soils. *FEMS Microbiol. Ecol.* **2015**, *91*, 1–9. [CrossRef] [PubMed]

biology

MDPI

Article

Leaf Nitrogen Allocation Trade-Offs Promote Efficient Utilization of Different Nitrogen Forms in *Hemarthria altissima*

Nan Xu [1,†], Xiaowei Wei [2,†], Ju Zhang [2], Mingyue Sun [2], Jinwei Zhang [3], Zihao Zhao [2] and Xuechen Yang [4,*]

1 Key Laboratory of Heilongjiang Province for Cold-Regions Wetlands Ecology and Environment Research, Harbin University, Harbin 150086, China; xunan0451@126.com

2 Jilin Provincial Key Laboratory for Plant Resources Science and Green Production, Jilin Normal University, Siping 136000, China; weixiaowei@jlnu.edu.cn (X.W.); 13894155011@163.com (J.Z.); 18543444519@163.com (M.S.); 13397005645@163.com (Z.Z.)

3 Department of Grassland Science, College of Animal Science and Technology, Northeast Agricultural University, Harbin 150086, China; zhangjw133@nenu.edu.cn

4 State Key Laboratory of Ecological Safety and Sustainable Development in Arid Lands, Xinjiang Institute of Ecology and Geography, Chinese Academy of Sciences, Urumqi 830011, China

* Correspondence: yangxuechen@ms.xjb.ac.cn

† These authors contributed equally to this work.

Simple Summary

Global environmental change has intensified nitrogen deposition, resulting in alterations to the availability and balance of nitrate (NO_3^-), ammonium (NH_4^+), and their mixtures in the soil. These nitrogen forms are known to exert distinct effects on plant physiology, photosynthesis, and nutrient-use strategies. While nitrate is often associated with enhanced plant growth, ammonium may induce stress or toxicity, depending on the species and environmental conditions. A comprehensive understanding of plant responses to different nitrogen forms is considered essential for maintaining grassland productivity and ecological resilience. In this study, the effects of nitrate, ammonium, and their combination on *Hemarthria altissima*, a valuable forage species in grassland ecosystems, were investigated. The influence of different nitrogen forms on soil nitrogen availability, photosynthetic performance, nitrogen allocation within the photosynthetic system, and photosynthetic nitrogen-use efficiency (PNUE) were analyzed. Key adaptive mechanisms of *H. altissima* under varying nitrogen conditions were revealed, contributing to strategies for sustainable grassland management under the pressures of global environmental change.

Abstract

The sharp increase in atmospheric nitrogen deposition has had profound effects on nitrogen availability and the photosynthetic capacity of terrestrial plants. Consequently, understanding the intricate trade-off between nitrogen sources and their allocation within leaves is essential for unraveling the photosynthetic responses of grassland ecosystems to nitrogen deposition. In a series of field experiments, the effects of different nitrogen forms (ammonium and nitrate nitrogen) on nitrogen assimilation and allocation in the C_4 plant *Hemarthria altissima* were thoroughly investigated. Towards the end of the growing season, *H. altissima* was observed to exhibit high photosynthetic efficiency. Ammonium nitrogen treatment notably enhanced photosynthetic nitrogen use efficiency (PNUE) by modifying the nitrogen allocation within the leaf's photosynthetic apparatus and leaf area, leading to a significant improvement in photosynthetic efficiency and biomass accumulation. Under ammonium nitrogen treatment, *H. altissima* directed more nitrogen toward its carboxylation process and other protein-related functions to increase carboxylation efficiency, thereby facilitating the accumulation of photosynthetic products. In contrast, under nitrate nitrogen treatment, the plant balanced growth and light absorption by allocating nitrogen to leaf light-capturing proteins. The application of both ammonium and nitrate nitrogen resulted in increased nitrogen content in the soil, as ammonium nitrogen is converted to nitrate nitrogen through nitrification. The net photosynthetic rate (A_n), nitrogen allocation to photosynthetic components (N_{psn}), and chlorophyll content per unit area (Chl_{area}) were all significantly and positively correlated with photosynthetic nitrogen use efficiency (PNUE). Notably, under the sole NH_4^+ treatment, nitrogen allocation to the photosynthetic components increased, which enhanced the NPQ and ETR in *H. altissima* leaves. These findings suggest that *H. altissima* preferentially utilizes ammonium nitrogen from the soil, optimizing its PNUE and biomass accumulation through a strategic allocation of nitrogen within its leaves. Further investigation is needed to explore how these nitrogen allocation strategies may vary under different environmental conditions and how they influence ecosystem-level productivity.

Keywords: ammonium nitrogen; nitrate nitrogen; photosynthetic nitrogen-use efficiency; leaf nitrogen allocation; *Hemarthria altissima*

1. Introduction

Nitrogen (N) plays a pivotal role in supporting plant growth and development, with a profound influence on ecosystem productivity. For plants, nitrogen is absorbed predominantly in the forms of ammonium (NH_4^+) and nitrate (NO_3^-), which are not only vital for metabolic processes but also represent the primary forms of nitrogen loading resulting from atmospheric deposition [1]. Grasses, especially forage species, exhibit a high degree of plasticity in their ability to absorb and utilize these nitrogen forms. This ability allows them to thrive in a wide range of environmental conditions, adjusting their nitrogen acquisition strategies based on the relative availability of NH_4^+ and NO_3^- in the soil [2]. Such flexibility is crucial for maintaining high biomass production in grasslands, which is essential for both agricultural productivity and ecological balance.

The availability of ammonium nitrogen (NH_4^+) and nitrate nitrogen (NO_3^-) in the soil directly impacts several physiological processes, from nitrogen uptake to photosynthesis and overall plant biomass. Grasses often show a stronger growth response to NH_4^+, as it is energetically less expensive to assimilate compared to NO_3^-. This is particularly advantageous under conditions where nitrogen availability is low or sporadic. However, while NH_4^+ provides a readily available nitrogen source, its accumulation can have toxic effects, particularly at higher concentrations. Excess NH_4^+ can lead to disruptions in ionic homeostasis, resulting in oxidative stress and metabolic disturbances, which impair plant growth and development [3]. Such imbalances can affect cellular respiration, nutrient uptake, and even chlorophyll synthesis, resulting in stunted growth and reduced photosynthetic efficiency.

On the other hand, nitrate (NO_3^-), while requiring more energy to assimilate, offers a more stable nitrogen source for grasses when available in adequate amounts. NO_3^- uptake by grasses is generally regulated through high-affinity and low-affinity transporter systems in roots, which allow plants to efficiently exploit soil nitrogen at varying concentrations [4]. The presence of NO_3^- in the soil also enhances the assimilation of NH_4^+, particularly in the roots, as nitrate reductase activity increases in response to nitrate uptake, facilitating the processing of both nitrogen forms [5]. This cross-regulation between NH_4^+ and NO_3^- highlights the complex relationship between nitrogen forms and their collective impact on plant growth. Notably, grasses that have access to both nitrogen forms tend to exhibit improved nitrogen-use efficiency (NUE), enhancing their ability to allocate nitrogen toward biomass production and photosynthetic capacity [6].

The relationship between carbon and nitrogen metabolism is intricately connected at multiple scales, from leaf to whole-plant levels. In grasses, nitrogen influences both carbon assimilation and photosynthetic efficiency, which are critical for overall growth and productivity. Nitrogen plays a central role in the synthesis of enzymes involved in carbon fixation and electron transport. Specifically, nitrogen is a key component of Rubisco (ribulose-1,5-bisphosphate carboxylase/oxygenase), the enzyme responsible for carbon fixation in the C_4 photosynthetic pathway of grasses, which is highly efficient in tropical and subtropical regions [7]. Thus, nitrogen availability directly influences photosynthetic nitrogen-use efficiency (PNUE), a critical trait for the productivity of forage grasses. Even small fluctuations in nitrogen availability can have a profound impact on carboxylation efficiency and light-harvesting efficiency, with significant consequences for plant biomass accumulation [8].

In the absence of sufficient nitrogen, grasses experience impaired photosynthesis, leading to reduced growth and altered nitrogen allocation across plant tissues. For instance, during nitrogen deficiency, grasses may reallocate nitrogen away from photosynthetic proteins and into structural components, including cell walls and defensive molecules, as a stress response to optimize growth under limited nitrogen availability [9]. This shift in nitrogen allocation impacts leaf mass per area (LMA) and the overall leaf economics spectrum, with native species typically investing more nitrogen into defensive proteins than invasive species, which prioritize photosynthetic machinery [10].

Gramineous forage grasses, such as maize, sorghum, and ryegrass, adapt their nitrogen-use strategies to optimize growth under variable soil nitrogen conditions. For example, under low-nitrogen conditions, some grasses shift nitrogen allocation toward maintaining critical nitrogen enzymes and bioenergetic pathways, ensuring they can sustain electron transport and cellular respiration [11]. In contrast, excessive nitrogen supply, particularly from ammonium, can lead to a shift toward greater root growth and higher allocation to structural components such as cell walls, rather than photosynthetic tissues, which ultimately reduces the efficiency of carbon fixation and slows growth [12].

Understanding the interaction between soil nitrogen forms and grass growth is essential for agricultural management, particularly in the context of optimizing fertilizer applications for forage production. By balancing ammonium and nitrate inputs, farmers and land managers can enhance forage yield, improve nutrient-use efficiency, and minimize nitrogen leaching and environmental pollution. Furthermore, identifying species that are more resilient to fluctuations in nitrogen availability could lead to more sustainable farming practices, especially in regions where nitrogen inputs are becoming increasingly unpredictable due to environmental changes and agricultural practices. The following hypothesis is proposed: (1) Increasing nitrogen deposition (inorganic nitrogen forms) will lead to enhanced nitrogen assimilation and improved nitrogen-use efficiency in *H. altissima*, particularly in environments with high ammonium nitrogen availability. (2) The adaptive response of *H. altissima* to rising nitrogen deposition will be characterized by increased nitrogen allocation to photosynthetic components, improving photosynthetic efficiency and biomass production, but potentially leading to trade-offs in other physiological processes like growth and root development.

2. Materials and Methods

2.1. Study Area and Growth Conditions

The field experiment was conducted at the Jilin Songnen Grassland Ecosystem National Observation and Research Station (44°34′ N, 123°31′ E), China, within a temperate continental monsoon climate zone. The region has hot, rainy summers and cold, dry winters. Soil in the 0–20 cm layer exhibited a pH of 8.68, EC of 78.76 $\mu S \cdot cm^{-1}$, total N of 1.02 $g \cdot kg^{-1}$, total P of 0.66 $g \cdot kg^{-1}$, organic C of 6.37 $g \cdot kg^{-1}$, NH_4^+-N of 1.23 $mg \cdot kg^{-1}$, and NO_3^--N of 1.89 $mg \cdot kg^{-1}$. Annual mean temperature ranged from 4.6 to 6.5 °C, with total precipitation between 280 and 620 mm, mostly falling between June and September [13]. The pot experiment used wind-sand soil (3.5 kg pot^{-1}) and followed a completely randomized block design with six replicates.

Hemarthria altissima (Poir.) Stapf & C. E. Hubb. is a perennial C_4 forage grass in the Poaceae family, known for its long, horizontally creeping rhizomes and strong nutritional and reproductive capacities. It is widely distributed across the natural meadows of the Songnen Plain, where it often dominates or serves as a key companion species in plant communities [14]. On May 10th, shoots of *H. altissima* were transplanted into plastic pots, collected from the eastern Eurasian meadow steppe. Based on the population density during the green period (May 10th–May 30th), six individuals per pot were planted in monoculture, and plots were harvested on September 15th. Nitrogen treatments included: no nitrogen (N0), sole NH_4^+-N [as $(NH_4)_2SO_4$ (Sinopharm Chemical Reagent Co., Ltd., Shanghai, China)] (AN), sole NO_3^--N [as $Ca(NO_3)_2$ (Sinopharm Chemical Reagent Co., Ltd., Shanghai, China)] (NN), and a 1:1 mixture of NH_4^+-N and NO_3^--N (NAN), with 10 g N m^{-2} applied in two equal doses on May 30th and June 15th. Dicyandiamide (DCD, 98.0%) (Sinopharm Chemical Reagent Co., Ltd., Shanghai, China) was added to the AN and NAN treatments to inhibit nitrification, at 10 mg m^{-2} y^{-1} and 5 mg m^{-2} y^{-1}, respectively. Apply Hogland nutrient solution (KH_2PO_4 2.5 mmol L^{-1}, $MgSO_4 \cdot 7H_2O$ 2 $mmol \cdot L^{-1}$, H_3BO_3 40 $\mu mol \cdot L^{-1}$, $MnCl_2 \cdot 4H_2O$ 10 $\mu mol \cdot L^{-1}$, $ZnSO_4 \cdot 7H_2O$ 0.8 $\mu mol \cdot L^{-1}$, $CuSO_4 \cdot 5H_2O$ 0.4 $\mu mol \cdot L^{-1}$, $Na_2MoO_4 \cdot 2H_2O$ 0.2 $\mu mol \cdot L^{-1}$, EDTA-Fe 20 $\mu mol \cdot L^{-1}$) (Tianjin Huasheng Tianhe Chemical Trading Co., Ltd., Tianjin, China) once a month, 200 mL pot^{-1} each time, all treatments received additional fertilizers (P, K, S) and micronutrients (Zn, B, Mn, Mo, Cu, Fe) to avoid nutrient limitations. Weeds, insects, and diseases were controlled, and the plots were exposed to natural precipitation with minimal irrigation. Harvesting occurred at the post-fruiting stage on September 15th.

2.2. Gas Exchange Measurements and Chlorophyll Fluorescence

From August 2nd to 10th, leaf assimilation rate (A_n, μmol m^{-2} s^{-1}), stomatal conductance (G_s, mmol m^{-2} s^{-1}), internal CO_2 (C_i, μmol mol^{-1}), and water use efficiency (*WUE*, %) were measured using a CIRAS-3 portable photosynthesis system (PP Systems, Amesbury, MA, USA) at 25 °C, with a CO_2 concentration of 400 μmol mol^{-1} and a 500 μmol s^{-1} flow rate. The photosynthetic photon flux density (PPFD) was set to 2000 μmol m^{-2} s^{-1} [15]. For the rapid A/C_i response curve [16], CO_2 partial pressure was varied from 50 to 1600 μmol mol^{-1}. Gas exchange measurements were taken from the 2nd and 3rd leaves of the shoot, between 8:00 AM and 3:00 PM, with six replicates per pot.

The maximum rate of Rubisco carboxylation (V_{cmax}, μmol m^{-2} s^{-1}) and the maximum rate of electron transport (J_{max}, μmol m^{-2} s^{-1}) were derived from A_n/C_i curve data, and the values were fitted using the models of [17,18]. The calculations followed these formulas:

$$V_{cmax} = \frac{(R_d + A_n)\left[C_i + K_C\left(1 + \frac{O}{K_O}\right)\right]}{(C_i - \Gamma^*)} \tag{1}$$

$$J_{max} = \frac{4(R_d + A_n)(C_i + 2\Gamma^*)}{(C_i - \Gamma^*)} \tag{2}$$

where R_d represents the mitochondrial respiration rate in the light (μmol m^{-2} s^{-1}), K_c and K_o are the Michaelis constants for carboxylation and oxygenation, respectively, O is the intercellular oxygen concentration (approximately 210 mmol mol^{-1}), and Γ^* is the CO_2 compensation point in the absence of respiration (μmol mol^{-1}). Additionally, the values for *Kc*, *Ko* and Γ^* were determined using the temperature-dependent functions outlined [18].

The following day, chlorophyll fluorescence measurements were taken using the IMAGING PAM M-series (Walz, Effeltrich, Germany). Prior to the measurements, samples were dark-adapted for 30 min. The maximum quantum yield of PSII (F_v/F_m), the effective quantum yield of PSII (φPSII), the non-photochemical quenching coefficient (NPQ), and the electron transport rate (ETR, μmol e^{-1} s^{-1} m^{-2}) were referred to previous analytical methods [15]. After measuring the chlorophyll fluorescence parameters, leaf area was determined using a portable leaf area meter (AM350, ADC Bio Scientific Ltd., Herts, UK).

2.3. Sample Collection and Chemical Analyses

On September 15, soil samples were collected using the five-point method. After passing through a 100-micron sieve and removing the roots, the moist soil was separated into two parts using a 1-mm sieve. One part was kept fresh (4 °C) for the analysis of ammonium nitrogen and nitrate nitrogen in the soil. The other half was air-dried naturally for the determination of total nitrogen (TN) in the soil. The total nitrogen content was determined using an elemental analyser (vario EL cube, Elementar, Langenselbold, Germany). A 10 g fresh soil sample was placed in a 100-mL flask containing 50 mL of 2 M KCL to measure extractable ammonium and nitrate. The sample was shaken at 180 rpm for 1 h on an orbital shaker. Afterward, the extract was filtered through a 0.45 μm nylon net Millipore filter (prewashed with 2 M KCL). The filtered extracts were stored in plastic vials and frozen at −20 °C for no longer than one week before analysis. Soil ammonium and nitrate concentrations were measured using an Alliance Flow Analyser (Alliance Flow Analyser, Futura, Frépillon, France). Dissolved inorganic nitrogen (DIN) was calculated as the sum of the extractable ammonium and nitrate in the soil.

Two leaves from each plant were collected, immediately frozen in liquid nitrogen, and stored at −80 °C for biochemical analysis. Additionally, two leaves were subjected to enzyme activity inhibition by heating at 105 °C for 30 min, then dried to a constant weight at 65 °C. These leaves were used for biomass measurement and total nitrogen content analysis (N_m, mg g^{-1}) using an Elementar Vario EL Cube (Elementar, Langenselbold, Germany). Leaf mass per unit leaf area (*LMA*, g m^{-2}) and leaf nitrogen content per unit leaf area (N_{area}, g m^{-2}) were calculated using the formula: $N_{area} = N_m \times LMA$. Chlorophyll content per leaf mass (Chl_m, mg g^{-1}) was quantified by extracting 0.1 g leaf tissue in ethanol, and absorbance was measured at 645 nm and 663 nm using a spectrophotometer (UVmini-1240, Shimadzu, Kyoto, Japan), following the method of Wellburn (1994) [19]. Chlorophyll content was calculated as follows:

$$Chl_a = 12.56 \times A_{665} - 2.71 \times A_{645} \quad (3)$$

$$Chl_b = 22.65 \times A_{645} - 4.35 \times A_{663} \quad (4)$$

$$Chl_m = Chl_a + Chl_b \quad (5)$$

Chlorophyll content per leaf area (Chl_{area}) was determined by multiplying Chl_m by *LMA*.

To quantify nitrate nitrogen and ammonium nitrogen content, 2.0 g of lyophilized leaf samples were incubated with 10 mL distilled water, boiled for 1 h, and filtered to obtain the crude extract. Nitrate concentration (NO_3^-) was measured using the salicylic acid chromogenic method [20], while ammonium concentration (NH_4^+) was determined by the phenol-hypochlorite method [21]. Free amino acids were analyzed using the ninhydrin colorimetric method [22].

For the analysis of different nitrogen forms, the procedure described by Takashima et al. (2004) and Onoda et al. (2017) was followed with some modifications [7,8]. Leaves were powdered in liquid nitrogen and homogenized in 2 mL of Na-phosphate buffer (pH 7.5, 100 mmol L^{-1}), followed by washing in a centrifuge tube. This procedure was repeated three times. The homogenates were centrifuged at 12,000× *g* for 10 min at 4 °C, and the supernatant was collected as soluble protein. The pellet was washed with 1 mL of phosphate buffer containing 3% sodium dodecyl sulfate (SDS), and after heating at 90 °C for 5 min, it was centrifuged again. This step was repeated six times, collecting the SDS-soluble protein. The residue was regarded as cell wall protein, washed with ethanol, and filtered onto quantitative filter paper. The supernatant was precipitated with 10% trichloroacetic acid (TCA) by heating at 85 °C for 5 min. The precipitate was filtered, washed with ethanol, and dried at 85 °C before being analyzed for nitrogen content by the Elementar Vario EL Cube.

Finally, the enzyme activities of ribulose-1,5-bisphosphate carboxylase/oxygenase (Rubisco), phosphoenolpyruvate carboxylase (PEPC), nitrate reductase (NR), nitrite reductase (NiR), glutamine synthetase (GSI), and Glutamate synthase (GOGAT)were determined in frozen leaves using specific ELISA kits (Shanghai Enzyme Biotechnology Co., Ltd., Shanghai, China), following the manufacturer's instructions.

2.4. Estimation of Nitrogen Allocation in the Photosynthetic Machinery and Its Efficiency in Photosynthetic Nitrogen Utilization

According to the LUNA model [23,24], leaf photosynthetic nitrogen is divided into three main components: the fractions allocated to the carboxylation system (PN_C, g g^{-1}), electron transport components (PN_B, g g^{-1}), and light-harvesting components (PN_L, g g^{-1}). These components were calculated as follows:

$$PN_C = \frac{V_{cmax}}{6.25 \times V_{cr} \times N_{area}} \quad (6)$$

$$PN_B = \frac{J_{max}}{8.06 \times J_{mc} \times N_{area}} \quad (7)$$

$$PN_C = \frac{C_c}{C_B \times N_{area}} \quad (8)$$

where 6.25 (g Rubisco g^{-1} N) is the conversion factor for Rubisco N at 25 °C [25], and V_{cr} = 20.78 (μmol CO_2 g^{-1} Rubisco s^{-1}) [23]. The factor 8.06 is the N conversion coefficient for cytochrome [26], and J_{mc} = 155.65 (μmol e^{-1} μmol cytochrome f s^{-1}) at 25 °C [23,24]. C_c refers to leaf chlorophyll content (mmol g^{-1}), and C_b is the chlorophyll binding to light-harvesting components (2.15 mmol g^{-1} N) [27].

The fractions of leaf N allocated to the thylakoid ($PN_B + PN_L$, g g^{-1}) and to the photosynthetic apparatus (PN_{PSN}, g g^{-1}) are the sums of PN_B and PN_L, and of PNc, PN_B, and PN_L, respectively. The N contents in the carboxylation system (Nc, g m^{-2}), bioenergetics (N_B, g m^{-2}), light-harvesting system (N_L, g m^{-2}), and the entire photosynthetic apparatus (N_{psn}, g m^{-2}) were calculated by multiplying PN_C, PN_B, PN_L, and PN_{psn} by N_{area}. The remaining leaf N was considered as other N. Photosynthetic N use efficiency (PNUE, μmol g N^{-2} s^{-1}) was determined as the ratio of A_n to N_{area} [28].

2.5. Statistical Analysis

All data were examined for a normal distribution (Kolmogorov–Smirnov test) and homogeneity of variance (Levene' s test) and conducted using R version 4.5.1 (R Core Team, 2025). Analyses were performed using the "Fisher's Least Significant Difference" function from "agricolae" package, differences were considered significant for $p < 0.05$. For correlation analysis, the "pearson" function in the "gpairs" and "ggpmisc" packages was utilized, and the "ggplot2" package was employed for creating graphics.

3. Results

3.1. Soil Nitrogen Characteristics

The effects of N0, AN, NN, and ANN on soil total nitrogen, soil NO_3^- -N, soil NH_4^+-N and soil DIN were significant ($p < 0.05$) (Figure 1). The soil total nitrogen of the AN, NN and ANN treatments were significantly higher than N0 treatment ($p < 0.05$) (Figure 1A). The soil NO_3^- -N, soil NH_4^+-N and soil DIN of the AN treatment were significantly higher than those of the N0, NN and ANN treatments ($p < 0.05$) (Figure 1B–D).

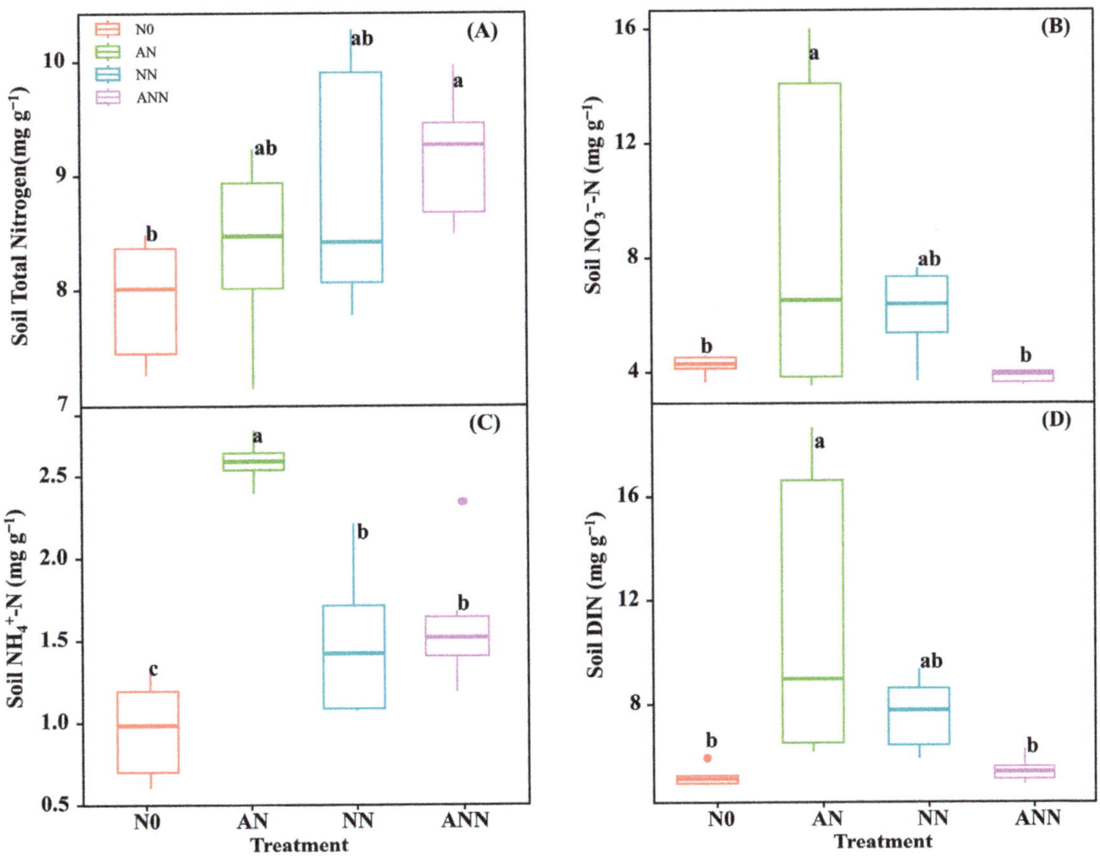

Figure 1. Effect of nitrogen forms treatments on soil total nitrogen (**A**), soil NO_3^--N (**B**), soil NH_4^+-N (**C**) and soil DIN (**D**) in *H. altissima*. Different lower-case letters indicate significant differences between the measuring dates under the unfertilized (N0) treatment and the fertilized (AN, NN, ANN) treatment, respectively ($p < 0.05$) (n = 6).

3.2. Leaf Nitrogen Assimilation Enzyme Activity

Nitrogen absorption from the soil depends on the activity of enzymes involved in nitrogen metabolism. NR and NiR activities were stimulated in the NN and ANN treatments; conversely, they were inhibited in the AN treatment ($p < 0.05$) (Figure 2A,B). The GSI activity of the AN treatment was significantly higher than N0 and NN treatments ($p < 0.05$) (Figure 2C). The GOGAT activity of the AN, NN and ANN treatments were significantly higher than N0 treatment ($p < 0.05$), but the GOGAT activity in the AN, NN, and ANN treatments showed no significant differences (Figure 2D).

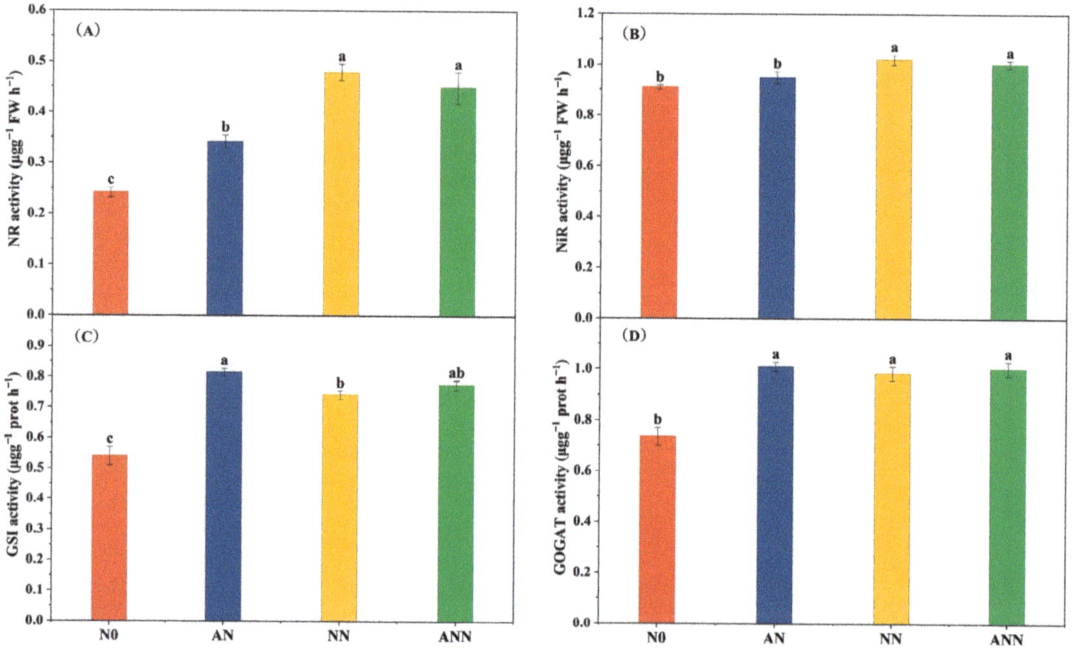

Figure 2. Effect of nitrogen forms treatments on NR activity (**A**), NiR activity (**B**), GSI activity (**C**) and GOGAT activity (**D**) in *H. altissima*. Different lower-case letters indicate significant differences between the measuring dates under the unfertilized (N0) treatment and the fertilized (AN, NN, ANN) treatment, respectively ($p < 0.05$) (n = 6).

3.3. Leaf Gas Exchange Parameters and Morphological Characteristics

The A_n and G_s of the AN and NN treatments were significantly higher than N0 and ANN treatments, and the A_n and G_s of the ANN were significantly higher than N0 ($p < 0.05$) (Figure 3A,B). The C_i of the AN and ANN treatments were significantly lower than N0 and NN treatments ($p < 0.05$) (Figure 3C). The *WUE* of the AN treatment was significantly higher than N0, NN and ANN treatments ($p < 0.05$) (Figure 3D). The leaf area and LMA of the AN, NN and ANN treatments were significantly higher than N0 treatment ($p < 0.05$) (Figure 3E,F).

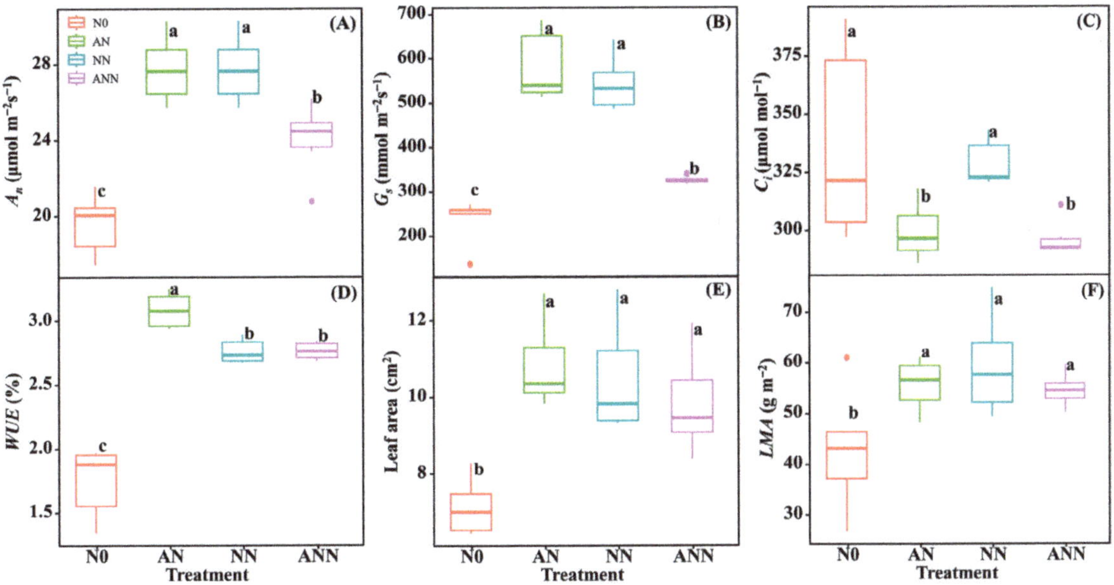

Figure 3. Effect of nitrogen forms treatments on net CO_2 assimilation rate (A_n) (**A**), stomatal conductance (G_s) (**B**), internal CO_2 (C_i) (**C**), water use efficiency (*WUE*) (**D**), leaf area (**E**) and leaf mass area (*LMA*) (**F**) in *H. altissima*. Different lower-case letters indicate significant differences between the measuring dates under the unfertilized (N0) treatment and the fertilized (AN, NN, ANN) treatment, respectively ($p < 0.05$) (n = 6).

3.4. Leaf Photosynthetic Pigment

The Chl_a of the AN, NN and ANN treatments were no significant difference (Table 1). The Chl_b of the AN treatment was 86.53% and 40.58% higher than N0 and NN treatments ($p < 0.05$). The Chl_m of the AN treatment were 29.55%, 15.54% and 20.42% higher than N0, NN and ANN treatments ($p < 0.05$). The Chl_{area} of the AN, NN and ANN treatments were significantly higher than N0 treatment ($p < 0.05$) (Table 1).

Table 1. Effect of nitrogen forms treatments on the content of chlorophyll in *H. altissima*.

N Form Treatments	Chl_a (mg g^{-1})	Chl_b (mg g^{-1})	Chl_m (mg g^{-1})	Chl_{area} (g m^{-2})
N0	0.80 ± 0.06 a	0.52 ± 0.01 c	1.32 ± 0.06 c	0.55 ± 0.04 b
AN	0.73 ± 0.07 ab	0.97 ± 0.06 a	1.71 ± 0.05 a	0.95± 0.05 a
NN	0.79 ± 0.05 ab	0.69± 0.05 bc	1.48 ± 0.05 b	0.88± 0.07 a
ANN	0.59 ± 0.09 b	0.88 ± 0.11 ab	1.42 ± 0.05 bc	0.81 ± 0.04 a

Different lower-case letters indicate significant differences between the measuring dates under the un-fertilized (N0) treatment and the fertilized (AN, NN, ANN) treatment, respectively ($p < 0.05$) (n = 6).

3.5. Leaf Photosynthetic Efficiency and Photosynthetic Nitrogen Utilization Efficiency

The V_{cmax} and J_{max} of the ANN treatment was significantly higher than other treatments, and the V_{cmax} and J_{max} of the AN and NN were significantly higher than N0 ($p < 0.05$) (Figure 4A,B). The total leaf biomass of the AN, NN and ANN treatments were significantly higher than N0 treatment ($p < 0.05$) (Figure 4C). The N_{area} of the N0 treatment was significantly higher than AN, NN and ANN treatments ($p < 0.05$) (Figure 4D). The PNUE of the AN treatment was significantly higher than N0, NN and ANN treatments, and the PNUE of the NN and ANN were significantly higher than N0 ($p < 0.05$) (Figure 4E).

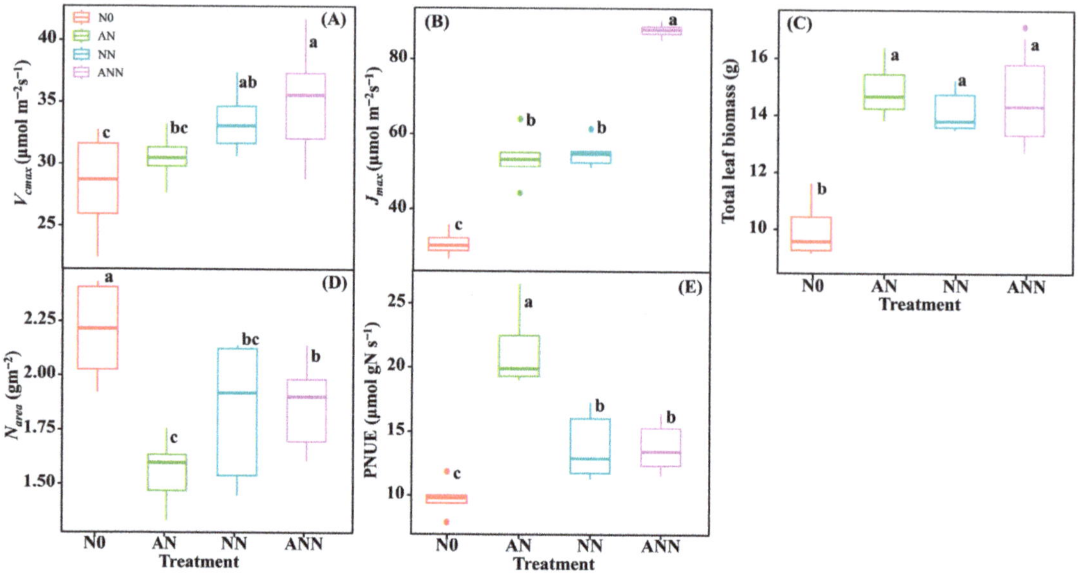

Figure 4. Effect of nitrogen forms treatments on maximum carboxylation rate (V_{cmax}) (**A**), maximum photoelectron transfer rate (J_{max}) (**B**), total leaf biomass (**C**), area-based nitrogen content (N_{area}) (**D**) and photosynthetic N use efficiency (PNUE) (**E**) in *H. altissima*. Different lower-case letters indicate significant differences between the measuring dates under the unfertilized (N0) treatment and the fertilized (AN, NN, ANN) treatment, respectively ($p < 0.05$) (n = 6).

3.6. *Leaf Photosynthetic Enzyme Activity*

The effects of N0, AN, NN, and ANN on Rubisco activity and PEPC activity were significant ($p < 0.05$) (Figure 5). The Rubisco activity and PEPC activity of the AN treatment was significantly higher than other treatments, and the Rubisco activity of the NN and ANN were significantly higher than N0 ($p < 0.05$) (Figure 5A). However, the PEPC activity in the N0, NN, and ANN treatments showed no significant differences (Figure 5B).

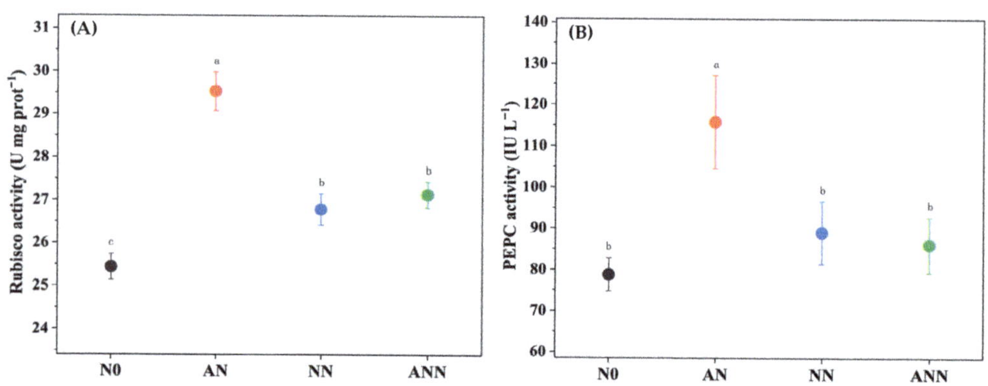

Figure 5. Effect of nitrogen forms treatments on Rubisco activity (**A**) and PEPC activity (**B**) in *H. altissima*. Different lower-case letters indicate significant differences between the measuring dates under the unfertilized (N0) treatment and the fertilized (AN, NN, ANN) treatment, respectively ($p < 0.05$) (n = 6).

3.7. Within-Leaf Nitrogen Allocation Estimate

The effects of different available nitrogen forms on the allocation of leaf nitrogen to different nitrogen components are shown in Figure 6. The rubisco, other soluble protein and carboxylation values expressed per unit leaf area were significantly higher under the AN treatment than under the N0, NN or ANN treatments ($p < 0.05$) (Table 1; Figure 6). Relative to the NN and ANN treatments, the AN treatment significantly increased the percentages of nitrogen allocated to rubisco (1.01% and −0.14%, respectively), other soluble protein (2.13% and 5.06%) and carboxylation (8.82% and 1.6%) proteins. Unexpectedly, no significant difference was found in N_B (Bioenergetics) between the NN and ANN treatments, but N_B was significantly higher in these treatments than in the N0 treatment ($p < 0.05$) (Table 1, Figure 6B). Relative to the AN and ANN treatments, the NN treatment significantly increased the percentages of nitrogen allocated to N_B (1.56% and 0.86%, respectively). Compared to the N0, AN and ANN treatments, N_L (Light-harvesting protein) increased under the NN treatment, while other nitrogen increased in ANN treatment ($p < 0.05$) (Table 2; Figure 6C,D).

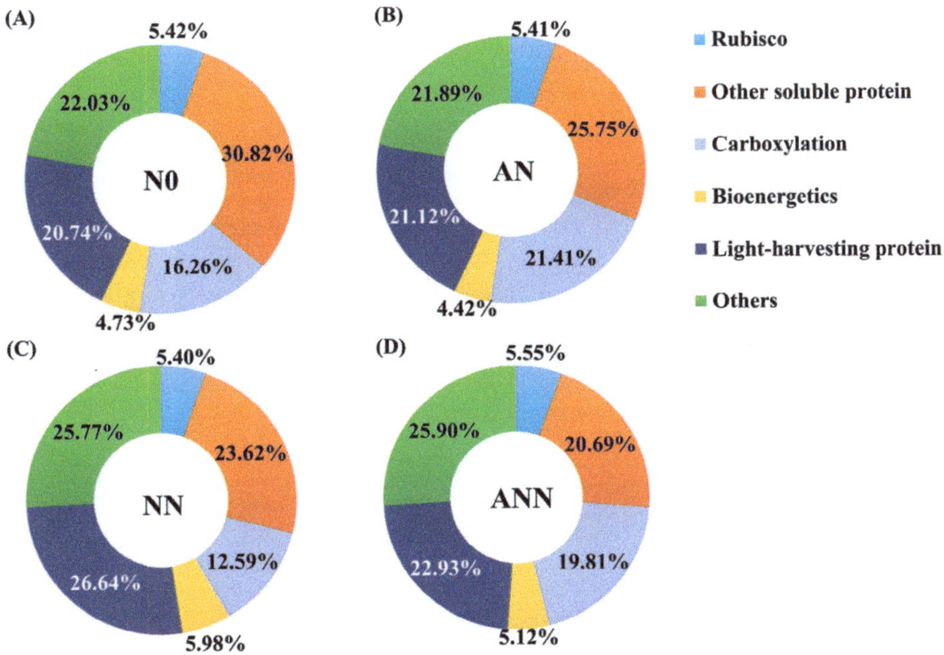

Figure 6. Effect of nitrogen forms treatments on the Nitrogen allocation in leaves of *H. altissima*. The data of percentages are the content of nitrogen in the corresponding components accounting for total leaf nitrogen content in *H. altissima*. (**A–D**) nitrogen partitioning in the unfertilized (N0) treatment and the fertilized (AN, NN, ANN) treatments. The size of pie chart indicates nitrogen content ($p < 0.05$) (n = 6).

Linear correlation analysis provided correlations of PNUE with photosynthetic responses of A_n and Chl_{area}, and nitrogen allocation of N_{psn} and N_{area} (Figure 7). Under different nitrogen forms treatments, PNUE are strongly shaped by A_n, Chl_{area} and N_{psn} ($p < 0.05$). Meanwhile, there is a negative correlation between PNUE and N_{area}, which will directly or indirectly affect the changes in PNUE.

Table 2. Effect of nitrogen forms treatments on the content of N compounds in *H. altissima*.

Parameters ($mg\ m^{-2}$)	Nitrogen Forms Treatments			
	N0	AN	NN	ANN
Rubisco	83.38 ± 0.38 c	105.53 ± 0.36 a	97.63 ± 0.34 b	107.72 ± 0.29 a
Other soluble protein	473.66 ± 1.86 b	502.66 ± 2.45 a	427.35 ± 4.26 c	401.33 ± 4.28 d
Carboxylation	249.86 ± 6.41 c	418.06 ± 20.32 a	227.81 ± 18.91 d	384.15 ± 23.19 b
Bioenergetics	72.75 ± 1.82 c	86.37 ± 1.29 b	108.23 ± 2.86 a	99.22 ± 4.16 ab
Light-harvesting protein	318.74 ± 5.18 d	412.23 ± 7.82 c	482.08 ± 12.94 a	444.80 ± 11.37 b
Other nitrogen	338.6 ± 9.16 d	427.35 ± 21.38 c	466.18 ± 22.35 b	502.34 ± 29.34 a

Different lower-case letters indicate significant differences between the measuring dates under the un-fertilized (N0) treatment and the fertilized (AN, NN, ANN) treatment, respectively ($p < 0.05$) (n = 6).

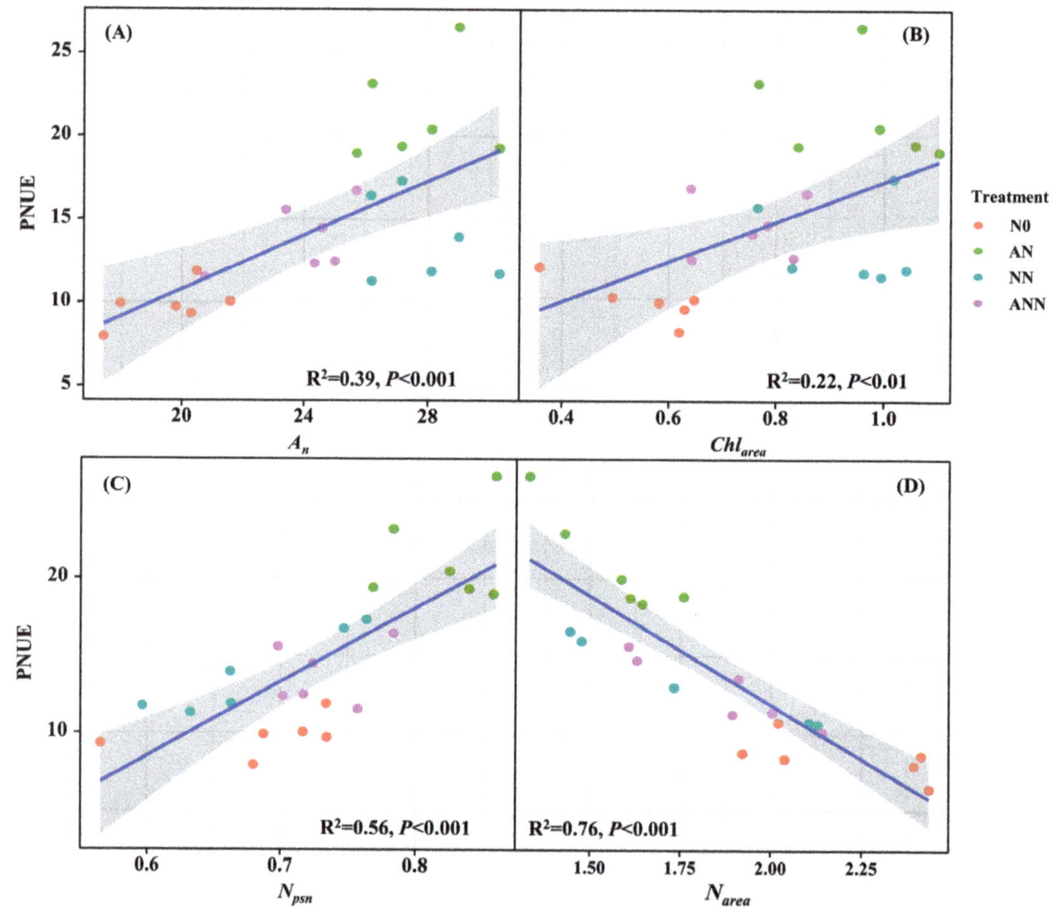

Figure 7. Relationships of photosynthetic N use efficiency (PNUE) with net CO_2 assimilation rate (A_n) (**A**), area-based chlorophyll content (Chl_{area}) (**B**), photosynthetic N (N_{psn}) (**C**) and area-based nitrogen content (N_{area}) (**D**) in *H. altissima*. Relationships between variables were assessed using linear regression analysis.

3.8. PSII Quantum Efficiencies

H. altissima plants demonstrated a clear advantage in allocating nitrogen to photosynthetic components in their leaves across various nitrogen treatments. To explore the potential effects of nitrate and ammonium on PSII quantum efficiencies, we examined the relationship between these nitrogen forms and the plants' photosynthetic performance. A strong, statistically significant positive linear correlation was observed between PNUE and Chl_{area} in *H. altissima* (Figure 7B). Additionally, measurements of Fv/Fm, φPSII, non-photochemical quenching (NPQ), and electron transfer rate (ETR) were significantly higher under the AN and ANN treatments compared to the N0 and NN treatments ($p < 0.05$) (Figure 8).

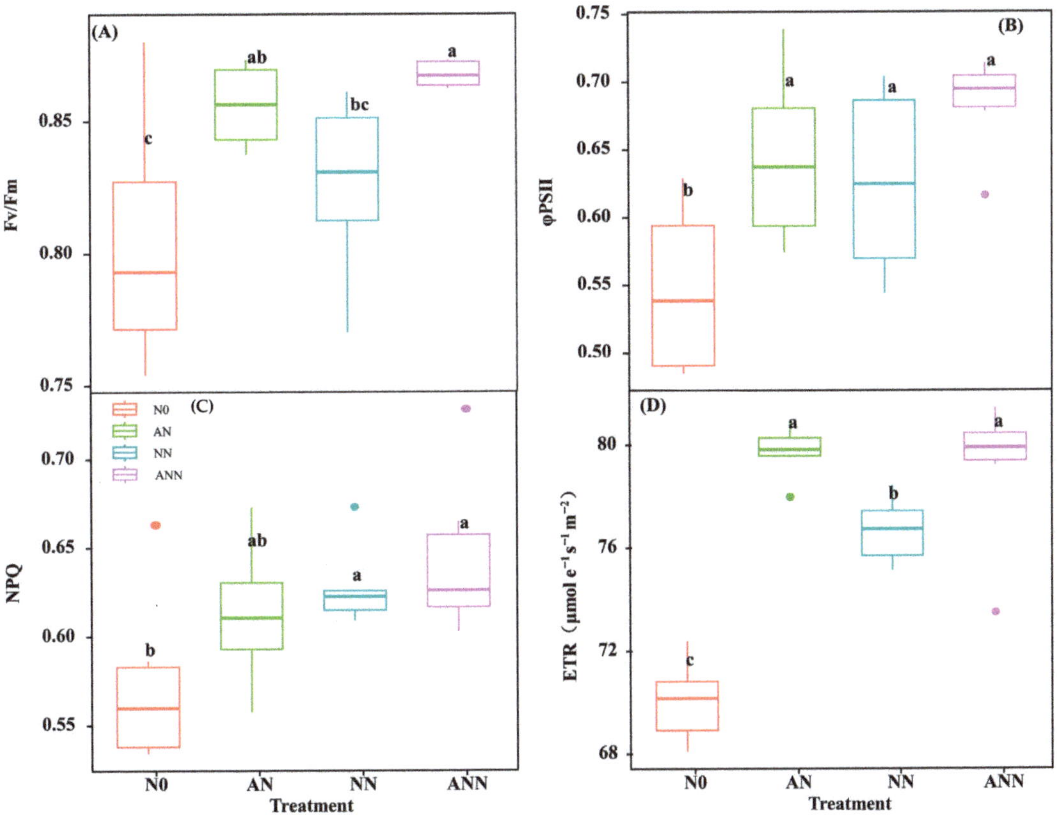

Figure 8. Effect of nitrogen forms treatments on the maximum quantum yield of PSII (Fv/Fm) (**A**), the effective quantum yield of PSII (φPSII) (**B**), non-photochemical quenching coefficient (NPQ) (**C**), and electron transport rate (ETR, μmol e^{-1} s^{-1} m^{-2}) (**D**) in *H. altissima*. Different lower-case letters indicate significant differences between the measuring dates under the unfertilized (N0) treatment and the fertilized (AN, NN, ANN) treatment, respectively ($p < 0.05$) (n = 6).

4. Discussion

4.1. Effects of Different Forms of Nitrogen on the Nitrogen Source and Nitrogen Metabolic Enzyme Activity in H. altissima

Nitrogen metabolism is crucial for the synthesis of key proteins required for photosynthesis. The different forms of nitrogen sources in the soil can significantly influence the activity of enzymes related to plant nitrogen assimilation. Nitrate reductase (NiR) and nitrate reductase (NR) are involved in the reduction of NO_3^- to NH_4^+, and they regulate this process through coupling mechanisms [29]. In this study, the inorganic nitrogen content in the soil significantly increased under AN treatment (Figure 1), which was directly related to changes in nitrogen metabolic enzyme activities. NN treatment significantly stimulated NR and NiR activities, consistent with previous research that indicates NR activity is mainly influenced by NO_3^- concentration, promoting nitrification [15]. As NO_3^- is converted into other forms of nitrogen, its availability decreases, but nitrogen in the soil continues to transfer to the leaves [30]. Our results indicate that AN and NN treatments significantly enhanced the activity of key nitrogen metabolic enzymes, such as glutamine synthetase (GS) and glutamate synthetase (GAGOT), with the AN treatment showing the most significant effect (Figure 2). This suggests that NH_4^+ play an important role in promoting nitrogen assimilation and improving overall plant metabolism, likely due to NH_4^+ preferentially entering the mesophyll cells to directly participate in nitrogen absorption and assimilation [31]. In particular, the increased GS activity under ammonium treatment may promote the synthesis of nitrogen in organic forms, thereby supporting the plant's nitrogen economy. In higher plants, GSI and GOGAT assimilate NH_4^+ into amino acids for plant uptake [5]. The concentration of NH_4^+ is closely related to GSI and GOGAT enzyme activity [32], but under different nitrogen supply treatments, GSI enzyme activity was significantly higher under AN treatment compared to other nitrogen forms, which is consistent with previous research on rice plants [33]. The results of this study reveal the relationship between NO_3^- and NH_4^+ supply and assimilation enzyme activity. According to our results, the activity of nitrogen isoenzymes was significantly increased after AN treatment.

4.2. *Effects of Different Forms of Nitrogen on Leaf Morphological Traits and Gas Exchange in H. altissima*

As is widely known, nitrogen is a vital nutrient for plant growth and development, and the form of nitrogen source directly affects leaf growth [34]. Leaf trait adjustments are often more important than biochemical characteristics in determining how a plant's leaves adapt to the environment for photosynthesis [8,24]. Nitrogen promotes leaf area growth and maintains the leaf's ability to absorb and utilize light energy, which helps increase net photosynthesis (A_n) and photosynthetic nitrogen use efficiency (PNUE) [8,15]. Different nitrogen sources had significant effects on the photosynthetic performance of *H. altissima*. Under AN and NN treatments, leaf A_n and stomatal conductance (G_s) significantly increased (Figure 3), indicating better nitrogen allocation to the photosynthetic apparatus. In addition to the biochemical and physiological processes of photosynthesis, G_s is also a key factor affecting CO_2 assimilation. Our results indicate that the G_s value of plants under AN treatment was higher than under other nitrogen source treatments (Figure 3B), which may be due to the role of ammonium and nitrate in regulating the opening and closing of stomata, as NH_4^+ is converted to nitrate through the anion transport system of nitrification, participating in the metabolism of guard cells [35]. The increase in G_s value facilitated more CO_2 absorption, and combined with better nitrogen allocation to photosynthetic proteins, this ultimately led to an increase in photosynthetic efficiency and PNUE. The accumulation of photosynthetic pigments in the leaves visually represented the photosynthetic efficiency, with Chl_m significantly higher under AN treatment compared to other treatments, indicating that ammonium nitrogen significantly improved the photosynthetic efficiency of *H. altissima*. This enhancement in photosynthetic efficiency may be closely related to the effective distribution of nitrogen in the leaves, which is crucial for optimizing the photosynthetic apparatus of C_4 plants. Key indicators of photosynthetic efficiency, such as Fv/Fm, φPSII, and ETR values [11], were significantly higher under NH_4^+ treatment, indicating an improvement in electron transport and light-harvesting efficiency (Figure 8). The high photosynthetic nitrogen use efficiency observed under nitrate conditions may be attributed to the optimized allocation of nitrogen in the photosynthetic apparatus, particularly the carboxylation system and the electron transport chain [36].

4.3. Effects of Different Forms of Nitrogen on Leaf Photosynthetic Performance and Photosynthetic Nitrogen Use Efficiency in H. altissima

The activity of key photosynthetic enzymes, such as Rubisco and PEP carboxylase (PEPC), is also affected by nitrogen sources. Under AN treatment, the activity of these enzymes, especially Rubisco and PEPC, was significantly increased, as they are core enzymes in carbon fixation in C_4 plants [37]. The increase in enzyme activity may be due to the effective distribution of nitrogen to the photosynthetic apparatus, where nitrogen plays a crucial role in the synthesis and function of photosynthetic proteins [38]. Under ammonium nitrogen conditions, the increase in enzyme activity in *H. altissima* suggests that nitrogen from ammonium salts is more effectively allocated to photosynthetic proteins than nitrogen from nitrates, thereby improving the photosynthetic rate. V_{cmax} as an indicator of Rubisco enzyme activity in the carbon fixation process of photosynthesis [35,36], was significantly increased by inorganic nitrogen sources. This indicates that ammonium nitrogen supply is closely related to the normal growth of *H. altissima* leaves. In this study, compared to N0, NN, and ANN treatments, the AN treatment showed higher N_{area}, A_n, and Chl_{area} in *H. altissima*, with PNUE increasing by 26.12%, 11.02%, and 10.82%, respectively. In addition, PNUE in *H. altissima* showed a significant positive correlation with A_n, N_{psn}, and Chl_{area}. Our results show that under AN treatment, nitrogen was preferentially allocated to the photosynthetic apparatus, especially to the chloroplasts, which is crucial for light absorption and energy transfer needed for photosynthesis [9,29]. This allocation promoted higher Chl_{area} and improved photosynthetic efficiency. Furthermore, the higher PNUE observed under AN treatment showed a strong negative correlation with Narea (Figure 7D), further emphasizing the role of nitrogen allocation in photosynthetic efficiency. The induction of higher photosynthetic efficiency under AN treatment may be due to the higher efficiency of nitrogen distribution in the leaves, which enhances the performance of photosynthesis. This result supports that NH_4^+ are more effective in nitrogen distribution within the C_4 photosynthetic apparatus, aiding in the photosynthetic optimization of *H. altissima*.

4.4. Effects of Different Forms of Nitrogen on Leaf Nitrogen Allocation and Trade-Offs in H. altissima

The allocation of nitrogen within the leaves is critical for optimizing photosynthetic performance because it determines how nitrogen is distributed to the photosynthetic apparatus and other cellular structures. As with many plant species, the allocation of nitrogen within the leaves reflects the trade-off between growth and defense. Leaf nitrogen allocation reflects the trade-off within the leaf economics spectrum, where fast-growing species tend to allocate more nitrogen to growth metabolism at the expense of structural components [32]. This shift in nitrogen allocation may reflect the plant's prioritization of photosynthesis while reducing nitrogen investment in structural functions. AN treatment led to a significant increase in the absolute content of soluble proteins and the proportion of Rubisco nitrogen (31.16%) (Table 2; Figure 6), a result consistent with previous research [4,6,10], who reported that 25–45% of leaf nitrogen is allocated to soluble proteins. Rubisco, a key enzyme in plant photosynthesis, accounts for 50% of soluble protein and 25% of leaf nitrogen [37]. In *H. altissima*, we found that AN treatment led to increased nitrogen allocation to Rubisco, reflected by lower bioenergetics and light-harvesting proteins, and higher carboxylation (Table 2; Figure 6). Compared to the NN and ANN treatments, N_B/N_{B+L} decreased under the ANN treatment, while N_L/N_{B+L} increased ($p < 0.05$) (Figure 9B). We found significant differences in nitrogen allocation to soluble proteins between AN treatment and other treatments, which is consistent with previous studies indicating that invasive species allocate more nitrogen to photosynthesis than native species, promoting growth and carboxylation [17,39]. The results of these studies suggest that more nitrogen is allocated to soluble proteins, sacrificing structural proteins [4,39]. These results suggest that under ammonium nitrogen conditions, nitrogen allocation helps improve nitrogen absorption and utilization, maximizing support for mesophyll cell photosynthesis. The changes in nitrogen investment strategies indicate that these components are crucial for ensuring the plant adapts to normal growth and physiological activities under inorganic nitrogen conditions.

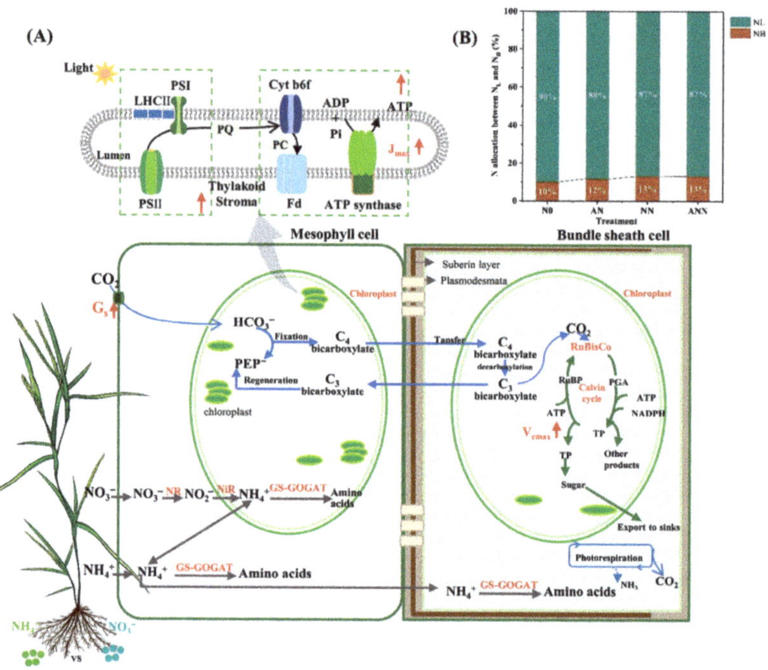

Figure 9. Effect of nitrogen form treatments on the change in nitrogen contents in the photosynthetic apparatus of leaves in *H. altissima*. (**A**) The percentage together indicates the increase (red arrows) on nitrogen in different photosynthetic apparatuses under AN compared to the N0, NN and ANN treatments. (**B**) The allocation of N between PN_B and PN_L within the thylakoid lumen under N0, AN, NN and ANN treatments.

Thus, we hypothesize that under AN treatment, *H. altissima* is usually in the "high return" zone of the leaf economics spectrum, with higher A_n, Chl_{area} and PNUE compared to other treatments, thereby allocating more nitrogen to leaf nitrogen pools related to photosynthesis and growth. Based on this analysis, species with larger investments in photosynthetic protein nitrogen typically show higher PNUE in many natural ecosystems [12,39,40]. In this study, *H. altissima* allocated 52.36% of its leaf nitrogen to the photosynthetic apparatus, consistent with previous studies on maize [36] and invasive plants [39]. According to our preliminary hypothesis, to determine if a plant is in the "high return" zone, the changes in leaf nitrogen allocation processes need to be assessed. In ecological models, nitrogen investment in the photosynthetic apparatus remains an important determinant of PNUE [15,41]. Photosynthesis is closely related to leaf nitrogen content, and nitrogen content is usually reflected by Calvin cycle proteins. Fast-growing plants allocate about two-thirds of their leaf nitrogen to the photosynthetic apparatus [6,12,14]. We found that the amount of nitrogen allocated to the photosynthetic apparatus was significantly positively correlated with PNUE ($R^2 = 0.56$, $p < 0.001$). The different forms of nitrogen sources play a crucial role in optimizing the photosynthetic efficiency and nitrogen utilization efficiency of *H. altissima* (Figure 9). AN treatment promoted the effective allocation of nitrogen to the photosynthetic apparatus, enhancing enzyme activity, photosynthetic efficiency, and PNUE. Under AN treatment, higher stomatal conductance and optimized leaf nitrogen allocation further facilitated CO_2 absorption and improved photosynthetic efficiency. The results highlight the importance of understanding nitrogen allocation dynamics for optimizing photosynthesis and plant growth, and suggest that increasing ammonium salt fertilizers in nitrogen-limited environments may be an effective strategy to optimize C_4 plant photosynthesis and productivity.

5. Conclusions

Our study reveals that inorganic nitrogen sources play a pivotal role in shaping the PNUE, nitrogen assimilation, and nitrogen allocation in *H. altissima* leaves. The plants exhibited a clear adaptive response, optimizing nitrogen distribution within the leaves to enhance photosynthesis. This resulted in increased PNUE and biomass production during the growing season, especially in environments with high ammonium nitrogen levels. Under AN and ANN treatments, *H. altissima* allocated more nitrogen to rubisco and the carboxylation apparatus, thereby improving ETR. In addition, Chl_m and NPQ increased, which helped to enhance the light protection capability. The treatment also shifted nitrogen allocation, with more nitrogen directed toward soluble proteins and the photosynthetic machinery. This reallocation suggests a trade-off between growth and absorption and utilization of nitrogen in *H. altissima*. Overall, our findings offer new insights into how inorganic nitrogen sources influence nitrogen dynamics in *H. altissima*, providing a deeper understanding of their nitrogen utilization strategies in the context of increasing nitrogen deposition. Understanding how *H. altissima* responds to inorganic nitrogen sources can guide the development of more targeted fertilization strategies. By knowing that ammonium nitrogen (AN) and ammonium nitrate nitrogen (ANN) enhance photosynthesis and biomass production, agriculturalists can optimize fertilizer types and application rates, especially in regions with high nitrogen deposition. This could improve crop yields without over-fertilizing, thus promoting both economic and environmental sustainability.

Author Contributions: Conceptualization, X.W., N.X. and X.Y.; methodology, X.W.; software, N.X. and J.Z. (Ju Zhang); validation, X.W., J.Z. (Ju Zhang) and M.S.; formal analysis, N.X. and Z.Z.; investigation, X.Y.; resources, X.Y.; data curation, X.W.; writing—original draft preparation, X.W.; writing—review and editing, N.X.; visualization, J.Z. (Jinwei Zhang); supervision, M.S.; project administration, X.Y.; funding acquisition, X.W. and X.Y. All authors have read and agreed to the published version of the manuscript.

Funding: This research was funded by the Fundamental Research Funds for the Science and Technology Project of the Jilin Provincial Education Department (JJKH20240565KJ). Xuechen Yang acknowledges support by the Innovation Leading Talents Project of the "Tianchi Talent" Recruitment Plan of Xinjiang Uygur Autonomous Region.

Institutional Review Board Statement: Not applicable.

Informed Consent Statement: Not applicable.

Data Availability Statement: The datasets generated for this study are available on request to the corresponding author.

Acknowledgments: We would like to express our gratitude to Lin Li and Bing Luo from Hansha Scientific Instruments Co., Ltd. for their technical support of the instruments.

Conflicts of Interest: The authors declare no conflicts of interest.

Abbreviations

A_n	Assimilation rate
G_s	Stomatal conductance
C_i	Internal CO_2 concentration
WUE	Water-use efficiency
V_{cmax}	Maximum rate of Rubisco carboxylation
J_{max}	Maximum rate of electron transport
LMA	A leaf mass per unit leaf area
N_{area}	A leaf nitrogen content per unit leaf area
N_C	Nitrogen contents in carboxylation
N_B	Nitrogen contents in bioenergetics
N_L	Nitrogen contents in light harvesting components
N_{psn}	Nitrogen contents in photosynthetic apparatus
PNUE	Photosynthetic nitrogen use efficiency
Fv/Fm	The maximum quantum yield of PSII
NPQ	The non-photochemical quenching coefficient
φPSII	The effective quantum yield of PSII
ETR	The electron transport rate

References

1. Liang, X.; Zhang, T.; Lu, X.; Ellsworth, D.S.; BassiriRad, H.; You, C.; Wang, D.; He, P.; Deng, Q.; Liu, H. Global response patterns of plant photosynthesis to nitrogen addition: A meta-analysis. *Glob. Change Biol.* **2020**, *26*, 3585–3600. [CrossRef]
2. Xia, J.; Wan, S. Global response patterns of terrestrial plant species to nitrogen addition. *New Phytol.* **2008**, *179*, 428–439. [CrossRef] [PubMed]
3. Hessini, K.; Hamed, K.B.; Gandour, M.; Mejri, M.; Abdelly, C.; Cruz, C. Ammonium nutrition in the halophyte Spartina alterniflora under salt stress: Evidence for a priming effect of ammonium? *Plant Soil* **2013**, *370*, 163–173. [CrossRef]
4. Yang, Z.; Yan, H.; Liu, H.; Yang, L.; Mi, G.; Wang, P. Enhancing Crop Nitrogen Efficiency: The Role of Mixed Nitrate and Ammonium Supply in Plant Growth and Development. *Biology* **2025**, *14*, 546. [CrossRef] [PubMed]
5. Prinsi, B.; Espen, L. Time-Course of Metabolic and Proteomic Responses to Different Nitrate/Ammonium Availabilities in Roots and Leaves of Maize. *Int. J. Mol. Sci.* **2018**, *19*, 2202. [CrossRef]
6. Li, Z.; Gao, W.; Jiang, C.; Lu, Y.; Kamran, M.; Yang, X. Appropriate nitrogen application rate with decreased basal/topdressing ratio improves yield, quality, water productivity, and N-use efficiency of forage maize in a rainfed region. *Agric. Water Manag.* **2025**, *317*, 109629. [CrossRef]
7. Cao, L.; Yan, C.; He, T.; Zhong, Q.; Yuan, Y.; Cao, L. Coordinated Regulation of Photosynthesis, Stomatal Traits, and Hormonal Dynamics in Camellia oleifera During Drought and Rehydration. *Biology* **2025**, *14*, 965. [CrossRef]
8. Onoda, Y.; Wright, I.J.; Evans, J.R.; Hikosaka, K.; Kitajima, K.; Niinemets, Ü.; Poorter, H.; Tosens, T.; Westoby, M. Physiological and structural tradeoffs underlying the leaf economics spectrum. *New Phytol.* **2017**, *214*, 1447–1463. [CrossRef]
9. Sugiura, D.; Terashima, I.; Evans, J.R. A decrease in mesophyll conductance by cell-wall thickening contributes to photosynthetic downregulation. *Plant Physiol.* **2020**, *183*, 1600–1611. [CrossRef]
10. Yuan, Y.; Xiao, J.; Liu, S.; He, T.; Rong, J.; Zheng, Y. Effect of Nitrogen Application Rate on Growth Physiology, Yield Quality, and Nitrogen Fertilizer Utilization Rate of Liriope muscari in Pots. *Biology* **2025**, *14*, 1104. [CrossRef]
11. Wei, X.; Han, L.; Xu, N.; Sun, M.; Yang, X. Nitrate nitrogen enhances the efficiency of photoprotection in *Leymus chinensis* under drought stress. *Front. Plant Sci.* **2024**, *15*, 1348925. [CrossRef]
12. Li, T.; Yang, L.; Zhao, D.; Li, T.; Liu, Y.; Wang, X. Induced photosynthesis significantly influences biomass in Betula platyphylla seedlings compared to steady-state photosynthesis under different nitrogen forms. *Plant Physiol. Biochem.* **2025**, *228*, 110225. [CrossRef] [PubMed]
13. Shi, Y.; Wang, J.; Ao, Y.; Han, J.; Guo, Z.; Liu, X.; Zhang, J.; Mu, C.; Le Roux, X. Responses of soil N_2O emissions and their abiotic and biotic drivers to altered rainfall regimes and co-occurring wet N deposition in a semi-arid grassland. *Glob. Change Biol.* **2021**, *27*, 4894–4908. [CrossRef] [PubMed]
14. Zhong, S.; Xu, Y.; Meng, B.; Loik, M.E.; Ma, J.-Y.; Sun, W. Nitrogen addition increases the sensitivity of photosynthesis to drought and re-watering differentially in C_3 versus C_4 grass species. *Front. Plant Sci.* **2019**, *10*, 815. [CrossRef]
15. Wei, X.; Yang, Y.; Yao, J.; Han, J.; Yan, M.; Zhang, J.; Shi, Y.; Wang, J.; Mu, C. Improved utilization of nitrate nitrogen through within-leaf nitrogen allocation trade-offs in *Leymus chinensis*. *Front. Plant Sci.* **2022**, *13*, 870681. [CrossRef]

16. Stinziano, J.R.; Morgan, P.B.; Lynch, D.J.; Saathoff, A.J.; McDermitt, D.K.; Hanson, D.T. The rapid A–Ci response: Photosynthesis in the phenomic era. *Plant Cell Environ.* **2017**, *40*, 1256–1262. [CrossRef]
17. Von Caemmerer, S. *Biochemical Models of Leaf Photosynthesis*; Csiro Publishing: Canberra, Australia, 2000.
18. Long, S.P.; Bernacchi, C.J. Gas exchange measurements, what can they tell us about the underlying limitations to photosynthesis? Procedures and sources of error. *J. Exp. Bot.* **2003**, *54*, 2393–2401. [CrossRef]
19. Li, X.; Zhang, W.; Niu, D.; Liu, X. Effects of abiotic stress on chlorophyll metabolism. *Plant Sci.* **2024**, *342*, 112030. [CrossRef] [PubMed]
20. Garmay, A.V.; Oskolok, K.V.; Monogarova, O.V.; Demidov, M.I. Determination of ammonium and nitrate in soils by digital colorimetry. *Environ. Monit. Assess.* **2024**, *196*, 948. [CrossRef]
21. de Castro, T.A.V.T.; Berbara, R.L.L.; Tavares, O.C.H.; da Graca Mello, D.F.; Pereira, E.G.; de Souza, C.d.C.B.; Espinosa, L.M.; García, A.C. Humic acids induce a eustress state via photosynthesis and nitrogen metabolism leading to a root growth improvement in rice plants. *Plant Physiol. Biochem.* **2021**, *162*, 171–184. [CrossRef]
22. Abramson, J.; Adler, J.; Dunger, J.; Evans, R.; Green, T.; Pritzel, A.; Ronneberger, O.; Willmore, L.; Ballard, A.J.; Bambrick, J.; et al. Accurate structure prediction of biomolecular interactions with AlphaFold 3. *Nature* **2024**, *630*, 493–500. [CrossRef]
23. Wang, J.; Wei, Y.; Chen, Y.; Jiang, S.; Ye, J.; Ding, P.; Shao, X. Alginate oligosaccharide alleviates chilling injury in postharvest peach fruit via regulating antioxidant and membrane lipid metabolisms. *Plant Physiol. Biochem.* **2025**, *228*, 110292. [CrossRef] [PubMed]
24. Ye, Z.; Hu, W.; Zhou, S.; Robakowski, P.; Kang, H.; An, T.; Wang, F.; Xiao, Y.; Yang, X. Limitations of the Farquhar–von Caemmerer–Berry Model in Estimating the Maximum Electron Transport Rate: Evidence from Four C_3 Species. *Biology* **2025**, *14*, 630. [CrossRef] [PubMed]
25. Nakanishi, K.; Fujiki, H.; Ozaki, K.; Yanahara, S.; Takeuchi, N.; Suzuki, Y.; Sugiyama, T.; Makino, A.; Ookawa, T.; Hirasawa, T.; et al. Decrease of cytokinin flux from roots enhances degradation of ribulose-1, 5-bisphosphate carboxylase/oxygenase: A mechanism of the accelerated decrease of leaf photosynthesis with senescence under soil moisture stress in rice (*Oryza sativa* L.). *Plant Soil* **2024**, *496*, 391–411. [CrossRef]
26. Rudenko, N.N. Photosynthesis and Carbon Metabolism in Higher Plants and Algae. *Plants* **2025**, *14*, 2161. [CrossRef]
27. Hikosaka, K.; Terashima, I. A model of the acclimation of photosynthesis in the leaves of C_3 plants to sun and shade with respect to nitrogen use. *Plant Cell Environ.* **1995**, *18*, 605–618. [CrossRef]
28. Poorter, H.; Evans, J.R. Photosynthetic nitrogen-use efficiency of species that differ inherently in specific leaf area. *Oecologia* **1998**, *116*, 26–37. [CrossRef]
29. Zayed, O.; Hewedy, O.A.; Abdelmoteleb, A.; Ali, M.; Youssef, M.S.; Roumia, A.F.; Seymour, D.; Yuan, Z.-C. Nitrogen journey in plants: From uptake to metabolism, stress response, and microbe interaction. *Biomolecules* **2023**, *13*, 1443. [CrossRef]
30. Hu, C.-C.; Liu, X.-Y.; Driscoll, A.W.; Kuang, Y.-W.; Brookshire, E.J.; Lü, X.-T.; Chen, C.-J.; Song, W.; Mao, R.; Liu, C.-Q. Global distribution and drivers of relative contributions among soil nitrogen sources to terrestrial plants. *Nat. Commun.* **2024**, *15*, 6407. [CrossRef]
31. Ding, S.; Shao, X.; Li, J.; Ahammed, G.J.; Yao, Y.; Ding, J.; Hu, Z.; Yu, J.; Shi, K. Nitrogen forms and metabolism affect plant defence to foliar and root pathogens in tomato. *Plant Cell Environ.* **2021**, *44*, 1596–1610. [CrossRef] [PubMed]
32. Li, H.; Wang, Y.; Wang, J.; Zhang, M.; Liu, W.; Li, X.; Lin, X. Improving Leaf GOGAT Activity After the Post-Silking Period Contributes to High Grain Yield with Reduced Nitrogen in N-Efficient Maize. *Agronomy* **2025**, *15*, 1379. [CrossRef]
33. Wu, Y.; Xu, Y.; Liu, Z.; Cai, M.; Pan, H.; Zhang, Q. Different responses of Lagerstroemia indica to varied supplies of ammonium and nitrate. *Sci. Hortic.* **2024**, *329*, 113001. [CrossRef]
34. Salesse-Smith, C.E.; Wang, Y.; Long, S.P. Increasing Rubisco as a simple means to enhance photosynthesis and productivity now without lowering nitrogen use efficiency. *New Phytol.* **2025**, *245*, 951–965. [CrossRef]
35. Xu, X.; Zhang, X.; Ni, W.; Liu, C.; Qin, H.; Guan, Y.; Liu, J.; Feng, Z.; Xing, Y.; Tian, G.; et al. Nitrogen-potassium balance improves leaf photosynthetic capacity by regulating leaf nitrogen allocation in apple. *Hortic. Res.* **2024**, *11*, uhad253. [CrossRef]
36. Peng, W.; Zhang, Y.; Yi, S.H. Increasing nitrogen absorption and assimilation ability under mixed NO_3^- and NH_4^+ supply is a driver to promote growth of maize seedlings. *J. Integr. Agric.* **2023**, *22*, 1896–1908.
37. Mu, X.; Chen, Q.; Chen, F.; Yuan, L.; Mi, G. Within-leaf nitrogen allocation in adaptation to low nitrogen supply in maize during grain-filling stage. *Front. Plant Sci.* **2016**, *7*, 699. [CrossRef]
38. Feng, H.; Guo, J.; Peng, C.; Kneeshaw, D.; Roberge, G.; Pan, C.; Ma, X.; Zhou, D.; Wang, W. Nitrogen addition promotes terrestrial plants to allocate more biomass to aboveground organs: A global meta-analysis. *Glob. Change Biol.* **2023**, *29*, 3970–3989. [CrossRef]
39. Feng, Y.L. Nitrogen allocation and partitioning in invasive and native Eupatorium species. *Physiol. Plant.* **2008**, *132*, 350–358. [CrossRef]

40. Feng, Y.L.; Lei, Y.B.; Wang, R.F.; Callaway, R.M.; Valiente-Banuet, A.; Inderjit; Li, Y.P.; Zheng, Y.L. Evolutionary tradeoffs for nitrogen allocation to photosynthesis versus cell walls in an invasive plant. *Proc. Natl. Acad. Sci. USA* **2009**, *106*, 1853–1856. [CrossRef]
41. Liu, T.; Ren, T.; White, P.J.; Cong, R.; Lu, J. Storage nitrogen co-ordinates leaf expansion and photosynthetic capacity in winter oilseed rape. *J. Exp. Bot.* **2018**, *69*, 2995–3007. [CrossRef] [PubMed]

biology

MDPI

Article

Responses of Soil Quality and Microbial Community Composition to Vegetation Restoration in Tropical Coastal Forests

Yuanqi Chen [1], Feifeng Zhang [1], Jianbo Cao [2], Tong Liu [1] and Yu Zhang [3,4,*]

[1] Institute of Geographic Environment and Carbon Peak & Neutrality, School of Earth Sciences and Spatial Information Engineering, Hunan University of Science and Technology, Xiangtan 411201, China; chenyq2016@163.com (Y.C.)
[2] Fujian Provincial Key Laboratory of the Development and Utilization of Bamboo Resources, Sanming University, Sanming 365000, China; caojianbo612@163.com
[3] School of Earth Sciences, Yunnan University, Kunming 650500, China
[4] Hunan Province Key Laboratory of Economic Crops Genetic Improvement and Integrated Utilization, School of Life and Health Sciences, Hunan University of Science and Technology, Xiangtan 411201, China
* Correspondence: anemone2016@163.com

Simple Summary

Assessing post-afforestation soil quality can identify the most effective vegetation restoration approaches, which is critical to the sustainable management of forest ecosystems. In this study, we evaluated how different vegetation restoration strategies (barren land control, disturbed short-rotation and undisturbed long-term *Eucalyptus* monocultures, a mixed native-species plantation, and a natural forest) affect soil quality and microbial communities in tropical ecosystems. The results showed that vegetation restoration significantly improved soil physicochemical properties and the overall soil quality index (SQI). Crucially, the SQI in the undisturbed long-term *Eucalyptus* monoculture and the mixed native-species plantation reached levels comparable to the natural forest, demonstrating the recovery potential of well-managed plantations. Microbial biomass (bacteria, fungi, arbuscular mycorrhizal fungi, and actinomycetes) increased from barren land to natural forest but remained lower in all plantations than in the natural forest, indicating incomplete microbial recovery. Strong positive correlations existed between microbial biomass and the SQI. The results indicate that intensive disturbances impede soil and microbial recovery, while microbial communities prove to be more sensitive restoration indicators than physicochemical properties alone. Collectively, afforestation with mixed native species offers rapid soil restoration, and undisturbed long-term monocultures can achieve similar soil quality outcomes over time. This work provides critical insights for optimizing tropical and subtropical afforestation practices.

Abstract

Afforestation substantially promotes vegetation restoration and modifies soil physical, chemical, and biological properties. The integrated effects of soil properties on soil quality, expressed via a composite soil quality index (SQI), remain unclear despite variations among individual properties. Here, five vegetation restoration treatments were selected as follows: (1) barren land (BL, control), (2) disturbed short-rotation *Eucalyptus* plantation (REP); (3) undisturbed long-term *Eucalyptus* plantation (UEP); (4) mixed native-species plantation (MF); and (5) natural forest (NF) following >50 years of restoration. Soil physicochemical properties and microbial community compositions were investigated, and soil quality was evaluated by an integrated SQI. Our results showed that vegetation restoration had strong effects on soil physicochemical properties, soil quality, and microbial communities. Most of the soil physicochemical properties exhibited significant differences among treatments.

 https://doi.org/10.3390/biology14091120

Soil dissolved organic carbon, total nitrogen, and ammonium nitrogen were the three key soil quality indicators. The SQI increased significantly with vegetation recovery intensity. In both UEP and MF, it reached levels comparable to NF, and was higher in UEP than in REP, implying that short-rotation practices impede soil restoration. In addition, microbial biomass (bacteria, fungi, arbuscular mycorrhizal fungi, actinomycetes, and total microbe PLFAs) increased from BL to NF. All plantations exhibited lower microbial biomass than NF, revealing incomplete recovery and a greater sensitivity to soil physicochemical properties. Conversely, the fungi-to-bacteria biomass ratio decreased sequentially (REP > BL > UEP > MF > NF). Strong positive correlations between microbial biomass and the SQI were observed. These results collectively indicate that afforestation with mixed tree species is optimal for rapid soil restoration, and undisturbed long-term monocultures can achieve similar outcomes. These findings highlight that tree species mixtures and reducing disturbance should be taken into consideration when restoring degraded ecosystems in the tropics.

Keywords: soil restoration; forest ecosystem; afforestation; vegetation types; forest conservation; short rotation

1. Introduction

Anthropogenic activities such as deforestation have led to a severe degradation of forest ecosystems, resulting in widespread soil erosion, nutrient depletion, reduced productivity, and biodiversity loss [1,2]. In response, extensive ecological restoration projects have been implemented in recent decades to rehabilitate degraded terrestrial ecosystems [3,4]. These efforts have achieved notable success, especially in enhancing environmental quality and reinforcing carbon sequestration [5,6]. Among the key restoration strategies, afforestation has played a pivotal role in reversing ecosystem degradation, significantly promoting vegetation recovery and, in turn, modifying soil physical, chemical, and biological properties [7]. Conversely, these soil characteristics are intrinsically linked to plant productivity and ecosystem services [8–10]. Therefore, assessing post-afforestation soil quality has emerged as a critical component in ensuring the sustainable management of forest ecosystems.

Afforestation approaches modulate vegetation establishment trajectories and soil restoration efficacy in degraded ecosystems via plant–soil feedback loops [11]. Different afforestation strategies, such as monoculture, mixed-species planting, or natural regeneration, can lead to markedly divergent outcomes in soil quality due to variations in tree species composition and their functional traits [12]. Specifically, tree species richness and diversity influence nutrient cycling, rhizosphere chemical properties, mycorrhizal associations, and microbial community dynamics, all of which collectively shape soil restoration trajectories [13–15]. In recent years, numerous studies have been conducted to evaluate the effects of afforestation on soil properties, and significant improvements in ecosystem functionality have been frequently recorded. Vegetation restoration enhances soil nutrient cycling and enzyme activities, particularly in degraded landscapes [16]. In southern China's erosion-prone regions, afforestation can also elevate soil fertility and increase the abundance and diversity of the soil bacterial community [16]. Based on the long-term monitoring in Ghana, it was found that both plantations and secondary forests exhibited soil carbon storage and key physicochemical properties comparable to those of primary forests in analogous climatic zones [17]. However, the outcome of afforestation is not universally positive, and contrasting effects have been documented. For example, the bacterial biomass represented by phospholipid fatty acids (PLFAs) declined with the successional stage [18]. Similarly, in the dry–hot valley of China, afforestation with *Eucalyptus camaldulensis* reduced the population of fungi and total microbial community, urease activity, and the soil quality index [19]. These discrepancies highlight that the effects of afforestation on soil properties are influenced by plant species composition, soil type, and environmental conditions, and different soil properties could produce varied responses [20,21]. For instance, soil total nitrogen content was the highest in coniferous-mixed plantations, but total phosphorus content was the highest in broad-leaved mixed plantations [22]. A meta-analysis concluded that afforestation increased soil carbon and nitrogen but not phosphorus accumulation [23]. Thus, a comparative analysis of soil quality under different afforestation strategies is essential, as it can not only identify the most effective vegetation restoration approaches but also refine soil management frameworks to maximize soil health and ecological functions [24].

Plantations in subtropical and tropical regions are predominantly monocultures dominated by a single tree species. Among these, eucalyptus (*Eucalyptus* spp.) has become a primary plantation due to its fast growth and short rotation (5–7 years), now covering over 5.40 million hectares across subtropical and tropical China [25]. However, successive planting generations under intensive management have caused severe soil degradation and productivity decline [26]. To mitigate these issues, both extending rotation periods and reducing management intensity are proposed as forest management practices [25,27]. Such measures may enhance biodiversity [28,29] and subsequently improve carbon sequestration and soil nutrient retention [27]. Nevertheless, the mechanistic effects of these adjustments on soil physicochemical and biological properties remain poorly resolved. Concurrently, multi-species afforestation has been advocated for superior soil fertility preservation [30]. Moreover, increasing the diversity of plantations is also a promising approach to adapt forests to climate change, which can be a viable and economically accessible strategy for sustainable wood production and reconciling economic and environmental benefits [31,32]. Some studies reported significantly greater soil quality improvements in mixed species plantations compared to monocultures [15,24], yet others documented negligible differences between monocultures and mixed plantations [33,34]. This contradiction underscores the idea that the context-dependent efficacy of forest management strategies on soil rehabilitation remains inadequately quantified.

The soil quality index (SQI) is regarded as an essential instrument for assessing changes in soil quality [35]. The utilization of the SQI can overcome the complexity of soil assessment [36]. To evaluate soil quality changes following vegetation restoration, a chrono sequence study incorporating five distinct land-use types was conducted, which included a barren land (as control), two pure *Eucalyptus* plantations over 50 years (one has undergone a short rotation every 5–7 years and the other is undisturbed), a mixed native-species plantation over 50 years, and a nearby undisturbed natural forest (over 200 years). We hypothesized the following: (1) Vegetation restoration will significantly improve soil quality across all forested sites relative to the barren land, primarily through enhanced soil fertility and microbial abundance, and the mixed native-species forest will have the highest soil quality. (2) Soil quality in the undisturbed long-term *Eucalyptus* plantation will exceed that of the disturbed short-rotation *Eucalyptus* plantation due to reduced management intensity. This work will advance the mechanistic understanding of soil restoration and provide critical guidelines for restoring degraded ecosystems.

2. Materials and Methods

2.1. Site Description

The study was conducted at the Xiaoliang Research Station of Tropical Coastal Ecosystems, Chinese Academy of Sciences (110°54′ E, 21°27′ N), located in Maoming city of

Guangdong Province, China. This region has a typical tropical monsoon climate with a mean annual precipitation of 2000 mm and a mean annual temperature of 23 °C. The soil is classified as a granite-derived Latosol that has been experiencing heavy erosion since the 1950s under harsh hydrothermal conditions [37]. In the degraded barren land, the monthly mean soil temperature at a 0–20 cm depth peaked at 47.5 °C, whereas soil total organic carbon and total nitrogen contents were only 6.0 g kg^{-1} and 0.3 g kg^{-1}, respectively [38]. Historically, the native vegetation in this region was evergreen broad-leaved seasonal rainforests. However, extensive deforestation has occurred [38]. The remaining native forests at the site are classified as tropical secondary forests, which have been preserved for over 200 years. Dominating tree species include *Cinnamomum camphora, Sterculia lanceolate,* and *Cryptocarya chinensis* [39].

Afforestation practices on the barren land (BL) have been implemented since 1959, although the harsh habitat severely limited natural vegetation recovery. Approximately 3.7 ha of BL was selected and assigned as a control representing the baseline condition prior to restoration. Due to complete topsoil erosion, only few herbaceous plants or xeric shrubs (e.g., *Dicranopteris linearis* and *Eriachne pallescens*) are to be found in the control areas [38]. *Eucalyptus exserta* were planted on the other BL in the early 1960s. Management subsequently diverged: half of the *E. exserta* plantation underwent short-rotation harvesting every 5–8 years (short-rotation *Eucalyptus* plantation, REP), while the other half remained undisturbed since planting (undisturbed *Eucalyptus* plantation, UEP). In 1974, one distinct catchment was clear-cut and reforested with multiple native tree species to create a mixed forest (MF). The vegetation surveys performed in 2015 showed that this MF supported an average of 14.6 native tree species in each 400 m^2 quadrate, and the dominant tree species are *Aphanamixis polystachya, Schefflera octophylla, Carallia brachiate, Symplocos chunii, Acacia auriculaeformis, Photinia benthamiana,* and *Cinnamomum burmannii* [39]. This MF stand is developing structural and compositional similarity to the undisturbed secondary natural forest (NF) [38]. Our study includes sites that experienced one of five restoration treatments: (1) barren land (control, BL), (2) disturbed short-rotation *Eucalyptus* plantation (REP), (3) undisturbed long-term *Eucalyptus* plantation (UEP), (4) mixed native-species plantation (MF), and (5) natural forest (NF). The detailed information about soil and vegetation was provided in previous studies [37,38]. Four replicate sampling plots per treatment were established, randomly located at a distance of >20 m apart from each other.

2.2. Soil Sampling and Physicochemical Properties

Surface soil samples (0–15 cm depths) were collected in May 2014 from all five vegetation restoration treatments. Within each replicate plot, soils were sampled from five randomly selected microsites using a stainless-steel core (3.0 cm diameter). Visible plant residues and roots were manually removed. The composite soil samples were sieved by a sieve with a 2 mm bore diameter and divided into two subsamples: one was preserved in a field-moist state for immediate analysis of soil moisture content and related parameters, while the other was air-dried for the subsequent determination of soil organic carbon and its associated physicochemical properties.

Soil moisture content (SMC) was measured by oven-drying at 105 °C to constant weight. Soil pH was measured potentiometrically at a soil-to-water ratio of 1:2.5 (w/v). Soil organic carbon (SOC) was determined by the traditional wet oxidation with potassium dichromate method [40]. Soil total nitrogen (total N) was analyzed by the micro-Kjedahl digestion method, and soil total phosphorus (total P) was digested with a sulfuric acid solution and quantified by the molybdenum–antimony (Mo-Sb) anti-spectrophotometer method. Soil dissolved organic carbon (DOC) was extracted with 0.5 M K_2SO_4, filtered (0.45 μm), and analyzed at a high temperature on a TOC analyzer (TOC-VCSH, Shimadzu,

Japan). Soil NH_4^+-N and NO_3^--N were measured by a flow injection analyzer (AA3, Bran Luebbe) [41].

2.3. Soil Microbial Biomass and Community Composition

Soil microbial biomass C (MBC) and N (MBN) were determined by the chloroform fumigation–extraction method [42]. Phospholipid fatty acid (PLFA) analysis was applied to characterize soil microbial community composition, and concentrations of individual PLFAs were quantified based on the internal standard concentration of 19:0 methylester [43]. The PLFAs i14:0, i15:0, a15:0, i16:0, a16:0, i17:0, a17:0, a18:0, i18:0, a19:0, 16:1ω7c, 16:1ω9c, 17:1ω8c, 18:1ω7, cy17:0, and cy19:0 [44] were used as bacterial (B) biomarkers. The PLFAs 18:1ω9c, 18:2ω6,9c [44], and 18:3ω6,9,12c were applied to denote fungal (F) biomarkers. The PLFA 16:1ω5c was considered as an arbuscular mycorrhizal fungal (AMF) biomarker [45]. The PLFAs 10Me 16:0, 10Me 17:0, and 10Me 18:0 were used as actinomycetes biomarkers. Total microbial biomass was represented by the sum of identified bacterial, fungal, AMF, and actinomycetes PLFAs. Soil microbial community structure represented by the fungal-to-bacterial ratio (F:B ratio) was calculated as the sum of fungal biomarker PLFAs divided by the sum of bacterial biomarker PLFAs [44].

2.4. Soil Quality Index (SQI) Evaluation

Twelve soil physicochemical and biological properties (SMC, pH, SOC, TN, TP, DOC, DN, NH_4^+-N, NO_3^--N, MBC, MBN, and MBC/MBN) were evaluated to identify a minimum dataset (MDS) for soil quality assessment. We employed principal component analysis (PCA) followed by Pearson's correlation analysis to select the most suitable indicators.

MDS selection procedure: According to Andrews et al. [46], to be MDS potentials, principal components (PCs) must have eigenvalues not less than 1.0 that explain more than 5% of the total variation. Within each retained principal component, those with an absolute value within 10% of the highest loading factor were selected as the important indicators. In addition, if multiple indicators were retained within a single PC and exhibited pairwise Pearson correlation coefficients > $|0.6|$, the indicator with the smallest absolute loading value in that PC was removed [47].

Scoring and SQI calculation: After MDS indicators were selected, a nonlinear scoring function was employed to convert the soil indicators into scores ranging from 0 to 1. Equation (1) for the soil indicator score was given in Andrews et al. [46]:

$$S = 1/[1 + (X/X_0)^b] \quad (1)$$

where S is the indicator score, X is the value of the soil indicator, X_0 is the mean value of each indicator, and b is the value of the equation's slope. Slope values (b) of −2.5 and 2.5 were used to illustrate a 'more is better' and a 'less is better' curve, respectively [47,48]. After scoring and weighting all MDS indicators, the integrated SQI was calculated by Equation (2) as follows:

$$SQI = \sum_{i=1}^{n} S_i \times W_i \quad (2)$$

where S_i is the score of the selected indicators, W_i is the weighting of the selected indicators, and n is the number of selected indicators [49].

2.5. Data Analysis

SMC, soil pH, the concentrations of SOC, TN, TP, NH_4^+-N, NO_3^--N, DOC, MBC, MBN, and microbial PLFAs, and the SQI were analyzed by a one-way ANOVA, and then a multiple comparison analysis (LSD) was employed to test the difference between vegetation restoration treatments. A Pearson correlation analysis was applied to test the relationships between the SQI on soil microbial biomass and community structure. Data were reciprocally or square-root-transformed when required to meet the assumptions of normality and homogeneity of variance. Statistical significance was determined at $p < 0.05$. All analyses were performed with SPSS 18.0 software (SPSS Inc., Chicago, IL, USA).

3. Results

3.1. Soil Physicochemical Properties

Vegetation restoration treatments significantly affected soil physicochemical properties (Table 1). Afforestation on barren land (BL) significantly increased SMC except REP. The SMC was highest in NF, followed by the MF and UEP, all of which were significantly greater than those in BL and REP ($p < 0.01$). Soil pH exhibited an inverse trend, with the highest value in BL and the lowest in NF. The differences between REP, UEP, and MF were not significant ($p > 0.05$; Table 1). SOC concentration increased progressively with restoration intensity ($p < 0.01$). SOC rose significantly from 3.0 g kg^{-1} in BL to 20.5 g kg^{-1} in MF. However, in MF, it remained significantly lower than in NF (26.7 g kg^{-1}; $p < 0.01$). SOC concentration in UEP (19.0 g kg^{-1}) was nearly double that of REP (9.8 g kg^{-1}) and statistically equivalent to MF ($p = 0.28$), and lower than in NF ($p < 0.01$). Interestingly, soil total N recovery varied notably: in MF, it exceeded that of NF ($p = 0.03$), while in UEP, it returned to levels comparable to NF ($p = 0.32$). Soil total N in REP was marginally lower than UEP ($p = 0.25$) but showed no significant difference from that in BL ($p = 0.73$). Soil total P mirrored the response pattern of soil total N to vegetation restoration treatments (Table 1). Soil DOC exhibited similar patterns to SMC, increasing from 159 mg kg^{-1} (BL) to 884 mg kg^{-1} (NF). DOC in REP (181 mg kg^{-1}) did not differ significantly from BL ($p = 0.72$). The soil N availability components showed treatment-specific responses. For instance, NH_4^+-N was the highest in UEP (p <0.01), with no significant differences among BL, REP, MF, and NF ($p > 0.05$), while NO_3^--N peaked in NF and it was significantly higher than the other four treatments (p <0.01). The NO_3^--N in UEP was significantly higher than REP ($p = 0.02$).

Table 1. Soil physiochemical properties in five vegetation restoration treatments.

Parameters	BL	REP	UEP	MF	NF
SMC (%)	10.1 ± 0.5 c	9.5 ± 0.3 c	19.6 ± 0.8 b	19.9 ± 1.1 b	23.3 ± 0.4 a
pH value	4.7 ± 0.05 a	4.4 ± 0.02 b	4.4 ± 0.04 bc	4.3 ± 0.1 bc	4.2 ± 0.04 c
SOC (mg g^{-1})	3.0 ± 0.2 d	9.8 ± 0.5 c	19.0 ± 1.6 b	20.5 ± 0.8 b	26.7 ± 0.7 a
TN (mg g^{-1})	0.01 ± 0.01 c	0.09 ± 0.04 c	0.35 ± 0.24 bc	1.16 ± 0.11 a	0.58 ± 0.22 b
TP (mg g^{-1})	0.02 ± 0.01 c	0.01 ± 0.01 c	0.08 ± 0.03 bc	0.16 ± 0.03 a	0.11 ± 0.03 ab
DOC (mg kg^{-1})	159 ± 26 d	181 ± 13 d	407 ± 62 c	694 ± 35 b	884 ± 50 a
NH_4^+-N (mg kg^{-1})	4.1 ± 0.6 b	7.0 ± 2.1 b	28.1 ± 7 a	9.0 ± 2.3 b	6.1 ± 1.9 b
NO_3^--N (mg kg^{-1})	3.5 ± 0.3 bc	1.8 ± 0.5 c	4.6 ± 0.5 b	3.7 ± 0.7 bc	11.2 ± 1.4 a

Note: BL, REP, UEP, MF, and NF represent barren land, short-rotation *Eucalyptus* plantation, undisturbed *Eucalyptus* plantation, mixed native-species plantation forest, and natural forest, respectively. SMC, SOC, TN, TP, DOC, NH_4^+-N, and NO_3^--N stand for soil moisture content, soil organic carbon, total nitrogen, total phosphorus, dissolved organic carbon, ammonium nitrogen, and nitrate nitrogen, respectively. Values are means ± SE; n = 4 plots. Different lowercase letters indicate significant differences between different restoration treatments at the $p = 0.05$ level.

3.2. Soil Microbial Properties

Restoration treatments enhanced MBC and MBN (Figure 1). Both MBC and MBN reached peak concentrations in NF, confirming that all restoration treatments facilitated microbial recovery. Specifically, MBC was significantly higher in UEP and MF than in BL and REP ($p < 0.05$), and no significant difference occurred between UEP and MF. Also, MBC in REP remained statistically indistinguishable from BL. For MBN, it was significantly higher in MF than in UEP, and UEP was higher than in REP and BL. There was no significant difference between REP and BL (Figure 1).

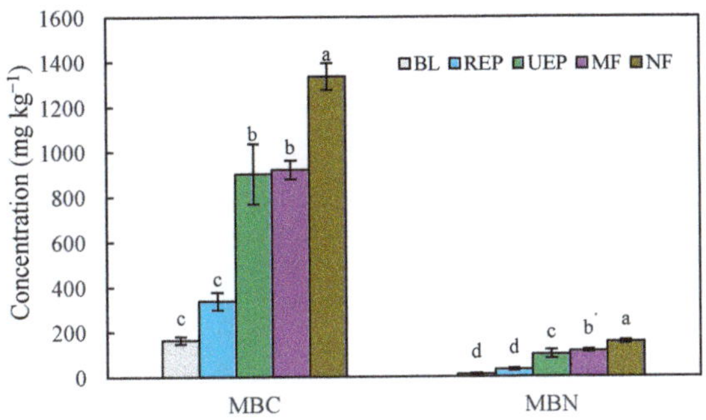

Figure 1. Soil microbial biomass carbon (MBC) and nitrogen (MBN) concentrations in all treatments. BL, REP, UEP, MF, and NF stand for barren land, short-rotation *Eucalyptus* plantation, undisturbed *Eucalyptus* plantation, mixed native-species plantation forest, and natural forest, respectively. Values are means ± SE; n = 4 plots. Different lowercase letters indicate significant differences between different restoration treatments at $p < 0.05$.

Total microbial biomass (summed PLFAs) and individual biomarker groups (bacteria, fungi, AMF, and actinomycetes PLFAs) increased with vegetation recovery treatments. They were the lowest in BL and the highest in NF for the biomass of bacteria, fungi, total microbes, AMF, and actinomycetes (Figure 2A,C). This meant that microbial biomass in the studied plantations did not reach the NF level. Microbial biomass (bacterial, fungal, total, AMF, and actinomycetes PLFAs) was significantly higher in MF than in UEP, and UEP was significantly higher than in REP. Meanwhile, the microbial biomass in REP was significantly higher than in BL, except for AMF biomass (Figure 2B). Conversely, the fungi-to-bacteria ratio (F:B) decreased along the restoration gradient: BL ≈ REP > UEP > MF > NF. Specifically, F:B ratios in BL, REP, and UEP exceeded those in MF and NF ($p < 0.05$), with no differences among BL, REP, and UEP. The F:B ratio in MF was marginally higher than in NF (Figure 2D).

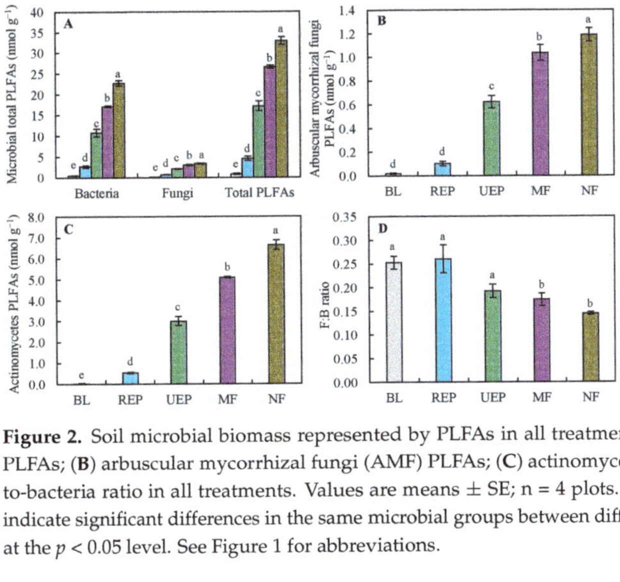

Figure 2. Soil microbial biomass represented by PLFAs in all treatments: (**A**) soil total microbial PLFAs; (**B**) arbuscular mycorrhizal fungi (AMF) PLFAs; (**C**) actinomycetes (Act) PLFAs; (**D**) fungi-to-bacteria ratio in all treatments. Values are means ± SE; n = 4 plots. Different lowercase letters indicate significant differences in the same microbial groups between different restoration treatments at the $p < 0.05$ level. See Figure 1 for abbreviations.

3.3. Soil Quality Index and Its Relationships with Microbial Community

PCA yielded three significant components (eigenvalues $\geq$ 1.0) explaining 89.2% of the total variance (Table 2). SMC, SOC, MBC, DOC, MBN, and DN were highly weighted indicators in PC-1 and were also significantly correlated with each other. DOC had the highest PC-1 weighting (0.978), so it was only retained in the MDS. TN and NH_4^+-N had the highest weighting in PC-2 and PC-3, respectively, securing their MDS inclusion. Thus, DOC, TN, and NH_4^+-N were the three key soil quality indicators and comprised the final MDS. The SQI was calculated using PCA-derived weighting factors. Vegetation restoration treatments promoted soil quality. The SQI increased from 0.05 to 0.66 with vegetation recovery (Figure 3). It was the lowest in BL and the highest in MF. The SQI in MF is equivalent to the NF levels. The SQI increased marginally in the REP compared with BL, and it was significantly higher in UEP than in REP. Meanwhile, the difference in the SQI between UEP and NF was not significant, while in UEP, it was significantly lower than in MF (Figure 3).

Table 2. Results of principal component analysis (PCA) of soil quality indicators in the 0–15 cm soil layer of the vegetation restoration chronosequence.

Principal Components	**PC-1**	**PC-2**	**PC-3**
Eigenvalues	8.20	1.38	1.13
Variance (%)	68.35	11.46	9.38
Cumulative (%)	68.35	79.82	89.20
Weighting value	0.346	0.34	0.314
Factor loading			
SMC	0.962	−0.093	−0.088
pH	−0.771	0.206	−0.021
SOC	0.968	−0.051	−0.028
TN	0.645	0.734	0.053
TP	0.743	0.630	0.118
MBC	0.967	−0.149	0.017
DOC	0.978	−0.059	0.117
MBN	0.976	−0.039	0.035
DN	0.955	−0.067	0.082
MBC/MBN	−0.682	−0.132	0.410
NH_4^+-N	0.260	−0.162	−0.889
NO_3^--N	0.698	−0.556	0.345

Note: SMC, pH, SOC, TN, TP, MBC, DOC, MBN, DN, MBC/MBN, NH_4^+-N, and NO_3^--N, stand for soil moisture content, pH value, soil organic carbon, total nitrogen, total phosphorus, microbial biomass carbon, dissolved organic carbon, microbial biomass nitrogen, dissolved nitrogen, ratio of microbial biomass carbon to nitrogen, ammonium nitrogen, and nitrate nitrogen, respectively.

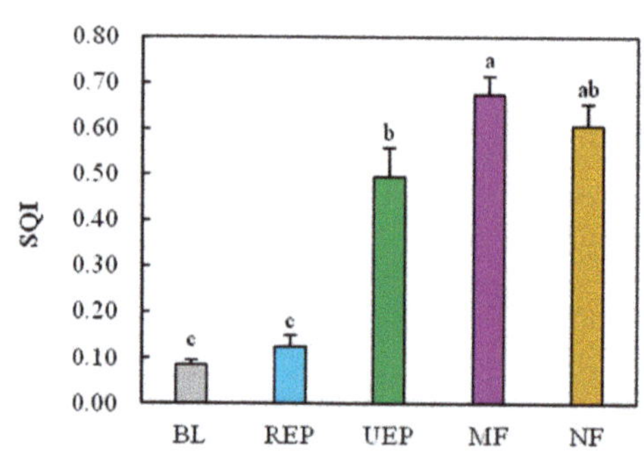

Figure 3. Soil quality index (SQI) for all treatments. BL, REP, UEP, MF, and NF stand for barren land, short-rotation *Eucalyptus* plantation, undisturbed *Eucalyptus* plantation, mixed native-species plantation forest, and natural forest, respectively. Values are means $\pm$ SE; n = 4 plots. Different lowercase letters indicate significant differences between different restoration treatments at $p < 0.05$.

Pearson correlation analysis showed that microbial biomass and community composition characteristics, such as bacteria, fungi, AMF, actinomycetes, and the F:B ratio, exhibited strong SQI linkages. Microbial biomass was significantly and positively correlated with the SQI, while the F:B showed a significantly negative correlation with the SQI (Table 3).

Table 3. Pearson correlation coefficients (r) indicating the direction and strength of relationships between soil physicochemical properties (microbial biomass) and the soil quality index (SQI) (** p <0.01, *** $p < 0.001$).

Parameter	SMC (B)	pH (F)	SOC (TMB)	TN (Act)	TP (AMF)	-- (F:B)
SQI	0.94 *** (0.89 ***)	−0.68 ** (0.92 ***)	0.89 *** (0.90 ***)	0.75 *** (0.90 ***)	0.79 *** (0.91 ***)	-- (−0.73 ***)

Note: SMC, pH, SOC, TN, TP, B, F, TMB, Act, AMF, and F:B stand for soil moisture content, pH value, soil organic carbon, total nitrogen, total phosphorus, the PLFAs of bacteria, fungi, total microbes, actinomyces, and arbuscular mycorrhizal fungi, and the ratio of fungi to bacterial PLFAs, respectively.

4. Discussion

4.1. Effects of Vegetation Recovery on Soil Quality

Vegetation recovery improves soil quality [50]. Soil quality index increased from barren land to natural forest in this study (Figure 3). Plant biomass generally drives soil improvement [51,52]. In these studied forests, the mixed plantation and natural forest had comparably high aboveground biomass C stocks and root biomass, which were much higher than those in the short-rotation *Eucalyptus* plantation [53]. Unfortunately, plant biomass in the undisturbed *Eucalyptus* plantation was not investigated, but the higher height and larger diameter at breast height were observed relative to the short-rotation *Eucalyptus* plantation. Therefore, the increase in plant biomass could be responsible for the soil quality improvement in these vegetation restoration treatments. However, compared to barren land (no plants), the greater carbon storage of plant biomass in the short-rotation *Eucalyptus* plantation did not result in a higher SQI in this study, which corresponded to 4.0% to 5.1% of those in the mixed forest and natural forest [53]. This finding suggests that plant biomass cannot fully explain the changes in soil quality in this study, which was consistent with other studies [20,54]. A recent study reported that root morphological traits affected soil quality [36], whereas they were absent in this study. Vegetation types also affected soil quality [19]. The deciduous broad-leaved forest had the highest soil quality index, which was higher than the natural forest, and the disturbed forest had the lowest SQI in the Karst areas of southwest China [12]. Furthermore, the minor SQI discrepancy between barren land and the short-rotation *Eucalyptus* plantation could be ascribed partly to their non-significant differences in DOC, TN, and NH_4^+-N in this study.

Soil physicochemical properties affect soil quality [36,55], which was supported by the significant correlations between soil physicochemical properties (SMC, pH, SOC, TN, and TP) and the SQI in this study. Ren et al. [52] found that soil organic matter and available P were the primary factors impacting soil quality. The trends in the SQI were the same as TN and TP among different vegetation types. As the TN was one of the three key soil quality indicators calculating the SQI, the correlations between them could be explained by the inclusion in the index, while the weighted value of TN in the SQI was not the greatest among the three key soil quality indicators (Table A1). Thus, we speculated that TN and TP could be the most important factors mediating soil quality.

Soil physicochemical properties showed different responses to vegetation restoration types [56]. The changes in SOC were more sensitive to vegetation recovery than TN and TP, which was consistent with the previous study [53]. High-intensity management like short-rotation cutting did not increase soil N and P [57]; thus, the improvement in soil quality was not significant in our study. Forest conservation or reducing disturbance facilitates the improvement of soil quality in plantations [58]. Brown et al. [17] reported that after 40 years of restoration, the soil carbon stocks and key soil physicochemical properties in plantations and secondary forests reached similar levels to those in the primary forests in the wet and moist climatic zones of Ghana. The significant improvement in soil physicochemical properties under protected management likely enhances soil quality, potentially explaining the SQI discrepancy between undisturbed long-term and disturbed short-rotation *Eucalyptus* plantations observed in this study.

4.2. Effects of Vegetation Recovery on Soil Microbial Community

Vegetation recovery significantly increased microbial biomass, regardless of MBC or PLFAs (Figures 1 and 2), aligning with the findings by Zhang et al. [59]. However, microbial communities responded differently. Notably, Zhang et al. [59] found that there was no significant difference in fungal biomass between the mixed forest and natural forest in the topsoil. No significant difference in AMF biomass between the barren land and short-rotation *Eucalyptus* plantation was also detected in the previous study [37], which was in accordance with our study. The shifts in microbial communities could probably be driven by vegetation recovery and the associated changes in soil physicochemical properties [60,61]. The significant correlations between soil physicochemical properties and microbial biomass supported these findings (Table A2). Soil microbial community composition was a more sensitive indicator to reflect the restoration of the ecosystem relative to soil physicochemical properties. Mixed plantations had higher microbial biomass than pure plantations, possibly due to higher biodiversity [62]. The positive effects of plant diversity on microbial biomass were shown across terrestrial ecosystems [63]. Litter as a substrate for microbes may mediate microbial communities [64]. However, there was no significant difference in litter biomass among the short-rotation *Eucalyptus* plantation, mixed forest, and natural forest [39]. Thus, the root-derived inputs rather than litter mass may exert a stronger influence on microbial biomass dynamics.

Elevated microbial biomass may likely contribute to enhanced soil quality through a greater accumulation of microbial residues. Previous studies have found that the accumulation of glomalin-related soil protein and amino sugars accelerated with vegetation recovery [37,65]. This link was further supported by significantly positive correlations between microbial biomass and the SQI (Table 3). The F:B ratio decreased as the restoration process progressed, which was in accordance with a previous study [65,66]. This could be explained by a faster increase in bacteria than fungi. Bacteria were more sensitive to local environmental drivers than fungi in forest ecosystems, particularly soil pH [66].

Beyond microbial biomass, soil microbiota functional recovery was also observed in the same plots with this research. Microbial enzyme activities and soil biodiversity in the mixed plantation reached comparable levels to the natural forest in terms of soil microbes and mite diversity [39,59]. This indicates significantly faster recovery of soil biodiversity under native-species mixed plantations than under disturbed *Eucalyptus* monocultures on these degraded tropical coastal terraces after over 60 years.

For the MDS-selected soil quality indicators, their ability to represent integrated soil quality remains constrained by the absence of key soil physical properties and plant parameters. Furthermore, the unmeasured plant biomass in the UEP hindered a direct assessment of its influence on soil quality. Additionally, although this study indicated comparable soil quality among UPE and NF, *Eucalyptus* monocultures are known to weakly support biodiversity [67], demonstrating lower bird functional diversity in relatively young plantations than in natural forests [68]. It thus raises concerns about broader ecosystem function beyond soil quality alone, necessitating urgent long-term monitoring of undisturbed *Eucalyptus* plantations.

5. Conclusions

Vegetation restoration treatments strongly influence soil quality and microbial communities. The SQI in the undisturbed long-term *Eucalyptus* plantation and mixed native-species forest reached levels comparable to the natural forest, indicating the recovery potential of well-managed plantations. Microbial biomass (bacteria, fungi, arbuscular mycorrhizal fungi, and actinomycetes) progressively increased from barren land to natural forest, but remained lower in all plantations than in the natural forest, suggesting incomplete microbial recovery and more sensitive responses relative to soil physicochemical properties alone. In addition, soil dissolved organic carbon, total nitrogen, and NH_4^+-N were three key soil quality indicators. Soil microbial biomass positively affected the SQI. Moreover, an undisturbed long-term *Eucalyptus* plantation has higher SQI and microbial communities than a disturbed short-rotation *Eucalyptus* plantation, indicating that intensive disturbance impeded the recovery of soil quality. Collectively, these findings suggest that afforestation with mixed native tree species is more favorable, and long-term forest protection implemented in pure plantations also proves effective for restoring degraded landscapes. These two restoration approaches enable soil quality to approach levels observed in the secondary native forest over time. This study highlights that tree species mixtures and reducing disturbance should be taken into consideration for ecological restoration in tropical and subtropical ecosystems.

Author Contributions: Conceptualization, Y.C.; methodology, Y.C. and Y.Z.; software, Y.C.; validation, Y.C., J.C., and Y.Z.; investigation, Y.C. and J.C.; data curation, F.Z.; writing—original draft preparation, Y.C., F.Z., and T.L.; writing—review and editing, Y.C., F.Z., J.C., T.L., and Y.Z.; visualization, F.Z. and T.L.; supervision, Y.C. and Y.Z.; project administration, Y.C.; funding acquisition, Y.C. All authors have read and agreed to the published version of the manuscript.

Funding: This research was funded by the National Natural Science Foundation of China (32271729 and U21A20189), the Hunan Provincial Natural Science Foundation of China (2024JJ5140), the Scientific Research Fund of Hunan Provincial Education Department, China (23A0385), and the Natural Science Foundation of Fujian Province of China (2023J011036).

Institutional Review Board Statement: Not applicable.

Informed Consent Statement: Not applicable.

Data Availability Statement: The data supporting this study's findings are available from the first author and the corresponding author upon reasonable request.

Acknowledgments: The authors would like to thank Weixin Zhang, Shenglei Fu, and Faming Wang for their help in conceptualization, and Bi Zou, Yingwen Li, and Yongxing Li for assistance in soil sampling in the field. We gratefully acknowledge constructive comments from anonymous reviewers that improved the manuscript.

Conflicts of Interest: The authors declare no conflicts of interest.

Abbreviations

The following abbreviations are used in this manuscript:

SQI	Soil quality index
BL	Barren land
REP	Disturbed short-rotation *Eucalyptus* plantation

UEP	Undisturbed long-term *Eucalyptus* plantation
MF	Mixed native-species plantation
NF	Natural forest
AMF	Arbuscular mycorrhizal fungi

Appendix A

Table A1. Normalization equation of scoring curves.

Parameter	Average (x0)	Curve Type	Slope (b)	Normalization Equation	Weighting Value (W)
DOC	0.40	More is better	−2.5	$S = 1/(1 + (X/0.4)^{-2.5})$	0.346
TN	452.81	More is better	−2.5	$S = 1/(1 + (X/452.81)^{-2.5})$	0.340
NH_4^+-N	10.75	More is better	−2.5	$S = 1/(1 + (X/10.75)^{-2.5})$	0.314

Note: DOC, TN, and NH_4^+-N stand for soil dissolved organic carbon, total nitrogen, and ammonium nitrogen, respectively.

Table A2. Pearson correlation coefficients (r) indicating the direction and strength of relationships between soil physicochemical properties and microbial biomass (* $p < 0.05$; ** $p < 0.01$; *** $p < 0.001$).

	SMC	pH	SOC	TN	TP	DOC	DN	NH_4^+-N	NO_3^--N
B	0.95 ***	−0.71 **	0.94 ***	0.64 **	0.72 ***	0.97 ***	0.94 ***	0.51	0.74 ***
F	0.94 ***	−0.72 **	0.96 ***	0.69 **	0.74 ***	0.93 ***	0.90 ***	0.30	0.64 **
TPLFAs	0.95 ***	−0.71 **	0.95 ***	0.66 **	0.74 ***	0.96 ***	0.94 ***	0.24	0.71 **
Act	0.94 ***	−0.71 **	0.94 ***	0.67 **	0.75 ***	0.96 ***	0.95 ***	0.22	0.76 ***
AMF	0.94 **	−0.65 **	0.93 ***	0.71 **	0.77 ***	0.94 ***	0.91 ***	0.25	0.66 **
F:B	−0.83 ***	0.56 *	−0.76	−0.52 *	−0.63 **	−0.84 ***	−0.85 ***	−0.20	−0.65 **

Note: SMC, pH, SOC, TN, TP, DOC, DN, NH_4^+-N, NO_3^--N, B, F, TPLFAs, Act, AMF, and F:B stand for soil moisture content, pH value, soil organic carbon, total nitrogen, total phosphorus, dissolved organic carbon, dissolved nitrogen, ammonium nitrogen, and nitrate nitrogen, the PLFAs of bacteria, fungi, total microbes, actinomyces, and arbuscular mycorrhizal fungi, and the ratio of fungi to bacterial PLFAs, respectively.

References

1. Bodo, T.; Gimah, B.G.; Seomoni, K.J. Deforestation and habitat loss: Human causes, consequences and possible solutions. *J. Geogr. Sci.* **2021**, *4*, 22–30. [CrossRef]
2. Qu, X.; Li, X.; Bardgett, R.D.; Kuzyakov, Y.; Revillini, D.; Sonne, C.; Xia, C.; Ruan, H.; Cao, F.; Reich, P.B.; et al. Deforestation impacts soil biodiversity and ecosystem services worldwide. *Proc. Natl. Acad. Sci. USA* **2024**, *121*, e2318475121. [CrossRef]
3. Chen, S.; Wen, Z.; Zhang, S.; Huang, P.; Ma, M.; Zhou, X.; Liao, T.; Wu, S. Effects of long-term and large-scale ecology projects on forest dynamics in Yangtze River Basin, China. *For. Ecol. Manag.* **2021**, *496*, 119463. [CrossRef]
4. Shao, Q.; Liu, S.; Ning, J.; Liu, G.; Yang, F.; Zhang, X.; Niu, L.; Huang, H.; Fan, J.; Liu, J. Remote sensing assessment of the ecological benefits provided by national key ecological projects in China during 2000–2019. *J. Geogr. Sci.* **2023**, *33*, 1587–1613. [CrossRef]
5. Lu, F.; Hu, H.; Sun, W.; Zhu, J.; Liu, G.; Zhou, W.; Zhang, Q.; Shi, P.; Liu, X.; Wu, X.; et al. Effects of national ecological restoration projects on carbon sequestration in China from 2001 to 2010. *Proc. Natl. Acad. Sci. USA* **2018**, *115*, 4039–4044. [CrossRef] [PubMed]
6. Zhang, Y.; Yuan, J.; You, C.; Cao, R.; Tan, B.; Li, H.; Yang, W. Contributions of National Key Forestry Ecology Projects to the forest vegetation carbon storage in China. *For. Ecol. Manag.* **2020**, *462*, 117981. [CrossRef]
7. Gatica-Saavedra, P.; Echeverría, C.; Nelson, C.R. Ecological indicators for assessing ecological success of forest restoration: A world review. *Restor. Ecol.* **2017**, *25*, 850–857. [CrossRef]
8. Mirghaed, F.A.; Souri, B. Contribution of land use, soil properties and topographic features for providing of ecosystem services. *Ecol. Eng.* **2023**, *189*, 106898. [CrossRef]
9. Paré, D.; Bognounou, F.; Emilson, E.J.; Laganière, J.; Leach, J.; Mansuy, N.; Martineau, C.; Norris, C.; Venier, L.; Webster, K. Connecting forest soil properties with ecosystem services: Toward a better use of digital soil maps-A review. *Soil Sci. Soc. Am. J.* **2024**, *88*, 981–999. [CrossRef]
10. Sandra, P.R.; Shankar, A.; Garkoti, S.C.; Adarsh, C.K. Understanding the effects of forest types, vegetation structural diversity, and soil properties on above-and below-ground carbon stock of moist deciduous forest ecosystems in Western Ghats, India. *Catena* **2025**, *257*, 109198. [CrossRef]

11. Oraon, P.R.; Sagar, V.; Beauty, K. Ecological restoration of degraded land through afforestation activities. In *Land and Environmental Management Through Forestry*; Raj, A., Jhariya, M.K., Banerjee, A., Nema, S., Bargali, K., Eds.; Scrivener Publishing LLC: Beverly, MA, USA, 2023; pp. 201–216. [CrossRef]
12. Ou, H.B.; Liu, X.S.; Wei, S.X.; Jiang, Y.; Gao, F.; Wang, Z.H.; Fu, W.; Du, H. The effects of different vegetation restoration models on soil quality in karst areas of southwest China. *Forests* **2024**, *15*, 1061. [CrossRef]
13. Marron, N.; Epron, D. Are mixed-tree plantations including a nitrogen-fixing species more productive than monocultures? *For. Ecol. Manag.* **2019**, *441*, 242–252. [CrossRef]
14. Guo, J.; Feng, H.; McNie, P.; Liu, Q.; Xu, X.; Pan, C.; Yan, K.; Feng, L.; Goitom, E.A.; Yu, Y. Species mixing improves soil properties and enzymatic activities in Chinese fir plantations: A meta-analysis. *Catena* **2023**, *220*, 106723. [CrossRef]
15. Zhang, J.; Zhu, S.; Liu, Y.; Yao, B.; Yu, M.; Ma, J.; Yang, X.; Xue, J.; Xiang, Y.; Li, Y.; et al. Impact of mixed plantations on soil physicochemical properties: Variations and controlling factors in China. *For. Ecol. Manag.* **2024**, *568*, 122107. [CrossRef]
16. Wang, X.; Zhuo, Z.; Zhou, M.; Li, S.; Lin, G.; Zhang, Y.; Jiang, F.; Huang, Y.; Lin, J. Response of the soil bacterial community to soil fertility during vegetation restoration in soil and water loss areas in south China. *J. Soil Sci. Plant Nutr.* **2024**, *24*, 3687–3698. [CrossRef]
17. Brown, H.C.A.; Appiah, M.; Quansah, G.W.; Adjei, E.O.; Berninger, F. Soil carbon and bio-physicochemical properties dynamics under forest restoration sites in southern Ghana. *Geoderma Reg.* **2024**, *38*, e00838. [CrossRef]
18. Dong, R.; Wang, X.; Wang, Y.; Ma, Y.; Yang, S.; Zhang, L.; Zhang, M.; Qin, J.; Quzha, R. Differences in soil microbial communities with successional stage depend on vegetation coverage and soil substrates in alpine desert shrublands. *Plant Soil* **2023**, *485*, 549–568. [CrossRef]
19. Peng, S.; Chen, A.; Fang, H.; Wu, J.; Liu, G. Effects of vegetation restoration types on soil quality in Yuanmou dry-hot valley, China. *Soil Sci. Plant Nutr.* **2013**, *59*, 347–360. [CrossRef]
20. Riestra, D.; Noellemeyer, E.; Quiroga, A. Soil texture and forest species condition the effect of afforestation on soil quality parameters. *Soil Sci.* **2012**, *177*, 279–287. [CrossRef]
21. Augusto, L.; Boča, A. Tree functional traits, forest biomass, and tree species diversity interact with site properties to drive forest soil carbon. *Nat. Commun.* **2022**, *13*, 1097. [CrossRef]
22. Xie, H.; Tang, Y.; Yu, M.; Wang, G.G. The effects of afforestation tree species mixing on soil organic carbon stock, nutrients accumulation, and understory vegetation diversity on reclaimed coastal lands in Eastern China. *Glob. Ecol. Conserv.* **2021**, *26*, e01478. [CrossRef]
23. Luo, X.; Hou, E.; Zhang, L.; Kuang, Y.; Wen, D. Altered soil microbial properties and functions after afforestation increase soil carbon and nitrogen but not phosphorus accumulation. *Biol. Fertil. Soils* **2023**, *59*, 645–658. [CrossRef]
24. Guo, Y.; Abdalla, M.; Espenberg, M.; Hastings, A.; Hallett, P.; Smith, P. A systematic analysis and review of the impacts of afforestation on soil quality indicators as modified by climate zone, forest type and age. *Sci. Total Environ.* **2021**, *757*, 143824. [CrossRef]
25. Xu, Y.; Du, A.; Wang, Z.; Zhu, W.; Li, C.; Wu, L. Effects of different rotation periods of *Eucalyptus* plantations on soil physiochemical properties, enzyme activities, microbial biomass and microbial community structure and diversity. *For. Ecol. Manag.* **2020**, *456*, 117683. [CrossRef]
26. Bose, T.; Hammerbacher, A.; Slippers, B.; Roux, J.; Wingfield, M.J. Continuous replanting could degrade soil health in short-rotation plantation forestry. *Curr. For. Rep.* **2023**, *9*, 230–250. [CrossRef]
27. Zhou, X.; Wen, Y.; Goodale, U.M.; Zuo, H.; Zhu, H.; Li, X.; You, Y.; Yan, L.; Su, Y.; Huang, X. Optimal rotation length for carbon sequestration in *Eucalyptus* plantations in subtropical China. *New For.* **2017**, *48*, 609–627. [CrossRef]
28. Chen, Y.; Cai, X.A.; Zhang, Y.; Rao, X.; Fu, S. Dynamics of understory shrub biomass in six young plantations of southern subtropical China. *Forests* **2017**, *8*, 419. [CrossRef]
29. Başkent, E.Z.; Kašpar, J. Exploring the effects of various rotation lengths on the ecosystem services within a multiple-use management framework. *For. Ecol. Manag.* **2023**, *538*, 120974. [CrossRef]
30. Furey, G.N.; Tilman, D. Plant biodiversity and the regeneration of soil fertility. *Proc. Natl. Acad. Sci. USA* **2021**, *118*, e2111321118. [CrossRef]
31. Cagnoni, L.B.; Weidlich, E.W.; Guillemot, J.; Morselo, C.; Weih, M.; Adler, A.; Brancalion, P.H. Stakeholders' perspectives of species diversity in tree plantations: A global review. *Curr. For. Rep.* **2023**, *9*, 251–262. [CrossRef]
32. Yao, X.; Hui, D.; Xing, S.; Zhang, Q.; Chen, J.; Li, Z.; Yang, X.; Deng, Q. Mixed plantations with N-fixing tree species maintain ecosystem C: N: P stoichiometry: Implication for sustainable production. *Soil Biol. Biochem.* **2024**, *191*, 109356. [CrossRef]
33. Shao, G.; Ai, J.; Sun, Q.; Hou, L.; Dong, Y. Soil quality assessment under different forest types in the Mount Tai, central Eastern China. *Ecol. Indic.* **2020**, *115*, 106439. [CrossRef]
34. Li, X.; Liu, Y.; Wu, G.; Lie, Z.; Sheng, H.; Aguila, L.C.R.; Khan, M.S.; Liu, X.; Zhou, S.; Wu, T.; et al. Mixed plantations do not necessarily provide higher ecosystem multifunctionality than monoculture plantations. *Sci. Total Environ.* **2024**, *914*, 170156. [CrossRef]

35. Bastida, F.; Zsolnay, A.; Hernández, T.; García, C. Past, present and future of soil quality indices: A biological perspective. *Geoderma* **2008**, *147*, 159–171. [CrossRef]
36. Chen, Y.; Chen, Z.; Zhang, W.; Tang, Z.; Zhang, Y. How forest types shape soil quality: The evidence from eastern China's north sub-tropical ecosystems. *Ecol. Indic.* **2025**, *175*, 113583. [CrossRef]
37. Zhang, J.; Li, J.; Ma, L.; He, X.; Liu, Z.; Wang, F.; Chu, G.; Tang, X. Accumulation of glomalin-related soil protein benefits soil carbon sequestration: Tropical coastal forest restoration experiences. *Land Degrad. Dev.* **2022**, *33*, 1541–1551. [CrossRef]
38. Ren, H.; Li, Z.; Shen, W.; Yu, Z.; Peng, S.; Liao, C.; Ding, M.; Wu, J. Changes in biodiversity and ecosystem function during the restoration of a tropical forest in South China. *Sci China Ser. C* **2007**, *50*, 277–284. [CrossRef]
39. Wu, W.; Kuang, L.; Li, Y.; He, L.; Mou, Z.; Wang, F.; Zhang, J.; Wang, J.; Li, Z.; Lambers, H.; et al. Faster recovery of soil biodiversity in native species mixture than in *Eucalyptus* monoculture after 60 years afforestation in tropical degraded coastal terraces. *Glob. Change Biol.* **2021**, *27*, 5329–5340. [CrossRef]
40. Lu, R.K. *Method of Analysis in Soil and Agrochemistry*; Agricultural Press: Beijing, China, 1999; pp. 31–33. (In Chinese)
41. Bao, S. *Analysis in Soil and Agrochemistry, 3rd ed.*; Agricultural Press: Beijing, China, 2000; pp. 25–76. (In Chinese)
42. Witt, C.; Gaunt, J.L.; Galicia, C.C.; Ottow, J.C.; Neue, H.U. A rapid chloroform-fumigation extraction method for measuring soil microbial biomass carbon and nitrogen in flooded rice soils. *Biol. Fertil. Soils* **2000**, *30*, 510–519. [CrossRef]
43. Bossio, D.A.; Scow, K.M. Impacts of carbon and flooding on soil microbial communities: Phospholipid fatty acid profiles and substrate utilization patterns. *Microb. Ecol.* **1998**, *35*, 265–278. [CrossRef]
44. Frostegård, A.; Bååth, E. The use of phospholipid fatty acid analysis to estimate bacterial and fungal biomass in soil. *Biol. Fertil. Soils* **1996**, *22*, 59–65. [CrossRef]
45. Joergensen, R.G. Phospholipid fatty acids in soil—Drawbacks and future prospects. *Biol. Fertil. Soils* **2022**, *58*, 1–6. [CrossRef]
46. Andrews, S.S.; Karlen, D.L.; Mitchell, J.P. A comparison of soil quality indexing methods for vegetable production systems in northern California. *Agr. Ecosyst. Environ.* **2022**, *90*, 25–45. [CrossRef]
47. Bastida, F.; Moreno, J.L.; Hernández, T.; García, C. Microbiological degradation index of soils in a semiarid climate. *Soil Biol. Biochem.* **2006**, *38*, 3463–3473. [CrossRef]
48. Chen, L.; Xiang, W.; Ouyang, S.; Wu, H.; Xia, Q.; Ma, J.; Zeng, Y.; Lei, P.; Xiao, W.; Li, S.; et al. Tight coupling of fungal community composition with soil quality in a Chinese fir plantation chronosequence. *Land Degrad. Dev.* **2021**, *32*, 1164–1178. [CrossRef]
49. Masto, R.E.; Chhonkar, P.K.; Singh, D.; Patra, A.K. Alternative soil quality indices for evaluating the effect of intensive cropping, fertilization and manuring for 31 years in the semi-arid soils of India. *Environ. Monit. Assess.* **2008**, *136*, 419–435. [CrossRef]
50. Guan, H.; Fan, J. Effects of vegetation restoration on soil quality in fragile karst ecosystems of southwest China. *PeerJ* **2020**, *8*, e9456. [CrossRef]
51. Panico, S.C.; Memoli, V.; Esposito, F.; Maisto, G.; De Marco, A.D. Plant cover and management practices as drivers of soil quality. *Appl. Soil Ecol.* **2018**, *129*, 34–42. [CrossRef]
52. Ren, Q.; Qiang, F.; Liu, G.; Liu, C.; Ai, N. Response of soil quality to ecosystems after revegetation in a coal mine reclamation area. *Catena* **2025**, *257*, 109038. [CrossRef]
53. Wang, F.; Ding, Y.; Sayer, E.J.; Li, Q.; Zou, B.; Mo, Q.; Li, Y.; Lu, X.; Tang, J.; Zhu, W.; et al. Tropical forest restoration: Fast resilience of plant biomass contrasts with slow recovery of stable soil C stocks. *Funct. Ecol.* **2017**, *31*, 2344–2355. [CrossRef]
54. Shen, Y.; Li, J.; Chen, F.; Cheng, R.; Xiao, W.; Wu, L.; Zeng, L. Correlations between forest soil quality and aboveground vegetation characteristics in Hunan Province, China. *Front. Plant Sci.* **2022**, *13*, 1009109. [CrossRef] [PubMed]
55. Schoenholtz, S.H.; Van Miegroet, H.; Burger, J.A. A review of chemical and physical properties as indicators of forest soil quality: Challenges and opportunities. *For. Ecol. Manag.* **2000**, *138*, 335–356. [CrossRef]
56. Lv, X.; Tang, Q.; Han, C.; Song, M.; Yuan, C.; Yang, Q.; Wei, J.; He, X.; Collins, A.L. Farmland abandonment and vegetation succession mediate soil properties but are determined by the duration of conversion. *Catena* **2024**, *238*, 107877. [CrossRef]
57. Xiang, W.; Xu, L.; Lei, P.; Ouyang, S.; Deng, X.; Chen, L.; Zeng, Y.; Hu, Y.; Zhao, Z.; Wu, H.; et al. Rotation age extension synergistically increases ecosystem carbon storage and timber production of Chinese fir plantations in southern China. *J. Environ. Manag.* **2022**, *317*, 115426. [CrossRef]
58. Sasanifar, S.; Alijanpour, A.; Shafiei, A.B.; Rad, J.E.; Molaei, M. Forest conservation mediating soil quality relationship with diversity of various plant layers in the biosphere of Arasabran, Iran. *Sci. Total Environ.* **2024**, *928*, 172475. [CrossRef]
59. Zhang, H.; Xiong, X.; Wu, J.; Zhao, J.; Zhao, M.; Chu, G.; Hui, D.; Zhou, G.; Deng, Q.; Zhang, D. Changes in soil microbial biomass, community composition, and enzyme activities after half-century forest restoration in degraded tropical lands. *Forests* **2019**, *10*, 1124. [CrossRef]
60. Zhao, C.; Long, J.; Liao, H.; Zheng, C.; Li, J.; Liu, L.; Zhang, M. Dynamics of soil microbial communities following vegetation succession in a karst mountain ecosystem, Southwest China. *Sci. Rep.* **2019**, *9*, 2160. [CrossRef]
61. Liu, Y.; Zhu, G.; Hai, X.; Li, J.; Shangguan, Z.; Peng, C.; Deng, L. Long-term forest succession improves plant diversity and soil quality but not significantly increase soil microbial diversity: Evidence from the Loess Plateau. *Ecol. Eng.* **2020**, *142*, 105631. [CrossRef]

62. Qiang, W.; He, L.; Zhang, Y.; Liu, B.; Liu, Y.; Liu, Q.; Pang, X. Aboveground vegetation and soil physicochemical properties jointly drive the shift of soil microbial community during subalpine secondary succession in southwest China. *Catena* **2021**, *202*, 105251. [CrossRef]
63. Chen, C.; Chen, H.Y.; Chen, X.; Huang, Z. Meta-analysis shows positive effects of plant diversity on microbial biomass and respiration. *Nat. Commun.* **2019**, *10*, 1332. [CrossRef] [PubMed]
64. Wang, C.; Lin, W.; Jia, S.; Chen, S.; Xiong, D.; Xu, C.; Yang, Z.; Liu, X.; Yang, Y. Effects of litter and root inputs on soil microbial community structure in subtropical natural and plantation forests. *Plant Soil* **2025**, 1–16. [CrossRef] [PubMed]
65. Li, T.; Yuan, Y.; Mou, Z.; Li, Y.; Kuang, L.; Zhang, J.; Wu, W.; Wang, F.; Wang, J.; Lambers, H.; et al. Faster accumulation and greater contribution of glomalin to the soil organic carbon pool than amino sugars do under tropical coastal forest restoration. *Glob. Change Biol.* **2023**, *29*, 533–546. [CrossRef]
66. Li, Q.; Feng, J.; Wu, J.; Jia, W.; Zhang, Q.; Chen, Q.; Zhang, D.; Cheng, X. Spatial variation in soil microbial community structure and its relation to plant distribution and local environments following afforestation in central China. *Soil Till. Res.* **2019**, *193*, 8–16. [CrossRef]
67. Lemessa, D.; Mewded, B.; Legesse, A.; Atinfau, H.; Alemu, S.; Maryo, M.; Tilahun, H. Do Eucalyptus plantation forests support biodiversity conservation? *For. Ecol. Manag.* **2022**, *523*, 120492. [CrossRef]
68. Melo, R.S.; Alexandrino, E.R.; de Paula, F.R.; Boscolo, D.; de Barros Ferraz, S.F. Promoting bird functional diversity on landscapes with a matrix of planted *Eucalyptus spp.* in the Atlantic Forest. *Environ. Manag.* **2024**, *73*, 395–407. [CrossRef] [PubMed]

biology

MDPI

Article

Effects of Different Afforestation Measures on Biological Soil Crust Properties and Microbial Communities in an Alpine Sandy Land

Shaobo Du [1,2,3], Huichun Xie [2,3,4,*], Gaosen Zhang [5], Feng Qiao [2,3,4], Guigong Geng [6] and Chongyi E [1,*]

1 College of Geographical Sciences, Qinghai Normal University, Xining 810008, China; wo827288809@163.com
2 Qilian Mountain Southern Slope Forest Ecosystem Research Station, Huzhu 810500, China; qiaofnm@163.com
3 Key Lab of Medicinal Animal and Plant Resources of Qinghai-Tibetan Plateau in Qinghai Province, Qinghai Normal University, Xining 810008, China
4 College of Life Sciences, Qinghai Normal University, Xining 810008, China
5 Key Laboratory of Extreme Environmental Microbial Resources and Engineering, Northwest Institute of Eco-Environment and Resources, Chinese Academy of Sciences, Lanzhou 730000, China; gaosenzhang@hotmail.com
6 Qinghai Academy of Agriculture and Forestry Sciences, Qinghai University, Xining 810016, China; genggg-298@163.com
* Correspondence: yezino.1@163.com (H.X.); echongyi@163.com (C.E.)

Simple Summary

Alpine sandy areas in the Gonghe Basin, which had been subjected to four afforestation measures, exhibited a gradual increase in water content, nutrient content, enzymatic activities, and bacterial community richness and diversity with succession from the bare sand stage to an algae crust and then to a moss crust. In particular, the above indicators for the two types of biological soil crusts had higher values at the *Populus simonii* plantation (YY) site than at other afforested sites. Effective phosphorus and organic matter were the primary environmental factors affecting bacterial community structure in algae and moss crusts, respectively, and the activities of four types of enzymes were significantly correlated with the relative abundance of most major bacterial phyla. The functional prediction results showed that YY preserved the balance of primary functions while offering precise support for the physiological characteristics and ecological needs of different crust types in the secondary functions. A comprehensive analysis of the indicators revealed that YY was more favorable for the development of biological soil crusts.

Abstract

A good understanding of the effects of different afforestation measures in alpine sandy land on the physicochemical properties, enzymatic activities, and bacterial community structure of such crusts enables elucidation of the succession patterns of biological soil crusts and provides a theoretical basis for precise optimization of desertification control programs in alpine sandy land. In the present study, four afforestation measures—*Salix cheilophila*+ *Populus simonii* (WLYY00), *S. cheilophila* (WL), *P. simonii* (YY), and *Caragana korshinskii* (NT00) plantations—were adopted. The physicochemical properties and enzymatic activities of bare sand, algae crust, and moss crust in the four afforested sites were analyzed using Illumina high-throughput sequencing and PICRUSt2 functional prediction to investigate the bacterial community structure and function. Results indicated the following: (1) Water content, nutrient content, enzymatic activities, and bacterial community richness and diversity increased stepwise with succession from the bare sand stage to algae crust and to moss crust. The enhancement effect of YY on the above indicators and fine particle content was most prominent. (2) The primary environmental factors affecting bacterial community structure in algae and moss crusts were adequate phosphorus and organic matter, respectively, and the correlations between the activities of the four enzymes and the bacterial community structure are also quite close. (3) Functional prediction indicated that metabolism was the main primary function of biological soil crusts at the various sample sites. YY maintained the balance of primary functions and provided precise support for the physiological characteristics and ecological needs of different crust types in the secondary functions. In conclusion, among the four types of afforestation measures with a restoration period of 24 years, YY provided a greater advantage in improving the nutrient content, bacterial community structure, and functional potentials of biological soil crusts. The results of this study can serve as a scientific reference for screening of afforestation measures and protecting and utilizing biological soil crusts during the ecological restoration of alpine sandy lands in the present study area and other regions.

Keywords: desertification control measures; alpine sandy land; biological soil crust; physicochemical properties; enzymatic activity; bacterial community structure

1. Introduction

Desertification is a severe ecological and environmental problem worldwide. It leads to a series of land degradation phenomena such as reduced soil water retention and productivity, resulting in the formation of large areas of sandy land [1,2]. Biological soil crusts (BSCs), formed by the cementation of algae, lichens, mosses, bacteria, and fungi to ground particles, are widely distributed on the surface layer of sandy land [3]. They are primarily divided into three successional stages: algae, lichen, and moss crusts. In particular, algae and moss crusts are the most common crust groups in sandy land due to their high biomass and wide coverage [4]. Algae can secrete high-molecular polymers such as polysaccharides to aggregate sand particles and form algae crusts on the soil surface, mainly including cyanobacteria, chlorophyta and diatoms. Among them, cyanobacteria play a key role in the formation and succession of crusts [5]. Moss crusts are organic complexes formed by the adhesion of rhizoids of Bryophyta plants to ground particles; they are in the advanced stage of the forward succession of BSCs. On sandy land, the main species that forms moss crusts is *Bryum argenteum* [6]. BSCs can serve various roles during the ecological restoration of sandy lands, including promoting vegetation succession, preventing wind erosion and stabilizing sand, retaining soil moisture, and enhancing soil nutrient content [7–9]. The strengths of these functions are closely related to the physicochemical properties and enzymatic activities of the crusts [10]. Relevant studies [11,12] have revealed significant differences in the physicochemical properties and enzymatic activities of BSCs subjected to different afforestation measures, as well as between BSCs at different developmental stages in identical environments. Therefore, elucidating the physicochemical properties and enzymatic activities of BSCs under different afforestation measures can provide a scientific basis for screening afforestation measures that effectively promote crust development and enhance its ecological function. This will be beneficial for the stable restoration and sustainable development of sandy land ecosystems. An understanding of changes in the physicochemical properties and enzymatic activities of BSCs across different developmental stages in the same environment can also shed light on the mechanisms of BSC succession and development. Bacteria are important in the formation and development of BSCs. Besides maintaining the structure and function of these crusts, bacteria can also drive ecosystem material cycling, influencing their physicochemical properties and enzymatic activities [13]. Therefore, an analysis of the composition and function of bacterial communities in BSCs enables an in-depth elucidation of the mechanisms of crust development and aids in assessing the restoration effect of afforestation measures through the analysis of differences in bacterial community function. This will contribute to the optimization of desertification control strategies.

With an extensive desertified land area of approximately 5573 km^2, the Gonghe Basin is one of the most severely desertified areas in China [14]. The high altitude and low temperature of the region make it a typical alpine sandy land [15]. Large-scale afforestation measures have been implemented in the alpine sandy land of the Gonghe Basin since the 1960s to control the spread of desertification effectively [16]. Researchers have also systematically studied and evaluated the effectiveness of specific representative management measures in soil improvement [17,18]. However, these studies have merely assessed the soil improvement effects of different afforestation measures from the perspective of soil physicochemical properties, while the degree of development of BSCs, an important indicator, has been neglected. Zhang et al. [19] compared the differences in the physicochemical properties among various BSCs in areas subjected to three different afforestation measures. However, an in-depth investigation of the enzymatic activity, bacterial community structure, and function of the crusts was not reported. Therefore, our understanding of the response characteristics of BSCs to different vegetation types is limited, thereby restricting the precise optimization of ecological restoration strategies in alpine sandy regions. We selected alpine sandy areas of the Gonghe Basin that had been separately subjected to four representative afforestation measures in 2000 to achieve an extensive understanding of the effects of different afforestation measures on BSCs in alpine sandy land. Samples of bare sand without crust cover (0–2 cm soil layer), algae crust, and moss crust were collected. The physicochemical properties and enzymatic activities of the samples were measured, and bacterial community structure analysis and functional prediction were performed using high-throughput sequencing and PICRUSt2 functional prediction. Correlations among the physicochemical properties, enzymatic activities, and bacterial community structure of algae and moss crusts were subsequently determined. Our results revealed the response characteristics of physicochemical properties, enzymatic activities, and bacterial community structure and function of BSCs under different afforestation measures, elucidated the succession and development patterns of BSCs in alpine sandy land, and provided scientific support for the screening of optimal afforestation measures that can effectively promote the development of BSCs and strengthen their ecological functions. This can serve as a theoretical basis for promoting stable restoration and sustainable management of the alpine sandy land ecosystem in the Gonghe Basin.

2. Materials and Methods

2.1. Study Area

The study area is located in a desertification control experimental station and afforestation base in Shazhuyu Township, Gonghe County, Hainan Prefecture, Qinghai Province, China (100°25′ E, 36°24′ N, Figure 1). Situated in the mid-western part of the Gonghe Basin, it belongs to the transition zone between alpine arid desert and semi-arid grassland [20]. The area is at an altitude of 2880 m and is characterized by an alpine climate, aridity, and intense radiation. Climatic characteristics include an annual precipitation of 264 mm, annual evaporation of 1528–1937 mm, average annual temperature of 2.4 °C, average annual wind speed of 2.7 m/s, and primarily northwesterly and westerly winds. The soil type is dominated by aeolian sandy soil, and BSCs are primarily algae and moss crusts, with the total coverage of other types of crusts being less than 1%.

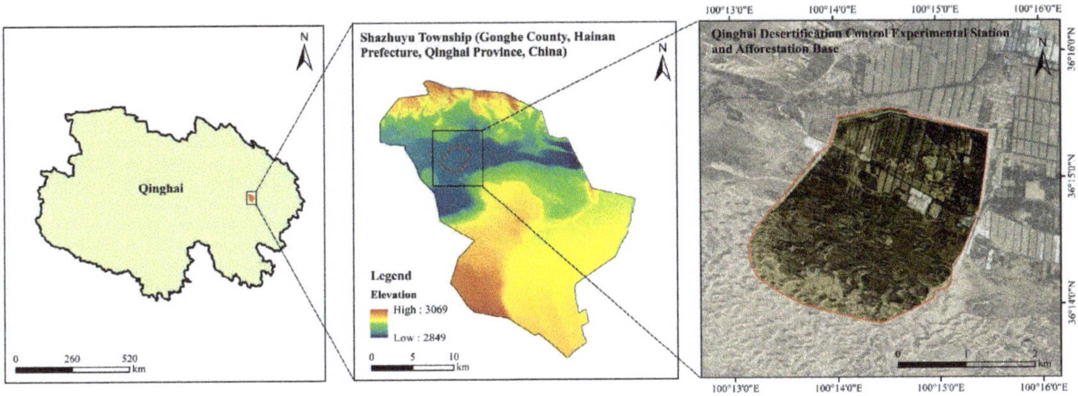

Figure 1. Overview of the study area.

2.2. Sample Site Selection and BSCs Sample Collection

Sample site selection and collection of BSC samples from the study area were performed in the second half of July 2024. Afforested sites on which the following four large-scale, representative afforestation measures had been implemented in 2000 were selected: *Salix cheilophila* + *Populus simonii* plantation (WLYY00), *S. cheilophila* plantation (WL), *P. simonii* plantation (YY), and *Caragana korshinskii* plantation (NT00) (Table 1). Prior to afforestation, all four sample sites were situated on mobile sand dunes on the alpine sandy land of the Gonghe Basin. Site conditions were generally consistent, with plant spacing being 1.5 m × 1.5 m. Afforestation was performed using pit sowing for NT00 and by planting seedlings in the remaining sites. Additionally, 3 large quadrats (50 m × 50 m) and 12 small quadrats (5 m × 5 m) were randomly established in the inter-dune area of each sample site. Based on the principles of random sampling and multi-point sampling, samples of bare sand (0–2 cm soil layer) without crust cover, algae crust, and moss crust were collected from each large quadrat. Samples of the same type were thoroughly mixed to form a single sample, and a total of 36 samples were obtained (3 large quadrats × 3 crust types × 4 sample sites). Each sample was divided into two parts: one part was used for the determination of physicochemical properties and enzymatic activities, and the other part was loaded into a 10 mL sterile centrifuge tube and stored in liquid nitrogen for subsequent bacterial sequencing. The coverage and thickness of crust samples within each small quadrat were recorded, and the average of the 12 small quadrats was calculated and used as the final value. During crust sample collection, a 9.0 cm diameter Petri dish was first used to delineate the sample collection area, and a sterile spatula was subsequently used to separate the crust layer from the underlying soil. The sampling tools were sterilized with 75% alcohol between sampling.

Table 1. Basic information of sample sites.

Sample Site	Longitude	Latitude	Elevation (m)	Area (m^2)	Coverage	Crust Type	Crust Thickness/mm	Crust Coverage/%
						Bare sand	nd	31.33 ± 11.47
WLYY00	100°14′27.23″ E	36°15′32.432″ N	2828	133,333	67%	Algae crust	7.12 ± 1.06	41.62 ± 14.38
						Moss crust	13.11 ± 2.11	27.04 ± 4.31
						Bare sand	nd	13.01 ± 2.74
YY	100°13′259.1″ E	36°15′32.713″ N	2827	46,666	81%	Algae crust	4.74 ± 0.37	22.73 ± 4.13
						Moss crust	9.22 ± 1.43	64.25 ± 11.74

Table 1. *Cont.*

Sample Site	Longitude	Latitude	Elevation (m)	Area (m^2)	Coverage	Crust Type	Crust Thickness/mm	Crust Coverage/%
WL	100°14′16.54″ E	36°14′50.678″ N	2830	40,000	64%	Bare sand	nd	31.12 ± 9.48
						Algae crust	5.16 ± 0.85	27.18 ± 3.04
						Moss crust	8.45 ± 1.39	41.71 ± 8.45
NT00	100°14′36.26″ E	36°15′5.3167″ N	2825	133,333	91%	Bare sand	nd	17.23 ± 7.51
						Algae crust	4.16 ± 0.77	21.08 ± 6.36
						Moss crust	8.68 ± 1.46	61.70 ± 14.54

Note: Mean ± standard error; nd = not determined; WLYY00: *Salix cheilophila* + *Populus simonii* plantation; WL: *S. cheilophila* plantation; YY: *P. simonii* plantation; NT00: *Caragana korshinskii* plantation.

2.3. Physicochemical Properties and Enzymatic Activities

Table 2 shows the methods used for measuring physicochemical properties and enzymatic activities.

Table 2. Methods for determining physicochemical properties and enzymatic activities.

Indicator	Method	Reference
Soil particle composition	Laser method	[21]
pH	Potentiometric method (water/soil ratio of 2.5:1)	[22]
Electrical conductivity (EC)	Conductometric method (water/soil ratio of 5:1)	[22]
Soil water content (SWC)	Drying method	[22]
Alkaline-dissolved nitrogen (AN)	Alkaline dissolution and diffusion method	[22]
Available phosphorus (AP)	Sodium bicarbonate leaching and molybdenum antimony colorimetric method	[22]
Total phosphorus (TP)	Sodium hydroxide fusion and molybdenum antimony colorimetric method	[22]
Soil organic matter (SOM)	Potassium dichromate–concentrated sulfuric acid external heating method	[22]
Total carbon (TC) and total nitrogen (TN)	Combustion method	[22]
Total potassium (TK) and available potassium (AK)	Flame photometry	[22]
Urease (URE)	Indophenolic acid colorimetric method	[23]
Alkaline phosphatase (ALP)	Disodium phenyl phosphate colorimetric method	[24]
Sucrase (SUC)	3,5-Dinitrosalicylic acid colorimetric method	[25]
Catalase (CAT)	Potassium permanganate titration method	[26]

Note: According to American standards, soil particles are categorized into three classes: clay (<0.002 mm in diameter), silt (0.002–0.05 mm), and sand (>0.05 mm).

2.4. 16S rDNA Extraction and Sequencing

Extraction and PCR amplification of 16S rDNA were performed according to the methods described by Du et al. [27]. DNA purification kit: MagaBio Soil Genomic DNA Purification Kit (Thermo Fisher Scientific, Shanghai, China). The V3—V4 variable region of the soil bacterial 16S rRNA gene was PCR amplified using the universal primers 338F and 806R. The 16S rDNA library was constructed using the TruSeqTM DNA Sample Prep Kit (Illumina Corporation, San Diego, CA, USA), and sequencing was performed on the MiSeq PE300 platform. Upon the completion of sequencing, the MiSeq paired-end sequencing data were merged into one sequence using FLASH 1.2.11 software. After quality control in Fastp 0.19.6, the samples were differentiated using barcodes and primers at both ends to obtain valid sequences.

2.5. Data Processing and Analysis

Analysis of variance (ANOVA) and Duncan's test for significant differences were performed in SPSS 27.0. Sequences were clustered into operational taxonomic units (OTUs) based on 97% similarity in UPARSE 7.0.1090, removing single sequences and chimeras during the clustering process, and compared to the Silva database to annotate species classification for each sequence. Alpha diversity indices (Coverage, Ace, Chao, Shannon, Simpson) were calculated in Mothur 1.30.2. Differences in the relative abundance of major bacterial phyla among different samples were assessed using the Kruskal–Wallis rank-sum test and one-way ANOVA. R software 3.3.1 was used to conduct inter-group difference tests with the alpha diversity index and to generate a principal coordinates analysis (PCoA) plot (calculate the sample distance matrix based on the distance algorithm of Bray–Curtis), soil bacterial community composition maps, correlation heatmaps, redundancy analysis (RDA) plots, and a Mantel test network heatmap.

PICRUSt2 functional prediction can be used to predict the functional information of bacterial communities in environmental samples, and to further understand some potential bacterial functional characteristics during environmental changes through functional composition and abundance. Functional prediction of the bacterial communities of samples was performed using the PICRUSt2 software based on the Kyoto Encyclopedia of Genes and Genomes (KEGG) database, and obtained the relative abundance of each function under the first-level and second-level function predictions

3. Results

3.1. Particle Composition of BSCs

The particle compositions of bare sand, algae crust, and moss crust were mainly dominated by sand particles, accounting for more than 45%, at the WLYY00, WL, and NT00 sites. In contrast, these were dominated by silt particles, accounting for more than 42%, at the YY site (Figure 2). Further comparison of the particle composition characteristics of the four afforested sites revealed that algae and moss crusts had lower sand content and a higher content of fine particles (clay + silt particles) than bare sand at all sites. This reflected the enrichment effect of the BSCs on fine particulate matter. Notably, there were significant differences in the particle composition of the two types of soil crusts in different afforested sites. The algae crust had a higher fine particulate matter content at the WLYY00 and NT00 sites, the moss crust had a higher fine particulate matter content at the YY site, and the fine particulate matter contents of algae and moss crusts were comparable at the WL site. The clay and silt contents for the same type of crust were significantly higher and the sand content was significantly lower at the YY site than at all other sites. This result indicated that YY exerted the most significant promoting effect on the enrichment of fine particulate matter by BSCs.

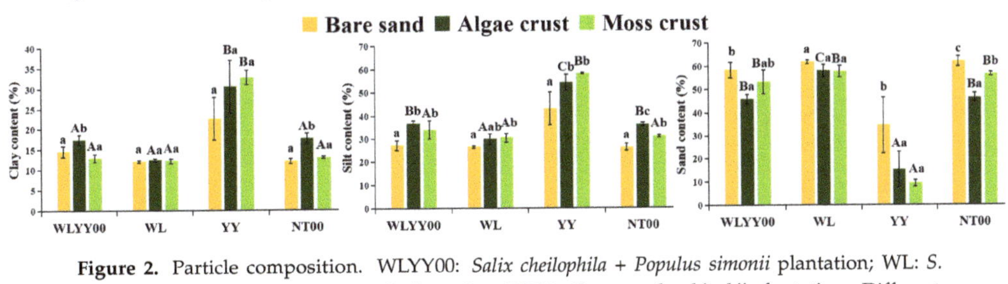

Figure 2. Particle composition. WLYY00: *Salix cheilophila* + *Populus simonii* plantation; WL: *S. cheilophila* plantation; YY: *P. simonii* plantation; NT00: *Caragana korshinskii* plantation. Different lowercase letters indicate significant differences in the indicators among bare sand, algae crust, and moss crust at the same site. Different uppercase letters indicate significant differences in the indicators of the same type of crust at different sites.

3.2. Physicochemical Properties and Enzymatic Activities of BSCs

Figure 3 shows that bare sand and two types of crusts in all four afforested sites were alkaline, with all pH values exceeding 8.22. The pH values of algae and moss crusts were significantly lower than that of bare sand, while there was no significant difference in pH between the two types of crusts. BSCs at the WLYY00 and YY sites had relatively lower pH values. Electrical conductivity (EC), soil water content (SWC), total carbon (TC), total nitrogen (TN), total phosphorus (TP), alkaline-dissolved nitrogen (AN), available phosphorus (AP), available potassium (AK), soil organic matter (SOM), catalase (CAT), sucrase (SUC), urease (URE), and alkaline phosphatase (ALP) values increased progressively from bare sand to algae crust to moss crust in each type of afforested site. At the WLYY00 and WL sites, the total potassium (TK) content of algae crust was significantly higher than that of moss crust, while moss crust had a significantly higher TK content than bare sand. At the YY and NT00 sites, TK content was highest in moss crust, but it was not significantly different from that of algae crust and was only significantly higher than that of bare sand. Among the four afforested sites, the YY site had higher contents of EC, SWC, and soil nutrients (TC, TN, TP, TK, AN, AP, AK, and SOM) than those of the same type of crust at the other sites. SUC, URE, and ALP activities of the same type of crust were also higher at the YY site than at the other sites, with only CAT activity being relatively low. These results indicated that YY was more conducive to promoting the improvement of physicochemical properties of BSCs on sandy land and also had a more pronounced effect on the enhancement of SUC, URE, and ALP activities, with only a relatively weak enhancement effect exerted on CAT activity.

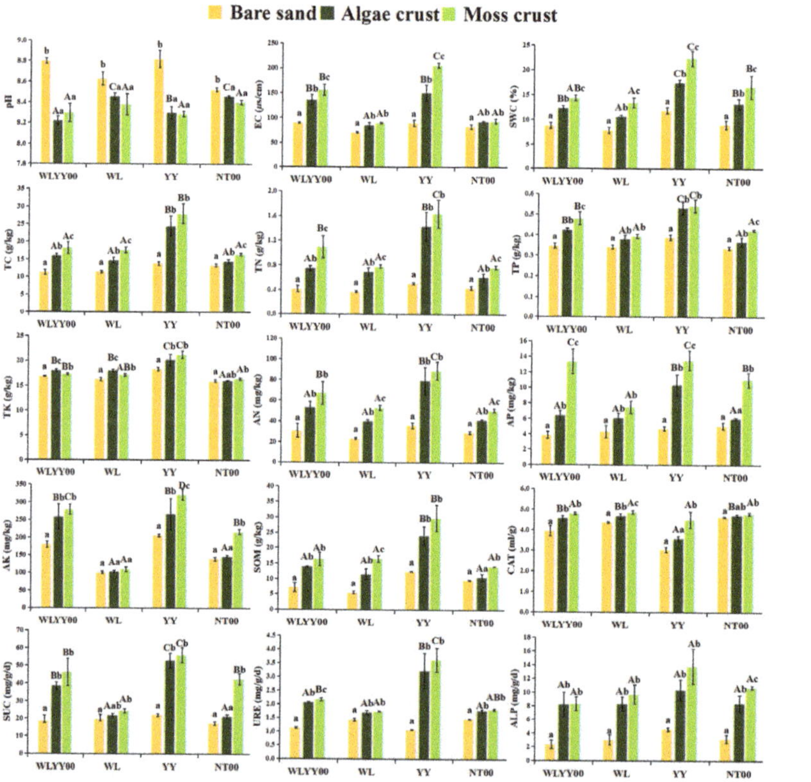

Figure 3. Physicochemical properties and enzymatic activities. WLYY00: *Salix cheilophila* + *Populus simonii* plantation; WL: *S. cheilophila* plantation; YY: *P. simonii* plantation; NT00: *Caragana korshinskii* plantation. Different lowercase letters indicate significant differences in the indicators among bare sand, algae crust, and moss crust at the same site. Different uppercase letters indicate significant differences in the indicators of the same type of crust at different sites.

3.3. Correlation Analysis of Physicochemical Properties and Enzymatic Activities

The majority of physicochemical properties and enzymatic activities were significantly correlated in pairs in both algae and moss crusts (Figure 4). In both types of crusts, ALP was significantly positively correlated with TC, TN, TP, TK, AN, and URE, and significantly negatively correlated with CAT. The ALP of algae crust showed a significant positive correlation with AP, while the ALP of moss crust was significantly positively correlated with SWC and significantly negatively correlated with pH. URE was significantly correlated with all indicators—except pH, where it was significantly negatively correlated only with CAT—in both types of crusts. The SUC of algae crust had significant correlations with all physicochemical properties, CAT, and URE, being significantly negatively correlated with CAT and pH and significantly positively correlated with the other indicators. In moss crust, SUC had no significant correlations with CAT and pH, while its correlations with the rest of the indicators were in agreement with those of algae crust. In both types of crusts, CAT was significantly negatively correlated with TC, TN, TK, AN, and SOM. The CAT of algae crust was significantly negatively correlated with EC, SWC, TP, AP, and AK. In the algae crust, pH was significantly negatively correlated with AK and EC, exhibited no significant correlations with SWC and TK, and showed significant positive correlations with the remaining physicochemical indicators. In moss crust, pH was significantly negatively correlated with SOM, TK, and TC, and had no significant correlation with the other indicators; AP was significantly positively correlated with AK, EC, TN, TP, and AN, and had no significant correlations with other indicators; and the remaining indicators were significantly positively correlated with each other. The results described above revealed that the correlations between physicochemical properties and enzymatic activities in the four afforested sites were closer and more extensive in algae crust than in moss crust.

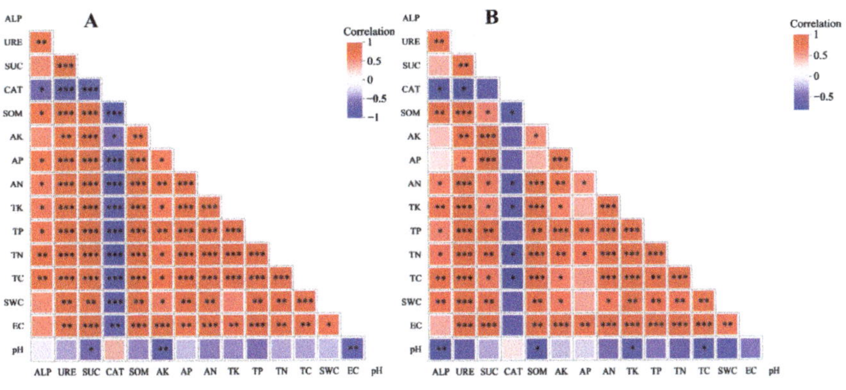

Figure 4. Correlations among physicochemical properties and enzymatic activities. (**A**,**B**) show the heatmaps of correlations between the physicochemical properties and enzymatic activities of algae crust and moss crust in different afforested sites, respectively; red denotes a positive correlation and blue denotes a negative correlation, with deeper colors indicating a higher degree of correlation. Asterisks within the color blocks indicate significance, *: $p \leq 0.05$, **: $p \leq 0.01$, ***: $p \leq 0.001$.

3.4. Structural Characteristics of Bacterial Communities in BSCs

3.4.1. Statistical Analysis of OTUs

A total of 1,677,744 valid sequences were obtained from 36 BSC samples after bacterial sequencing. After clustering and transformation, 9898 OTUs belonging to 1 kingdom, 50 phyla, 141 classes, 322 orders, 540 families, and 1092 genera were obtained using classification and annotation. The bacterial community coverage index exceeded 0.977 in the various samples (Figure S1), indicating that the current sequencing volume was sufficient to cover the vast majority of species in the bacterial communities of bare sand, algae crust, and moss crust at the four afforested sites.

3.4.2. Alpha Diversity Indices

The alpha diversity indices of bare sand and the two types of crusts in the four afforested sites were analyzed to assess the species richness and diversity levels of the bacterial communities (Figure 5). Specifically, the ACE and Chao indices were used to assess the richness of bacterial communities, while the Shannon and Simpson indices were used to characterize the diversity of bacterial communities. Figure 5 shows that the ACE, Chao, and Shannon indices gradually increased in all four afforested sites. In contrast, the Simpson index gradually decreased with succession from bare sand to algae crust and to moss crust. This indicated that both the richness and species diversity of the bacterial communities increased with the gradual maturation of crust development. Among the four afforested sites, the YY site had higher ACE, Chao, and Shannon indices and a lower Simpson index in the algae and moss crusts than the same crusts at the other sites. Therefore, it is evident that YY was more conducive for achieving an increase in richness and optimization of diversity of bacterial communities in the BSCs, as compared with the other three afforestation measures.

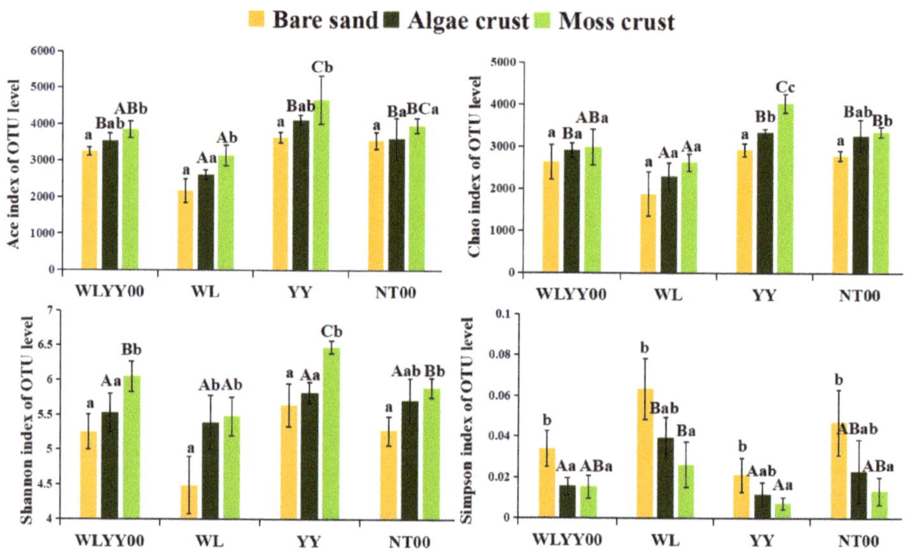

Figure 5. Alpha diversity indices of bacterial communities. WLYY00: *Salix cheilophila* + *Populus simonii* plantation; WL: *S. cheilophila* plantation; YY: *P. simonii* plantation; NT00: *Caragana korshinskii* plantation. Different lowercase letters indicate significant differences in the indices among bare sand, algae crust, and moss crust at the same site. Different uppercase letters indicate significant differences in the indices of the same type of crust at different sites.

3.4.3. Bacterial Community Composition

Bacterial communities of algae crusts and moss crusts of the four afforested sites were mainly comprised of eight bacterial phyla (Figure 6). Proteobacteria, Cyanobacteriota, Bacteroidota, and Actinomycetota were the dominant phyla, with the relative abundances being 21.77–31.56%, 8.47–35.13%, 7.40–19.20%, and 8.17–16.99%, respectively. The relative abundances of Gemmatimonadota and Cyanobacteriota in the algae crust were lower and higher, respectively, than those of moss crust at all four afforested sites. Notably, the relative abundances of Cyanobacteriota in algae crust of the WLYY00, NT00, and YY sites were 3.19, 3.18, and 2.21 times that of the corresponding moss crusts, respectively. Other major bacterial phyla did not exhibit similar patterns of differences between algae and moss crusts.

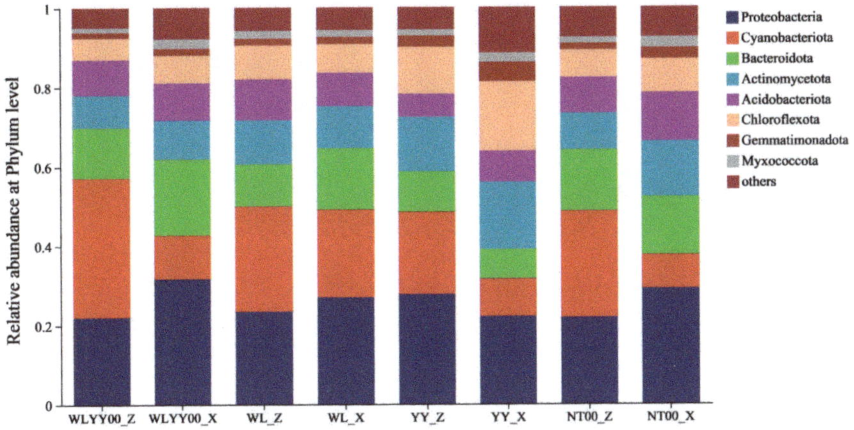

Figure 6. Bacterial community composition of BSCs. "Others" indicate all bacterial phyla with ranks higher than the 8th when sorted in descending order of relative abundance, sorted from high to low; WLYY00: *Salix cheilophila* + *Populus simonii* plantation; WL: *S. cheilophila* plantation; YY: *P. simonii* plantation; NT00: *Caragana korshinskii* plantation; _Z: algae crust; _X: moss crust.

3.4.4. PCoA

The PCoA plot shows the degree of similarity of bacterial community composition among different samples, with a smaller distance between sample points in the plot indicating higher similarity. PCoA performed in this study ($R = 0.785$, $p = 0.001$, Figure 7) revealed that the two types of crust samples each formed a relatively centralized distribution, suggesting higher similarity between bacterial community compositions of the same type of crust. The algae crust and moss crust sample points of the WL and YY sites were close to each other, which indicated that the bacterial community composition of the two types of crusts had greater similarity in these two afforested sites. However, at the WLYY00 and NT00 sites, the distributions of the two types of crusts were located further apart, demonstrating the presence of greater differences in bacterial community composition. This may be related to the selective shaping of bacterial communities in different crusts by unique microenvironmental conditions (e.g., moisture) in these two afforested sites, leading to clear differentiation of bacterial community composition between the two types of crusts. The bacterial community compositions of the same type of crust samples at the WLYY00 and NT00 sites had greater similarity, suggesting that the bacterial community structure of the same crust type possessed higher internal consistency under the influence of these two afforestation measures.

3.4.5. Significance Testing of Intergroup Differences Among Major Bacterial Phyla

Significance testing of intergroup differences among eight major bacterial phyla in the algae and moss crusts of the four afforested sites was conducted (Figure 8). Results indicated that the relative abundances of Chloroflexota in algae crust and Proteobacteria, Bacteroidota, Actinomycetota, Gemmatimonadota, and Myxococcota in moss crust differed significantly among the four afforested sites. This indicated that on alpine sandy land, the effects of different afforestation measures with a restoration period of 24 years on the bacterial community of BSCs exhibited crust type specificity. The relative abundances of the major bacterial phyla in moss crust were more sensitive to responses to afforestation measures, and the structure of its major bacterial communities was susceptible to regulation by the afforestation measures. In contrast, the bacterial community structure of the algae crust was relatively stable.

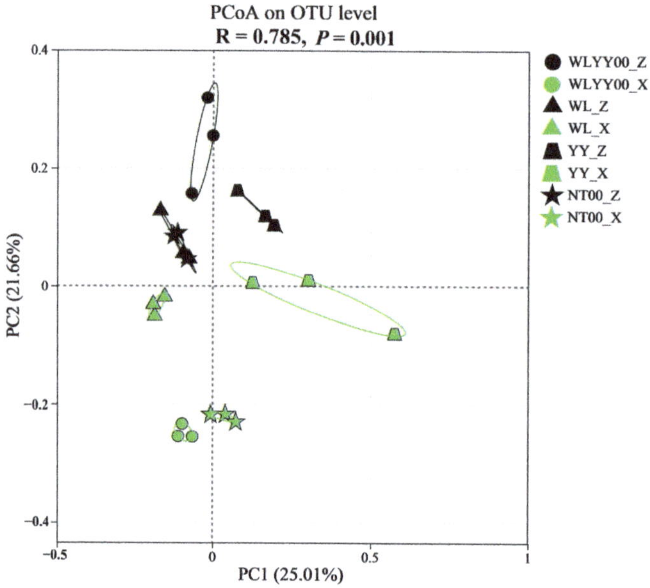

Figure 7. PCoA plot of bacterial communities. WLYY00: *Salix cheilophila* + *Populus simonii* plantation; WL: *S. cheilophila* plantation; YY: *P. simonii* plantation; NT00: *Caragana korshinskii* plantation; _Z: algae crust; _X: moss crust.

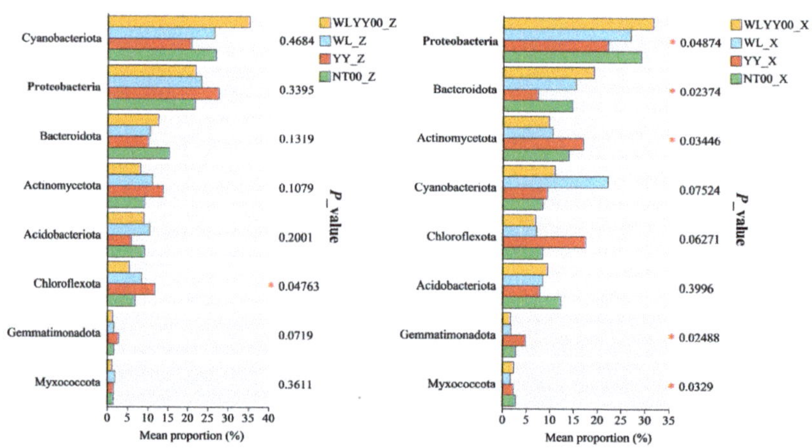

Figure 8. Significance testing of intergroup differences among major bacterial phyla. *: $p \leq 0.05$; WLYY00: *Salix cheilophila* + *Populus simonii* plantation; WL: *S. cheilophila* plantation; YY: *P. simonii* plantation; NT00: *Caragana korshinskii* plantation; _Z: algae crust; _X: moss crust.

3.5. Correlation Analysis of Physicochemical Properties, Enzymatic Activities, and Bacterial Community Structure

3.5.1. Algae Crust

Correlation analysis was performed on the physicochemical properties, enzymatic activities, and bacterial community structure of algae crust in the four afforested sites (Figure 9). Results revealed that most physicochemical properties and enzyme activity indicators are closely correlated with the bacterial community structure, but have no significant correlation with the relative abundance of multiple major bacterial phyla. Figure 9A shows that the relative abundances of Actinomycetota, Chloroflexota, Myxococcota, Cyanobacteriota, and Bacteroidota had no significant correlations with any physicochemical properties or enzymatic activities. The relative abundance of Acidobacteriota was not significantly correlated with pH but significantly correlated with all remaining indicators, with only CAT showing a significant positive correlation. Relative abundances of Proteobacteria and Gemmatimonadota were significantly positively correlated with URE, AP, TP, SOM, TC, and TN; the relative abundance of Proteobacteria was significantly positively correlated with ALP and TK; and the relative abundance of Gemmatimonadota was significantly positively correlated with SWC. From Figure 9B, it can be seen that AP was the most critical physicochemical factor affecting the community structure of algae crust bacteria in the four afforested sites of this study, followed by TC, TN, and TP. Figure 9C shows that the activities of the four enzymes are closely correlated with the community structure of algae crust, among which the correlation of URE is the most significant.

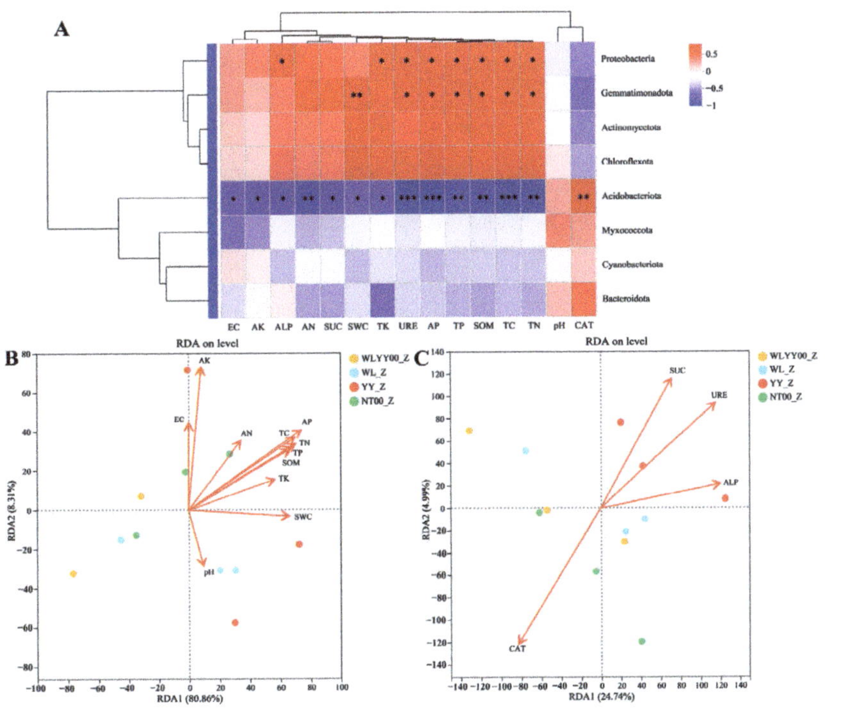

Figure 9. Correlations between environmental factors and bacterial community structure of algae crust. (**A**) Heat map of correlations of the physicochemical properties and enzymatic activities of algae crust with the relative abundances of major bacterial phyla. (**B**,**C**) RDA plots of bacterial community structure with the physicochemical properties and enzymatic activities of algae crust, respectively. WLYY00: *Salix cheilophila* + *Populus simonii* plantation; WL: *S. cheilophila* plantation; YY: *P. simonii* plantation; NT00: *Caragana korshinskii* plantation; _Z: algae crust; red denotes a positive correlation and blue denotes a negative correlation, with deeper colors indicating a higher degree of correlation. Asterisks within the color blocks indicate significance, *: $p \leq 0.05$, **: $p \leq 0.01$, ***: $p \leq 0.001$.

3.5.2. Moss Crust

Correlation analysis was performed on the physicochemical properties, enzymatic activities, and bacterial community structure of moss crust in the four afforested sites (Figure 10). Results indicated that the physicochemical properties and enzyme activity indicators of moss crusts are closely correlated with the bacterial community structure, and also have a significant correlation with the relative abundance of most major bacterial phyla. This differed from the findings for algae crust described above. Figure 10A shows that the relative abundances of Proteobacteria and Bacteroidota were significantly positively correlated with pH and CAT, and significantly negatively correlated with ALP, AN, TK, TC, SOM, and URE, while the relative abundance of Bacteroidota was significantly negatively correlated with TN, SWC, and TP. The relative abundance of Cyanobacteriota was significantly positively correlated with CAT and significantly negatively correlated with AP, AK, and SUC. Relative abundances of Actinomycetota, Chloroflexota, and Gemmatimonadota were significantly negatively correlated with CAT and significantly positively correlated with SUC, ALP, TK, TC, SOM, TN, URE and SWC. The relative abundances of Actinomycetota and Chloroflexota were significantly negatively correlated with pH, the relative abundances of Chloroflexota and Gemmatimonadota were significantly positively correlated with AN and TP, and the relative abundance of Chloroflexota was significantly positively correlated with AK and EC alone. The relative abundance of Myxococcota was significantly correlated with AP only, with the correlation being positive, while the relative abundance of Acidobacteriota was not significantly correlated with any of the indicators. Figure 10B shows that SOM was the most critical physicochemical factor affecting the community structure of moss crust bacteria in the four afforested sites of this study, followed by TC and TN. From Figure 10C, it can be observed that the activities of the four enzymes are also closely related to the community structure of moss skinning bacteria, among which ALP is the most significant.

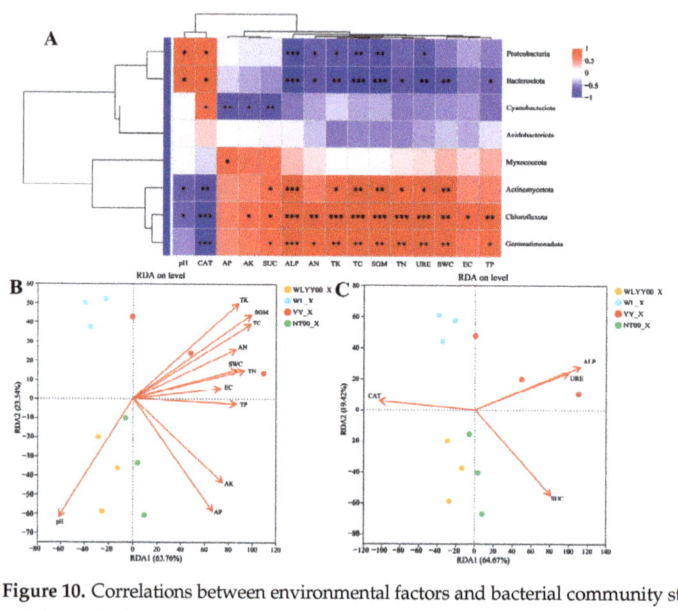

Figure 10. Correlations between environmental factors and bacterial community structure of moss crust. (**A**) shows the heat map of correlations of the physicochemical properties and enzymatic activities of moss crust with the relative abundances of major bacterial phyla; (**B**,**C**) show the RDA plots of bacterial

community structure with the physicochemical properties and enzymatic activities of moss crust, respectively; WLYY00: *Salix cheilophila + Populus simonii* plantation; WL: *S. cheilophila* plantation; YY: *P. simonii* plantation; NT00: *Caragana korshinskii* plantation; _X: moss crust; red denotes a positive correlation and blue denotes a negative correlation, with deeper colors indicating a higher degree of correlation. Asterisks within the color blocks indicate significance, * $p \leq 0.05$, ** $0.001 < p \leq 0.01$, *** $p \leq 0.001$.

3.6. Functional Prediction of Bacterial Communities in BSCs of Different Afforested Sites

3.6.1. Prediction of Primary Functions

Functional prediction indicated that the bacterial communities of BSCs and bare sand in the four afforested sites possessed six primary functions. All relative abundances were greater than 1%, with metabolism having the highest relative abundance and accounting for more than 76% (Figure 11). The relative abundance of metabolism was not significantly different among bare sand, algae crust, and moss crust at the WLYY00 and WL sites. However, the relative abundances of metabolism in both crust types were significantly higher than that of bare sand at the YY and NT00 sites, with that of moss crust being higher than that of algae crust. With the succession from bare sand to algae crust and to moss crust in the four afforested sites, the relative abundance of genetic information processing and human diseases gradually increased. In contrast, the relative abundance of environmental information processing and organismal systems gradually decreased. The relative abundance of cellular processes in bare sand was significantly higher than that in algae and moss crusts, but did not differ significantly between the two types of crusts. Cross-site comparisons revealed no significant differences in the relative abundances of metabolism and cellular processes among the same type of crusts in the four afforestation sites. The relative abundances of genetic information processing and environmental information processing were not significantly different among moss crust of the various sites, and the relative abundances of human diseases and organismal systems were not significantly different among algae crust of the various sites. The relative abundance of genetic information processing in algae crust was considerably higher at the NT00 site than in other afforested sites. Relative abundances of environmental information processing in algae crust and human diseases in moss crust were the highest at the WLYY00 site. The relative abundance of organismal systems in the moss crust was the highest at the WL site, followed by that at the NT00 and WLYY00 sites. These results demonstrated that, except for YY, the other three afforestation measures were significantly targeted at enhancing the primary functions of bacterial communities in soil crusts. NT00 was more conducive to the exertion of the genetic information processing function in algae crust, WLYY00 provided optimal effects in promoting the environmental information processing function of algae crust and human disease function of moss crust, and WL was more favorable for the exertion of the organismal systems function of moss crust.

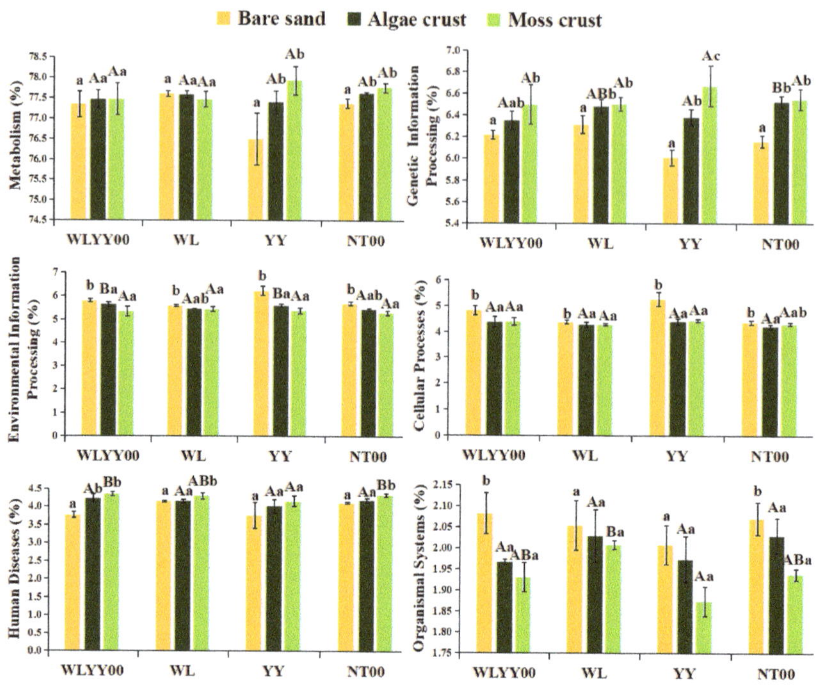

Figure 11. Relative abundances of primary functions of bacterial communities. Different uppercase letters indicate significant differences in the same type of crust across different afforested sites ($p < 0.05$), and different lowercase letters indicate significant differences among different types of crusts in the same afforested site ($p < 0.05$); WLYY00: *Salix cheilophila* + *Populus simonii* plantation; WL: *S. cheilophila* plantation; YY: *P. simonii* plantation; NT00: *Caragana korshinskii* plantation.

3.6.2. Prediction of Secondary Functions

Functional prediction indicated that the bacterial communities of BSCs and bare sand in the four afforestation sites possessed a total of 46 secondary functions. In the present study, only the top six bacterial community secondary functions in terms of relative abundance (accounting for more than 2.5% each) were retained. Global and overview maps had the highest relative abundance of approximately 40% (Figure 12). At the four afforested sites, the relative abundances of algae and moss crusts in global and overview maps were higher than that of bare sand, with the differences being statistically significant at the YY and NT00 sites. However, the relative abundance of this secondary function did not differ significantly between algae and moss crusts in the four afforested sites. The relative abundances of carbohydrate metabolism and amino acid metabolism in bare sand, algae crust, and moss crust of the four afforested sites exhibited the trend of bare sand > moss crust > algae crust, while the relative abundances of energy metabolism, metabolism of cofactors and vitamins, and membrane metabolism exhibited the trend of bare sand > algae crust > moss crust. This suggested that BSCs had a significant advantage in exerting the global and overview maps function, but showed a weaker performance than bare sand in other major secondary functions. The functional differences between the two crust types indicated that the algae crust was more likely to facilitate the exertion of functions such as energy and vitamin metabolism and membrane transport, while moss crust provided a greater advantage in functions such as carbohydrate and amino acid metabolism. Cross-site comparisons revealed that the relative abundance of membrane transport in algae crust was much higher at the YY site than at other sites, and the relative abundances of the remaining five secondary functions did not differ significantly among the four afforested sites. Among the moss crusts of the four sites, the relative abundances of global and overview maps and carbohydrate metabolism were highest at the YY site, the relative abundance of amino acid metabolism was highest at the WLYY00 site, and the relative abundances of energy metabolism, metabolism of cofactors and vitamins, and membrane transport were highest at the WL site. There was little variation in the relative abundances of multiple primary secondary functions of algae crust among the four afforested sites, which was similar to the results obtained from significance testing of intergroup differences among major bacterial phyla in algae crust. This again demonstrated that the structure of algae crust bacterial communities in alpine sandy land possessed high stability and was affected by the type of afforestation measure to a smaller extent. In contrast, the secondary functional characteristics of moss crust exhibited a clearer dependence on the type of afforestation measure. This suggested that functional differentiation was more significant in moss crust, and the expression of its related functions was more susceptible to regulation by afforestation measures. Therefore, targeted functional enhancement in moss crust may be achieved through the optimization of afforestation measures. When the functional advantages of the afforestation measures were examined, we observed that YY favored the promotion of the membrane transport function of algae crust and the global and overview maps and carbohydrate metabolism functions of moss crust. WLYY00 exceled in enhancing the amino acid metabolism function of moss crust, while WL improved the energy metabolism, metabolism of cofactors and vitamins, and membrane transport functions of moss crust.

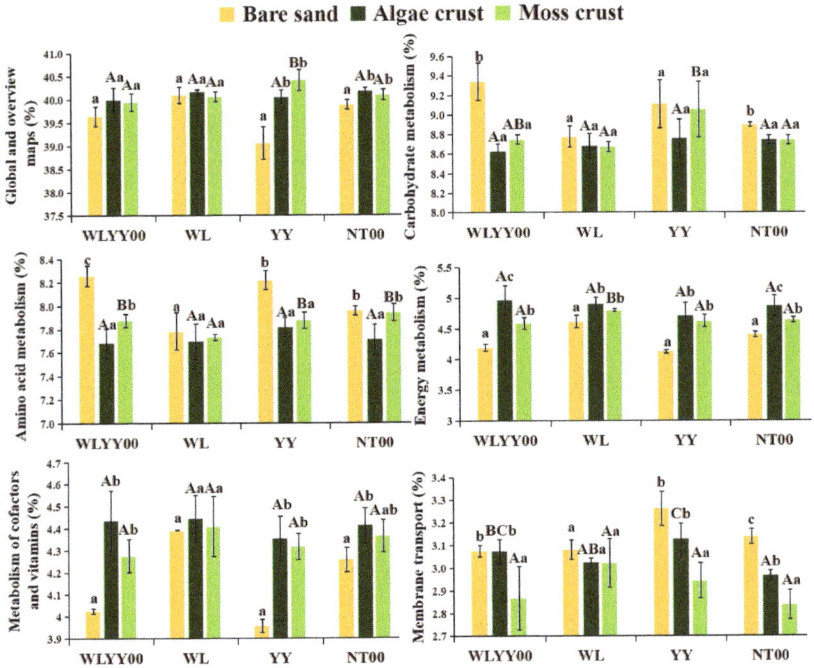

Figure 12. Relative abundances of secondary functions of bacterial communities. Different uppercase letters indicate significant differences in the same type of crust across different afforested sites ($p < 0.05$), and different lowercase letters indicate significant differences among different types of crusts in the same afforested site ($p < 0.05$); WLYY00: *Salix cheilophila* + *Populus simonii* plantation; WL: *S. cheilophila* plantation; YY: *P. simonii* plantation; NT00: *Caragana korshinskii* plantation.

4. Discussion

4.1. Characteristics of BSCs Particle Composition in Areas with Different Afforestation Measures

The alpine sandy land of the Gonghe Basin is prone to severe soil erosion. Wind and water erosion, the main forms of erosion in the region, directly affect soil particle composition. A lack of effective afforestation measures for soil protection will lead to continued erosion of the ground surface due to the lack of vegetation cover and difficulties in BSCs formation, resulting in a reduction in fine particle content in the soil [28–30]. In the present study, BSCs development in the study area achieved gradual maturation after 24 years of restoration under four different afforestation measures. The fine particle content of BSCs was higher than that of bare sand without crust cover. This indicated that BSCs reduced fine particle loss by wind transportation, which enabled the enrichment of fine particulate matter. However, at the WLYY00, WL, and NT00 sites, the particle composition of the two types of BSCs was still dominated by sand particles, which was similar to the results of a study by Du et al. [31] conducted in the Tengger Desert and a study by Cui et al. [12] conducted on the Erdos Plateau. This can be explained by the fact that the WLYY00, WL, and NT00 sites were located in the alpine sandy land of the Gonghe Basin, where the sand content in the original soil substrate was inherently high. The development of BSCs is primarily achieved by stabilizing the surface particles by cementation, making it challenging to fundamentally alter the basal particle composition of the regional soil [3]. At the YY site, the sand particle content of the two types of BSCs was significantly lower than that of the other afforested sites. The content of fine particles, such as clay and silt particles, was significantly higher than that of the other sites, with particle composition mainly dominated by silt particles. Such a result indicated that YY provided better effects in improving the particle composition of BSCs and promoting the enrichment of fine particulate matter (especially silt particles). YY also enabled effective retention and stabilization of fine particulate matter and promoted BSCs development towards finer granulation, which was more conducive to soil improvement. Differences in particle composition between the two types of crusts were relatively small. Significant differences were only observed at the NT00 site, where the algae crust had significantly higher fine particle and sand content than the moss crust. This suggests that the ability of algae crust to enrich fine particulate matter was superior to that of moss crust under the environmental conditions of the NT00 site.

4.2. Characteristics of Physicochemical Properties and Enzymatic Activities of BSCs in Different Afforested Sites

In sandy land ecosystems, afforestation measures serve as an important means of improving soil quality. Such measures can directly act on soil and regulate the basic soil environment by introducing litter and root exudates [32], while also indirectly improving soil quality by altering the physicochemical properties and enzymatic activities of BSCs [33,34]. Sandy lands are prone to imbalances in water-salt transport in soil due to scarce precipitation and high evaporation, a characteristic that makes them highly susceptible to soil salinization [35]. In the present study, the pH of algae and moss crusts was significantly lower than that of bare sand, and EC was significantly higher than that of bare sand. These differences in physicochemical properties were in agreement with the results obtained by Yan et al. [36] in the Tengger Desert. The heatmaps of correlations between physicochemical properties and enzymatic activities also demonstrated that pH was negatively correlated with EC. Therefore, BSCs can effectively alleviate further soil alkalization in sandy land, but may increase soil salinity to a certain extent through the retention or accumulation of salts in the crust layer. This represents a differential regulatory effect between the two key dimensions of salinization and alkalization. SWC was significantly higher in the algae and moss crusts than in bare sand, and the SWC of the same type of crust was significantly higher at the YY site than at other sites. Such a phenomenon can be explained by the fact that BSCs act as a special cover on land surfaces, stabilizing the soil. Their formation and development enable filling of degraded patches on land surfaces, thus influencing surface water circulation (e.g., infiltration and evaporation) and leading to improved water retention capacity in BSCs [37]. The high SWC level of BSCs at the YY site suggested the existence of microenvironments more suitable for crust development (e.g., vegetation cover and precipitation retention capacity) or higher developmental maturity of the crusts in this afforestation measure, which strengthened the soil water retention effect. We also found that moss crust had higher SWC and EC than algae crust. This demonstrated a superior water retention capacity and the ability to release more ion-containing organic or inorganic compounds in moss crust, thus further increasing the ionic concentration in the soil solution [38].

An examination of nutrient content and enzymatic activity characteristics in moss crust, algae crust, and bare sand revealed that, with the exception of TK content, which did not show consistent differences, all other nutrient and enzymatic activity indicators showed the trend of moss crust > algae crust > bare sand in the four afforested sites. This clearly reflected a trend of gradual enhancement in nutrient accumulation and enzymatic activity levels with an increase in the degree of BSCs development. This was consistent with the results obtained by Yang et al. [39] in a study on the differences in nutrient content between BSCs and bare sand in the Gurbantunggut Desert. Such a phenomenon is attributed to the presence of cryptogams within BSCs, which can fix carbon and nitrogen through photosynthesis and nitrogen fixation. This gives rise to more active carbon and nitrogen cycling processes compared with bare sand, which in turn promotes soil nutrient accumulation and enzymatic activities in BSCs [40–42]. Compared with algae crust, moss crust also possesses stronger photosynthetic and respiratory capacities and can secrete higher levels of enzymes through pseudo-roots and apoplasts, which contribute to higher nutrient content and enzymatic activities [11,40,43]. Our results also showed that the YY site had the highest nutrient content and ALP, URE, and SUC activities and the lowest CAT activity for the same type of crust across the four afforested sites. This demonstrated that YY was more conducive to promoting the accumulation of nutrients in BSCs and significantly enhancing the activities of most soil enzymes. Such a phenomenon can be ascribed to the use of *P. simonii* as the sand-fixing tree species at the YY measure. After 24 years of restoration, the root system of the trees had fully developed, with root length density being considerably higher than that of other sand-fixing plants. This enabled the secretion of higher levels of organic acids and organic substances through a larger root contact area, thereby enhancing soil nutrient content [44]. The weaker CAT activity suggested that the oxidative stress environment (e.g., reactive oxygen content) of BSCs at the YY site may differ from that of other sample sites, or that the antioxidant system may possess unique regulatory mechanisms.

4.3. Characteristics of Bacterial Community Structure and Functions of BSCs in Different Afforested Sites

Microorganisms are important components of BSCs. In particular, bacteria possess stronger adaptability in arid and semi-arid environments, and can promote soil material cycling and energy flow in sandy land ecosystems. Therefore, the environmental changes and functions of sandy land can be characterized by bacterial community structure [45,46]. In the present study, the dominant phyla of bacterial communities in BSCs of the study sites were Proteobacteria, Cyanobacteriota, Bacteroidota, and Actinomycetota. This finding is consistent with the community compositions of different types of BSCs in Mu Us sandy land reported by Li et al. [47]. Compared with the bacterial community composition of the surface soil of the alpine sandy land [18,27], the bacterial community composition of the BSCs had a higher relative abundance of Cyanobacteriota, which served as the dominant phylum. This is explained by the fact that Cyanobacteriota are the core builders of algae and moss crusts, and their secreted extracellular polysaccharides can provide a stable matrix for the attachment of other bacteria to form the basic crust framework. They also photosynthesize, promoting the conversion of carbon, nitrogen, and phosphorus, and primarily account for the higher nutrient content of the BSCs compared with that of bare sand [48]. Bacterial community richness and diversity levels of bare sand, algae crust, and moss crust followed the trend of sand < algae crust < moss crust in all afforested sites. Such a phenomenon can be explained as follows: BSCs in all four afforested sites had not yet fully developed. Therefore, nutrient saturation had not yet been reached, and competition among bacterial communities had not yet intensified. With a higher nutrient content, there was a greater abundance of resources, such as carbon, nitrogen, and phosphorus, available to bacteria, and a larger number of supporting bacterial communities. This contributed to higher community richness and diversity levels [49]. Significance testing of intergroup differences among major bacterial phyla revealed that different afforestation measures exerted more significant effects on the structure of major bacterial phyla of the bacterial community of moss crust. This was attributed to the significant improvement in soil physicochemical properties resulting from the different measures, combined with the greater susceptibility of the bacterial community structure of moss crust to the influence of soil environmental factors [50]. Bacterial community richness and diversity levels of the same type of biological crust were the highest at the YY site, due to the higher conduciveness of YY to improving the microhabitat conditions of biological crusts (e.g., enhancing nutrient supply, optimizing water retention, or stabilizing the physical structure). As a result, YY provided a suitable environment for the survival and reproduction of more types of bacteria, thus promoting the development of bacterial communities towards higher richness and complexity. AP and SOM were the primary environmental factors affecting the bacterial community structure of algae crust and moss crust, respectively, and TC and TN were the second and third environmental factors influencing both types of crusts. This was similar to the results of a study conducted in the Tengger Desert [51]. Therefore, despite differences in the alpine sandy land environment between the study area of the present study and the Tengger Desert, commonalities exist in the key environmental factors affecting the bacterial community structure of BSCs. The fact that both TC and TN were the second and third environmental factors affecting the bacterial community structure of both types of crusts suggests that the basic nutrients carbon and nitrogen served an important and universal regulatory role in the construction and stabilization of algae crust and moss crust bacterial communities in alpine sandy land, and were core common nutrient factors shaping the bacterial community composition in different types of BSCs. The correlations between the activities of the four enzymes and the bacterial community structure are all relatively close. This was due to the involvement of these enzymes in the cycling process of key elements such as carbon, nitrogen, and phosphorus in soil, which directly regulated the form and content of nutrients available to microorganisms in the soil and thus affected the bacterial community structure [52].

Functional prediction of bacterial communities in the BSCs revealed a total of six primary functions at each site, with metabolism being the main function. This was consistent with the results of functional prediction of soil bacterial community structures in the Hulun Buir Sandy Area by Du et al. [53], indicating that the distribution characteristics of primary functions are similar in other types of soils [54,55]. It can be deduced that the metabolic function serves as the core in bacterial communities, supporting the basic processes of soil ecosystems, and is not readily influenced by regional differences.

Among the various secondary functions, global and overview maps, a secondary function subordinate to the primary function of metabolism, was identified as the main function. This suggested the need for bacterial communities to rely on coordination and regulation by the global and overview maps function to utilize the metabolism function fully. Notably, YY, which exhibited the most optimal improvement effects towards particle composition, physicochemical properties, enzymatic activities, and bacterial community structure in BSCs, did not show specific enhancement advantages for any of the six primary functions. However, it demonstrated significant differences in the secondary functions. For instance, YY was more conducive to promoting the exertion of the membrane transport function in the bacterial community of algae crust and the global and overview maps and carbohydrate metabolism functions in the bacterial community of moss crust. Therefore, YY maintained the balance of primary functions in function regulation and precisely supported the secondary functions according to the physiological characteristics and ecological needs of different types of crusts.

The results of this study indicate that, in terms of the response characteristics of BSCs to different afforestation measures, it is recommended to give priority to the YY measure in high-cold sandy land. This study has three main limitations. First, the sampling was only conducted in late July 2024, lacking a repetitive design in the time dimension. This makes it difficult for the research results to fully represent the characteristics of different seasons, especially failing to reflect the potential impact of environmental factor changes during seasonal transitions on the research subjects, significantly restricting the generalization and application of the conclusions on a cross-seasonal scale. Second, the sampling design did not fully consider the heterogeneity of microhabitats, which may result in some sample point data failing to fully capture the characteristics of small-scale habitats. Thirdly, although the composite sample processing method reduces random errors to a certain extent and smooths out local extreme values, it may also mask the subtle variations within the sample points, compressing the variances within the sample points and failing to fully reflect the true characteristics of a single sub-sample. Based on this, future research can enhance temporal representativeness by increasing the sampling frequency across multiple seasons, refining the sampling plan to cover microhabitat heterogeneity, and conducting comparative analysis of single samples and composite samples to further clarify the impact of different sample processing methods on data variation, thereby providing more comprehensive and reliable support for related research conclusions.

5. Conclusions

Biological soil crusts of the various afforested sites enriched fine particle content, increased soil water content, and reduced pH, thus enhancing soil resistance to wind erosion, improving water retention, and inhibiting further soil alkalization. Nutrient content, enzymatic activities, and bacterial community richness and diversity exhibited a trend of increase from bare sand to algae crust and to moss crust. Among the four afforestation measures with a restoration period of 24 years, YY was the most effective at enhancing biological soil nutrient content and optimizing bacterial communities. AP and SOM were the primary environmental factors affecting the bacterial community structure of algae and moss crusts, respectively. TC and TN were the second and third environmental factors affecting both types of crusts, and all four measured enzymatic activities exerted significant effects on bacterial community structure. Metabolism and global and overview maps were the main primary function and secondary function of biological soil crusts in sandy land, respectively. YY, which exhibited the best effects in regulating biological soil crust development, did not show specific enhancement advantages for any of the six primary functions. However, it promoted the exertion of the membrane transport function in the bacterial community of the algae crust and the exertion of global and overview maps, as well as carbohydrate metabolism functions in the bacterial community of moss crust.

6. Patents

[1] Shaobo Du, Huichun Xie, Chongyi E., Tianyue Zhao, Shuang Ji, Zhiqiang Dong, Shaoxiong Zhang, Haokun Wu. A plant fixation device for desertification control in deserts. ZL202322873122.9, 26 July 2024.

[2] Shaobo Du, Huichun Xie, Chongyi E., Shujing Qi, Haokun Wu, Shuang Ji, Tianyue Zhao, Zhiqiang Dong. An invention relates to a portable spraying device for desert algae biological control of sand. ZL202421424816.2, 28 March 2025.

Supplementary Materials: The following supporting information can be downloaded at: https://www.mdpi.com/article/10.3390/biology14111530/s1, Figure S1: Coverage index of bacterial communities. WLYY00: *Salix cheilophila* + *Populus simonii* plantation; WL: *S. cheilophila* plantation; YY: *P. simonii plantation*; NT00: *Caragana korshinskii* plantation; _W: bare sand; _Z: algal crust; _X: moss crust.

Author Contributions: Conceptualization, S.D. and H.X.; methodology, S.D.; software, S.D.; validation, H.X. and C.E.; formal analysis, H.X. and C.E.; investigation, G.G.; resources, G.Z.; data curation, S.D.; writing—original draft preparation, S.D.; writing—review and editing, F.Q. and G.Z.; visualization, S.D.; supervision, H.X. and C.E.; project administration, H.X. and C.E.; funding acquisition, H.X. and C.E. All authors have read and agreed to the published version of the manuscript.

Funding: This research was funded by the National Natural Science Foundation of China, grant no. 42171011, the Kunlun Talent High end Innovation and Entrepreneurship Talent Program Featured Talent Foundation of China, grant no. 1003-005024011, Qinghai Provincial Natural Science Foundation General Foundation of China, grant no. 2023-ZJ-908M and Formation mechanism and utilization team of characteristic germplasm resources in the Qinghai Tibet Plateau, grant no. QHKLYC-GDCXCY-2024-597. The APC was funded by Kunlun Talent High end Innovation and Entrepreneurship Talent Program Featured Talent Foundation of China.

Institutional Review Board Statement: Not applicable.

Informed Consent Statement: Not applicable.

Data Availability Statement: The original contributions presented in the study are included in the article/Supplementary Material, further inquiries can be directed to the corresponding authors.

Acknowledgments: We are extremely grateful to the other members of the research group, apart from the author, for their assistance in the field collection work.

Conflicts of Interest: The authors have no conflicts of interest to declare.

Abbreviations

The following abbreviations are used in this manuscript:

BSCs	Biological soil crusts
WLYY00	*Salix cheilophila*+ *Populus simonii* plantation
WL	*Salix cheilophila* plantation
YY	*Populus simonii* plantation
NT00	*Caragana korshinskii* plantation

References

1. Xue, J.; Gui, D.; Lei, J.; Sun, H.; Zeng, F.; Mao, D.; Jin, Q.; Liu, Y. Oasification: An unable evasive process in fighting against desertification for the sustainable development of arid and semiarid regions of China. *Catena* **2019**, *179*, 197–209. [CrossRef]
2. Lu, H. New trend of studying desert evolution and desertification. *Sci. Sin. Terrae* **2025**, *55*, 2089–2093. [CrossRef]
3. Garcia-Pichel, F. The microbiology of biological soil crusts. *Annu. Rev. Microbiol.* **2023**, *77*, 149–171. [CrossRef]
4. Zhao, J.; Zheng, Y.; Zhang, B.; Chen, Y.; Zhang, Y. Progress in the study of algae and mosses in biological soil crusts. *Front. Biol. China* **2009**, *4*, 143–150. [CrossRef]
5. Zhang, Q.; Lv, J.; Ma, Y.; Li, E.; Shen, C.; Chen, J. Microbial community structure and potential function of algal crusts in different regions of Gurbantunggut Desert, Xinjiang, China. *Acta Ecol. Sin.* **2024**, *44*, 6317–6330. [CrossRef]
6. Yang, Y.; Zhang, K.; Yang, Y.; Bu, C. Field Artificial Cultivation Technology of Moss Dominated Crust in Mu Us Sandland. *J. Soil Water Conserv.* **2016**, *36*, 165–170. [CrossRef]
7. Su, Y.G.; Li, X.R.; Chen, Y.W.; Zhang, Z.S.; Yan, L. Carbon fixation of cyanobacterial-algal crusts after desert fixation and its implication to soil organic carbon accumulation in desert. *Land Degrad. Dev.* **2013**, *24*, 342–349. [CrossRef]
8. Lu, Q.; Xiao, Y.; Lu, Y. Employment of algae-based biological soil crust to control desertification for the sustainable development: A mini-review. *Algal Res.* **2022**, *65*, 102747. [CrossRef]
9. Elbert, W.; Weber, B.; Burrows, S.; Steinkamp, J.; Büdel, B.; Andreae, M.O.; Pöschl, U. Contribution of cryptogamic covers to the global cycles of carbon and nitrogen. *Nat. Geosci.* **2012**, *5*, 459–462. [CrossRef]
10. Jiao, B.; Zhang, B.; Zhao, K.; Yan, L.; Wu, Z. Promoting effect of biological soil crusts succession on soil nitrogen transformation and microbial activity in water-wind erosion crisscross region of Loess Plateau. *J. Desert Res.* **2023**, *43*, 191–199. [CrossRef]
11. He, H.; Liu, W.; Chang, Z.; Hou, C.; Sun, L.; Chi, X. Effects of biological soil crust succession on soil nutrients, microbial community composition in desert regions. *Arid Land Geogr.* **2024**, *47*, 1724–1734. [CrossRef]
12. Cui, Y.; Lv, Y.; Li, B. Physicochemical properties of soil microbiotic crusts on Erdos Plateau. *Soils* **2004**, *36*, 197–202. [CrossRef]
13. Zhou, H.; Wu, B.; Gao, Y.; Cheng, L.; Jia, X.; Pang, Y.; Zhao, H. Composition and influencing factors of the biological soil crust bacterial communities in the *Sabina vulgaris* community in Mu Us Sandy Land. *J. Desert Res.* **2020**, *40*, 130–141. [CrossRef]
14. Li, X.; Yao, Z.; Dong, Z. Driving mechanism of aeolian desertification in Gonghe Basin of Qinghai Province. *Bull. Soil Water Conserv.* **2018**, *38*, 337–344. [CrossRef]
15. Li, Q.; Jia, Z.; He, L.; Zhao, X.; Yang, X. The allocation and cycling characteristics of main nutrients for *Caragana intermedia* with different stand age on alpine sandy land. *For. Res.* **2023**, *36*, 119–128. [CrossRef]
16. Zhou, H.; Liu, Y. Effects of soil crusts on physicochemical properties of shallow soil in alpine sandy area. *J. Arid Land Resour. Environ.* **2022**, *36*, 154–160. [CrossRef]
17. Li, Q.; Yang, D.; Jia, Z.; Zhang, L.; Zhang, Y.; Feng, L.; He, L.; Yang, K.; Dai, J.; Chen, J.; et al. Changes in soil organic carbon and total nitrogen stocks along a chronosequence of *Caragana intermedia* plantations in alpine sandy land. *Ecol. Eng.* **2019**, *133*, 53–59. [CrossRef]
18. Du, S.; Xie, H.; E, C.; Zhang, G.; Qiao, F.; Geng, G. Effectiveness of different sand control measures in improving the surface soil in alpine sandy land. *Land Degrad. Dev.* **2025**, *36*, 5269–5286. [CrossRef]
19. Zhang, M.; Wang, Z.; Deng, L.; Zhou, H. Differences in the physical and chemical properties of biological soil crusts in different shrub communities in the Gonghe Basin. *Arid Zone Res.* **2023**, *40*, 1797–1805. [CrossRef]
20. Zhao, H.; Cheng, L.; Jia, X.; Wu, B.; Li, Y.; Yue, Y.; Zhou, H.; Zhao, X. The dynamics of soil carbon release covered with biological soil crusts in Alpine sand area. *Acta Ecol. Sin.* **2020**, *40*, 6396–6404. [CrossRef]
21. Yoshihara, Y.; Tatsuno, Y.; Miyasaka, K.; Sasaki, T. Can complementarity in water use help explain diversity-productivity relationships in semi-arid grasslands? *J. Arid Environ.* **2020**, *173*, 103994. [CrossRef]
22. Bao, S. *Soil Agrochemical Analysis*, 3rd ed.; China Agricultural Press: Beijing, China, 2000.
23. Zhang, N.; He, X.D.; Gao, Y.B.; Li, Y.H.; Wang, H.T.; Ma, D.; Zhang, R.; Yang, S. Pedogenic carbonate and soil dehydrogenase activity in response to soil organic matter in *Artemisia ordosica* community. *Pedosphere* **2010**, *20*, 229–235. [CrossRef]
24. Li, M.; Nie, H.; Zhang, S.; Zhou, F.; Han, D.; Zhan, L.; Tian, Y.; Shi, M.; Zhang, E. Correlation between FDA enzyme activity and soil fertility under combining application organic and nitrogen fertilizer in facility vegetable field. *Acta Hortic. Sin.* **2016**, *43*, 907–917. [CrossRef]
25. Hu, B.; Liang, D.; Liu, J.; Lei, L.; Yu, D. Transformation of heavy metal fractions on soil urease and nitrate reductase activities in copper and selenium co-contaminated soil. *Ecotoxicol. Environ. Saf.* **2014**, *110*, 41–48. [CrossRef]
26. Wang, L.; Mei, W.; Chen, X.; Fang, H.; Liu, X.; Yang, D.; Wang, W. Determination of water-soluble sugar in rice by 3,5-dinitrosalicylic acid colorimetric method. *J. Chin. Cereals Oils Assoc.* **2020**, *35*, 168–173.
27. Du, S.; Xie, H.; Zhang, G.; Qiao, F.; Geng, G.; E, C. Improvement effects of different afforestation measures on the surface soil of alpine sandy land. *Biology* **2025**, *14*, 144. [CrossRef]

28. Zuo, X.A.; Zhao, X.Y.; Zhao, H.L.; Li, Y.Q.; Guo, Y.R.; Zhao, Y.P. Changes of species diversity and productivity in relation to soil properties in sandy grassland in Horqin sand land. *Environ. Sci.* **2007**, *28*, 945–951. [CrossRef]
29. Li, Y.Q.; Zhao, H.L.; Yi, X.Y.; Zuo, X.A.; Chen, Y.P. Dynamics of carbon and nitrogen storages in plant-soil system during desertification process in Horqin sandy land. *Environ. Sci.* **2006**, *27*, 635–640. [CrossRef]
30. Su, Y.; Zhao, H. Influences of grazing and exclosure on carbon sequestration in degraded sandy Grassland, Inner Mongolia, North China. *Environ. Sci.* **2003**, *27*, 23–28. [CrossRef]
31. Du, J.; Li, Y.; Yang, X.; Li, Y.; Ma, X. Effects of biological soil crusts types on soil physicochemical properties in the southeast fringe of the Tengger Desert. *J. Desert Res.* **2018**, *38*, 111–116. [CrossRef]
32. Li, Y.; Brandle, J.; Awada, T.; Chen, Y.; Han, J.; Zhang, F.; Luo, Y. Accumulation of carbon and nitrogen in the plant-soil system after afforestation of active sand dunes in China's Horqin Sandy Land. *Agric. Ecosyst. Environ.* **2013**, *177*, 75–84. [CrossRef]
33. Du, S.; Xie, H.; Zhang, G.; Qiao, F.; Geng, G.; E, C. Response characteristics of biological soil crusts under different afforestation measures in alpine sandy land. *Biology* **2025**, *14*, 532. [CrossRef]
34. Yin, R.; Wu, Y.; Zhang, X.; Hasi, R.J.; Tian, X.; Wang, J.; Li, Z.; Miao, H. Study on physicochemical properties of biological crusts and subsurface sediments in southern edge of Mu Us desert. *J. Soil Water Conserv.* **2013**, *27*, 120–124. [CrossRef]
35. Liu, Z.; Xu, C.; Zhu, X.; Zhou, D.; Tian, J.; Gu, F.; Huang, S.; Li, Z.; Zhao, Z.; Wang, X. Comprehensive assessment and evolution analysis of soil salinization in artificial oasis in arid desert area. *China Environ. Sci.* **2022**, *42*, 367–379. [CrossRef]
36. Yan, P.; Qu, J.; Wang, L.; Xiao, J.; Zhang, Y.; Wang, X.; Guo, S. Effect of sand barrier fixation on the formation and development of biological soil crust. *Arid Zone Res.* **2023**, *40*, 1931–1937. [CrossRef]
37. Li, Y.; Li, S.; Xiao, B.; Wang, F. Evaporation characteristics of soil covered with moss crust in the wind-water erosion crisscross region of the Loess Plateau. *J. Soil Water Conserv.* **2020**, *34*, 208–215. [CrossRef]
38. Tian, S.; Liu, X.; Jin, B.; Chen, Y.; Wang, Y.; Tian, X.; Zhao, X. Effects of litter on soil organic carbon fixation in *Reaumuria soongorica* communities in the Sangong River basin. *Acta Ecol. Sin.* **2019**, *39*, 5339–5347. [CrossRef]
39. Yang, J.; Zhang, L.; Guo, X.; Lu, Y.; Guo, H.; Zhang, Y.; Zhou, X. Vertical distribution characteristics and influencing factors of soil organic carbon under biological soil crusts in the Gurbantunggut Desert. *Acta Ecol. Sin.* **2024**, *44*, 2946–2954. [CrossRef]
40. Belnap, J.; Büdel, B.; Lange, O.L. Biological soil crusts: Characteristics and distribution. In *Biological Soil Crusts: Structure, Function, and Management*; Belnap, J., Lange, O.L., Eds.; Springer: Berlin/Heidelberg, Germany, 2003; pp. 3–30. [CrossRef]
41. Li, X.R.; Zhang, P.; Su, Y.G.; Jia, R.L. Carbon fixation by biological soil crusts following revegetation of sand dunes in arid desert regions of China: A four-year field study. *Catena* **2012**, *97*, 119–126. [CrossRef]
42. Miralles, I.; Ladrón de Guevara, M.; Chamizo, S.; Rodríguez-Caballero, E.; Ortega, R.; van Wesemael, B.; Cantón, Y. Soil CO_2 exchange controlled by the interaction of biocrust successional stage and environmental variables in two semiarid ecosystems. *Soil Biol. Biochem.* **2018**, *124*, 11–23. [CrossRef]
43. Si, S.; Li, Y.; Hui, R.; Liu, L.; Xie, M.; Wang, Y. Effects of snow on the photosynthetic physiological characteristics of biological soil crusts in A desert region. *J. Desert Res.* **2018**, *38*, 560–567. [CrossRef]
44. Shi, K.; Jia, Z.; Zhang, H.; Zhang, Y.; Li, Q.; Feng, L.; He, L. Root distribution characteristics of typical sand-fixing plants in Gonghe Basin of Qinghai Province. *Sci. Soil Water Conserv.* **2016**, *14*, 78–85. [CrossRef]
45. Bao, J.; Sun, J.; Wang, J. A review on microbial community assembly in biological soil crusts. *J. Desert Res.* **2022**, *42*, 33–43. [CrossRef]
46. Zhang, Q.; Wang, Q.; Ouyang, H.; Lan, S.; Hu, C. Pyrosequencing reveals significant changes in microbial communities along the ecological succession of biological soil crusts in the Tengger Desert of China. *Pedosphere* **2018**, *28*, 350–362. [CrossRef]
47. Li, J.; Yang, R.; Duan, X.; Liu, J.; Liu, B. Effects of different biological soil crust types on microbial community structure and composition in Baijitan, China. *J. Ecol. Rural Environ.* **2023**, *39*, 97–106. [CrossRef]
48. Li, K.; Zhang, B.; Zhao, K.; Zhang, Y.; Liu, Y. Succession of carbon-fixing microbial community in different stages of biological soil crusts in the Mu Us Sandy Land. *Acta Ecol. Sin.* **2024**, *44*, 1177–1190. [CrossRef]
49. Fan, J.; Li, S.; Wang, R.; Yu, H.; Huang, J. Influence of biocrusts succession process on bacterial community structure of biocrustal layer and subsoils in desert steppe. *Chin. J. Ecol.* **2021**, *40*, 2033–2044. [CrossRef]
50. Ma, L.H.; Meng, X.C.; Wang, G.Q.; Ma, Z.F.; Li, Y.K.; Li, Y.M.; Zhou, H.K.; Zhang, F.W.; Lin, L. Effects of moss crust inoculation on soil properties and microbial communities in alpine meadow in Sanjiangyuan, China. *Chin. J. Plant Ecol.* **2025**, *49*, 173–188. [CrossRef]
51. Liu, L.; Liu, Y.; Zhang, P.; Song, G.; Hui, R.; Wang, Z.; Wang, J. Development of bacterial communities in biological soil crusts along a revegetation chronosequence in the Tengger Desert, northwest China. *Biogeosciences* **2017**, *14*, 3801–3814. [CrossRef]
52. Zhang, J.; He, J.; Wei, J.; Huang, C.; Wang, W.; Cai, Y. Effects of application period of microbial inoculants on rhizosphere soil bacterial diversity, enzyme activity and yield and quality of flue-cured tobacco. *Crops* **2025**, *2*, 162–171. [CrossRef]
53. Du, Y.; Gao, G.; Chen, L.; Ding, G.; Zhang, Y.; Cao, H. Soil bacteria community structure and function prediction in the Hulun Buir Sandy Area. *China Environ. Sci.* **2019**, *39*, 4840–4848. [CrossRef]

54. Yang, J.; Liu, F.; Han, X.; Lin, X.; Zhang, M.; Nie, X. Diversity and functional prediction of soil bacterial communities across different elevations in tropical rainforest of Diaoluo Mountain. *Southwest China J. Agric. Sci.* **2024**, *37*, 2270–2279. [CrossRef]
55. Yuan, Z.; Jiang, J.; Hu, L.; Duan, T.; Yang, J.; Zhu, H.; Zhang, H.; Zhang, X.; Zhu, L.; Fu, Q. Analysis of bacterial community structure and functional prediction in citrus rhizosphere soils from different regions. *Till. Cultiv.* **2025**, *45*, 8–18+36. [CrossRef]

biology

MDPI

Article

Improving Soil Properties and Microbiomes by Mixed *Eucalyptus–Cupressus* Afforestation

You-Wei Zuo [1,2], Yu-Ying Liu [1,2], Ya-Xin Jiang [3], Wen-Qiao Li [1,2], Yang Peng [1,2], Sheng-Mao Zhou [1,2], Shi-Qi You [1,2], Sheng-Qiao Liu [1,2] and Hong-Ping Deng [1,2,*]

1 Center for Biodiversity Conservation and Utilization, Key Laboratory of Eco-Environment in the Three Gorges Reservoir Region, Ministry of Education, School of Life Sciences, Southwest University, Beibei, Chongqing 400715, China

2 Chongqing Key Laboratory of Plant Resource Conservation and Germplasm Innovation, Institute of Resources Botany, School of Life Sciences, Southwest University, Beibei, Chongqing 400715, China

3 College of Pharmaceutical Sciences, Southwest University, Beibei, Chongqing 400715, China

* Correspondence: denghp@swu.edu.cn; Tel.: +86-13883395687

Simple Summary

Forests are important ecosystems that provide many benefits to the environment and human society, such as clean air, water regulation, and carbon storage. However, different types of forests, such as natural forests, plantations, and mixed forests, may affect the soil and its biological communities in different ways. In this study, we explored how various forest types in a subtropical mountain area influenced the soil's physical and chemical properties, microbial communities, and chemical compounds known as metabolites. We found that soil in mixed forests had higher levels of nutrients and supported a more diverse and beneficial group of microbes compared to monoculture plantations. These microbes were also linked with compounds that are important for nutrient cycling and plant health. By using advanced analysis techniques, we were able to understand how changes in forest type can impact the soil's ability to support healthy ecosystems. This research is valuable for forest managers, conservationists, and policymakers who aim to improve forest quality and sustainability. It shows that choosing more diverse forest types can promote healthier soil, which benefits the entire ecosystem and the services it provides to people.

Abstract

Monoculture plantations of *Eucalyptus* in China have raised ecological concerns due to water depletion, soil degradation, and fire risk. Integrating *Eucalyptus* with *Cupressus* offers a sustainable approach to improving forest ecosystem health. In this study, we established five forest treatments, pure *Eucalyptus* (1:0), mixed *Eucalyptus–Cupressus* at three ratios (2:1, 1:1, and 1:2), and pure *Cupressus* (0:1), to assess their effects on soil properties, microbial diversity, and metabolomic profiles. Laboratory analyses revealed significant differences in physicochemical soil properties (such as water content ($p < 0.05$), pH ($p < 0.001$), organic carbon ($p < 0.001$), and nitrogen ($p < 0.001$)) among various groups within the mixed forests. Microbial community investigations highlighted a unique microbial signature in *Eucalyptus–Cupressus* mixed forests, especially when the tree ratio was 1:2, characterized by a rich (Chao1, $p < 0.05$) and diverse (Shannon, $p < 0.05$) array of bacterial taxa. The mixed *Eucalyptus–Cupressus* forest also exhibited an uplift in microbial communities, bacterial genera such as *RB41*, and fungal genera including *Penicillium*, *Talaromyces*, and *Mortierella*, which are associated with enhanced organic matter decomposition and nutrient cycling. Interactive networks within microbial communities were revealed through co-occurrence and Spearman correlation analyses, highlighting potential symbiotic relationships and ecological complexities. Metabolomic analysis, coupled with pathway analysis, further illuminated metabolic shifts in the mixed forests, emphasizing alterations in key metabolic pathways such as phenylpropanoid biosynthesis, tyrosine metabolism, arachidonic acid metabolism, and isoquinoline alkaloid biosynthesis. Collectively, these results show that moderately mixed *Eucalyptus–Cupressus* forests improve soil fertility and microbial multifunctionality, providing a practical model for sustainable and resilient forest management in subtropical regions.

Keywords: comprehensive evaluation; ecological restoration; mixed forests; plant–soil–microbial interactions; plant and soil properties

1. Introduction

Eucalyptus trees (e.g., *E. grandis* and *E. urophglla*) are well-suited for cultivation in areas with abundant sunlight, such as plains, slopes, and roadsides [1]. These trees exhibit strong adaptability to soil conditions, possess well-developed root systems, fast growth, a short maturation period, and have wide applications in industries like construction and paper production [2]. Due to its distinctive biological characteristics, many countries and regions have introduced *Eucalyptus* for afforestation and timber production [3,4]. However, with the extensive and large-scale planting of *Eucalyptus*, ecological issues in *Eucalyptus* plantations have become increasingly prominent. Growing evidence has raised concerns that the extensive root system of *Eucalyptus*, which depletes surface and groundwater during its rapid growth, leading to rapid water evaporation and causing soil aridity and compaction in forested areas [5]. *Eucalyptus* is believed to deplete soil fertility significantly due to its rapid growth, leading to severe soil degradation, which subsequently hinders the growth of subsequent generations of *Eucalyptus* or alternative tree species [6]. Its high-water consumption leads to soil aridity and nutrient depletion, adversely affecting the growth of native plants [4]. Moreover, the allelopathic properties of *Eucalyptus* release toxins into the soil, inhibiting the germination and growth of other plant species and further exacerbating soil toxicity [7]. The dense root systems of *Eucalyptus* trees also compact the soil, reducing its porosity and disrupting the natural soil structure, which diminishes its ability to retain water and support diverse microbial communities. Furthermore, *Eucalyptus* branches, leaves, and trunks contain a significant amount of oil. In the continuous high temperatures of summer, they produce a highly flammable eucalyptus oil, and transform *Eucalyptus* forests into potential "gasoline bombs" [8,9]. It has been reported that the total *Eucalyptus* cultivation area in China spans 5,460,000 hectares, with a dominant presence in southwestern China, particularly in Sichuan and Chongqing Provinces [10,11]. This extensive cultivation of *Eucalyptus* highlights the urgency of exploring alternative strategies to address the challenges arising from the high-density planting of *Eucalyptus* and to establish a sustainable forest ecosystem.

Mixed forests of *Eucalyptus*, in combination with tree species like *Quercus* spp., *Parashorea* spp., *Acacia* spp., *Manglietia* spp., *Pinus* spp., *Erythrophleum* spp., and *Castanopsis* spp., have demonstrated substantial economic and ecological benefits compared to monoculture *Eucalyptus* forests [12,13]. These mixed forests are able to foster the healthy growth of *Eucalyptus* and improve the physicochemical properties of the forest soil in their early stages. Additionally, *Cupressus* spp. thrives in warm and humid conditions and exhibits adaptability to various soil types. When combined with slow-growing nitrogen-fixing tree species, *Cupressus* spp., such as *C. funebris*, contributes to the formation of robust artificial forest communities, making it suitable for afforestation in hilly and mountainous regions with poor soil quality [14]. *Eucalyptus–Cupressus* mixed forests have demonstrated advantages in terms of understory biodiversity and leaf litter composition compared to pure *Eucalyptus* and *Cupressus* forests [15,16], whereas the detailed information about soil nutrients cycling and microbial structures in the mixed forest ecosystems remains elusive.

Within the intricate microenvironment of forest soil, soil metabolites serve as crucial connectors linking soil, plants, and soil microorganisms. Secondary metabolites produced by plants are released into the soil through leaching and volatilization from aboveground plant parts. These compounds influence the growth and development of surrounding plants, soil organisms, and microorganisms [17,18]. In turn, soil microbial activity can affect nutrient absorption in plants and the synthesis and secretion of plant metabolites. It has been documented that soil microbiota can assist plants in mitigating abiotic stress and activating nutrient cycling in different mixed forest ecosystems [19]. Some metabolites produced by soil microorganisms play a critical role in regulating various biological activities of neighboring organisms. For instance, certain nitrogen-fixing bacteria may form associations with the roots of specific tree species in the mixed forest, leading to increased nitrogen inputs through biological nitrogen fixation [20]. The fixed nitrogen becomes available to all trees, such as *Eucalyptus* and *Cupressus* trees, helping to mitigate abiotic stress, particularly in nitrogen-deficient environments. However, the intricate relationships and interactions between various tree species, especially *Eucalyptus–Cupressus* mixed forests, and the associated soil microbiota remain an area of research that requires further exploration.

This study hypothesizes that mixed forests of *Eucalyptus* and *Cupressus* can significantly enhance soil physicochemical properties, microbial community composition, and soil metabolite profiles compared to monoculture systems. We further anticipate that different mixing ratios of these two species will lead to distinct shifts in microbial structure and metabolic expression patterns, reflecting complex ecological interactions. The objective of this research is to elucidate the tree–soil–microbe–metabolite interplay under varying plantation compositions, thereby providing a mechanistic understanding of mixed forest ecosystem functioning. By integrating multi-omics approaches with soil chemical analyses, this study aims to generate novel insights that will inform ecological restoration practices and support the development of more sustainable and resilient forest ecosystems, particularly in regions impacted by large-scale *Eucalyptus* afforestation.

2. Materials and Methods

2.1. Site Description

The research area is situated in the Yubei District of Chongqing, China, and experiences a pronounced subtropical humid climate characterized by distinct continental monsoon patterns. The region has an average temperature of 17.3 °C, and the annual precipitation averages around 1100 mm. Native vegetation includes subtropical evergreen and deciduous broad-leaved forests, mixed coniferous and broad-leaved forests, and evergreen broad-leaved forests, complemented by extensive artificial forest plantations.

The research area includes forest plots with similar site conditions and approximately 15-year-old *Eucalyptus* (*E. grandis* × *E. urophglla*) and *Cupressus* (*C. funebris*) artificial mixed forests or pure forests. Historically, the area was composed of anthropogenically influenced mixed woodlands, where early-stage plantations of *Eucalyptus* were interspersed with native shrubs and grasses; however, long-term dominance and allelopathic interference from *Eucalyptus* eventually transformed much of the landscape into monoculture eucalyptus forests. Five afforestation patterns were established, with each denoted as follows: *Eucalyptus* pure forest (named as group A, *Eucalyptus–Cupressus* 1:0), *Eucalyptus* and *Cupressus* mixed forests with tree ratios of 2:1 (group B), 1:1 (group C), 1:2 (group D), and *Cupressus* pure forest (group E, *Eucalyptus–Cupressus* 0:1). In 2021, 20 m × 20 m sample plots were randomly established within each type of forest plot. The individual trees within these sample plots were examined, and data on tree height, diameter at breast height (DBH), and other relevant measurements were collected and recorded (Table S1).

2.2. Soil Sampling

Soil samples were collected from each forest type using a serpentine five-point sampling method within each 20 m × 20 m plot. After carefully removing surface litter and fallen branches, bulk soil was sampled from the top 0–20 cm layer using a sterile auger at each point. In total, three biological replicates were collected for each forest treatment. Subsamples from each of the five points within a plot were thoroughly mixed to form one composite soil sample per replicate. Each composite sample was then divided into two portions. One portion was immediately weighed and oven-dried at 105 °C to determine soil water content. The other portion was transferred into sterile, self-sealing plastic bags, stored on ice in the field, and transported to the laboratory. Upon arrival, the samples were passed through a 2 mm sieve to remove debris and roots, and the processed soils were stored at −80 °C for subsequent physicochemical, microbial, and metabolomic analyses.

2.3. Soil Physical and Chemical Properties

A 10 g soil sample was sieved through a 1 mm mesh and combined with 25 mL of distilled water. After standing for 30 min, the suspension pH was measured using a PH meter (PHS-2F). For soil organic carbon (OC) measurement, 0.1–0.5 g (exact value recorded) of soil sample was sieved through a 60-mesh sieve (<0.25 mm) and then mixed with 10 mL of 0.36 mol/L potassium dichromate-sulfuric acid solution. The mixture was thoroughly shaken and boiled at 185–190 °C. After cooling, 3–4 drops of phenanthroline were added to the solution, followed by the addition of a standard solution of ferrous sulfate ($FeSO_4$) (0.2 mol/L). The content of the added $FeSO_4$ was recorded, and the soil organic matter was subsequently calculated using the previously established method. For soil water content (WC) detection, the fresh soil sample was precisely weighed and then subjected to a 12 h baking process in a preheated oven at 105 °C. Afterward, the treated soil sample was transferred to a dryer and allowed to cool to room temperature for 30 min before being weighed again. The calculation method used was consistent with the previous approach. Soil total nitrogen (TN) was analyzed using the semimicro-Kjeldahl (KDY-9820) digestion method. Soil total phosphorus (TP) was quantified colorimetrically using the molybdate method. Soil total potassium (TK) was measured through flame spectrophotometry. Soil available nitrogen (AN) was determined using the potassium dichromate external heating method. Soil available phosphorus (AP) was analyzed using the molybdenum blue method. Soil available potassium (AK) was extracted using 1 M ammonium acetate (pH 7.0) and quantified via flame photometry.

2.4. DNA Extraction and High-Throughput Sequencing

Initially, we evaluated soil microbiome composition and diversity of three main mixed types—*Eucalypt–Cypress* mixed forest, *Eucalypt–Ficus* mixed forest and *Eucalypt–Ginkgo* mixed forest—and *Eucalypt* pure forest. Soil DNA extraction from these four groups and different ratio of *Eucalyptus–Cupressus* mixed forests was conducted, with 0.5 g of soil utilized for each of the replicates. The extraction process was carried out using the PowerSoil DNA Isolation Kit, following the manufacturer's instructions. To ensure the purity and quality of the extracted DNA, a combination of PCR amplification and subsequent 2% agarose gel electrophoresis was employed. Specific primers were chosen for the amplification of soil bacterial and fungal structures. The bacterial 16S rDNA (V3 + V4) region was targeted by the primers 338F and 806R, while the fungal ITS sequences were amplified using the primers ITS1F and ITS2R. Each 25 μL PCR reaction contained

12.5 μL of 2× Taq PCR Master Mix (Takara, Japan), 1 μL of each primer (10 μM), 1 μL of DNA template (~10 ng), and 9.5 μL of nuclease-free water. The thermal cycling program consisted of an initial denaturation at 95 °C for 3 min; 30 cycles of 95 °C for 30 s, 55 °C for 30 s, and 72 °C for 45 s; and a final extension at 72 °C for 10 min. PCR products were verified by 2% agarose gel electrophoresis, and clear amplicons were purified using the GeneJET Gel Extraction Kit (Thermo Scientific, Waltham, MA, USA). A secondary 8-cycle PCR was performed to attach Illumina sequencing adapters and indices. PCR products displaying distinct bands were then subjected to purification using the GeneJET gel extraction kit, following the guidelines provided by Thermo Scientific. Subsequently, equal concentrations of PCR products from each sample were prepared for sequencing on the Illumina MiSeq platform, employing 300-bp paired-end reads. This sequencing was carried out at TinyGene Bio-Tech Co., Ltd. in Shanghai, China, using technology provided by Illumina, Inc., based in San Diego, CA, USA.

2.5. Sequence Processing and Bioinformatic Analysis

To assess the microbial species composition in each sample site, the effective data from all soil samples were clustered into Operational Taxonomic Units (OTUs) at a 97% identity threshold. For bacterial 16S rRNA gene sequences, the Greengenes database was used (v2022.10, https://ngdc.cncb.ac.cn/databasecommons, accessed on 18 November 2025), and for fungal ITS sequences, the UNITE database was employed (v10.0, https://unite.ut.ee/), with species taxonomy annotation performed using the classify-sklearn algorithm. Alpha diversity of microbial communities was comprehensively assessed using QIIME2 (2019.4) software. Alpha diversity differences between different mixed forest modes were visually presented through graph generation with Kruskal–Wallis rank-sum tests and Dunn's post hoc tests to validate the significance of differences. Utilizing QIIME2, a distance matrix was computed for differential OTU tables, followed by principal coordinates analysis (PCoA) analysis. Anosim and Permdisp algorithms were used to test the significance of differences among groups. The distance matrices were analyzed and visualized using R scripts, the Vegan package, and Ggplot2 package. Venn diagrams and heatmaps for microbial composition analysis were created using R scripts, the VennDiagram package (v1.7.3), and the Pheatmap package (v1.0.13). A co-occurrence network among different microbial taxa was constructed based on Spearman correlation analysis, with multiple hypothesis testing controlled using the Benjamini–Hochberg procedure to adjust for false discovery rate (FDR); only taxa pairs showing strong positive correlations (correlation coefficient >0.8) and adjusted *p*-values (FDR) below 0.01 were retained for network visualization. This network was created utilizing the "igraph" (v2.2.1) and "Hmisc" (v5.2.4) packages, which utilized unique correlations derived from pairwise comparisons of OTU abundance. To minimize potential bias from sequencing depth or sample sparsity, network stability and randomization tests were performed using permutation-based null models to differentiate true biological interactions from spurious correlations. To describe the network structure, various network topology metrics and features, such as modularity and eigenvector centrality, were applied. The network was then visualized using Gephi software (v0.10.1) with the Fruchterman Reingold layout. Linear discriminant analysis (LDA) was conducted using the R package "lefser" (v3.22) to identify microbial taxa that significantly differentiated the mixed forest types, thereby revealing potential biomarkers associated with variations in microbial community composition. Functional analysis and significance (p-value < 0.05) were conducted and presented using the "tidyverse" (v2.0.0) and "ggplot2" (v4.0.1) packages. Functional predictions for 16S rRNA and ITS sequences were conducted using PICRUSt2, and differential metabolites underwent functional pathway enrichment and topological analysis using the MetaboAnalyst (v4.0) package. Pathway enrichment results were visualized using KEGG Mapper (v5.1). Spearman rank correlation analysis was used to explore the associations between microbial communities and soil characteristics. Additionally, we employed structural equation modeling (SEM) to explore the relationships between soil properties (e.g., water content, organic matter, pH, and nutrients), microbial properties (such as gene abundance, diversity, and network complexity), and soil metabolomic multifunctionality. To evaluate model fit, we used various parameters including the chi-square (χ^2) test, *p*-value (>0.05), the root mean square error of approximation (RMSEA) ranging from 0 to 0.05, and a high goodness-of-fit index (GFI) exceeding 0.90. This analysis was conducted using Amos version 25.0.

2.6. Metabolome Sequencing and Differentially Expressed Metabolites (DEMs) Analysis

Soil samples from every group were used for metabolite extraction. To each homogenized sample, we added 500 mL of 80% methanol and vortexed them for 5 min. Subsequently, each sample was partitioned into 20 mL portions to ensure quality. Afterward, all the solutions underwent centrifugation at 15,000× *g* for 20 min at 4 °C. Following this, the supernatant was diluted with water and once more subjected to centrifugation at 15,000× *g* for 20 min at 4 °C. The resulting supernatant was utilized for the LC-MS analysis. For liquid chromatography and mass spectrometry, we employed the Thermo Ultimate 3000 system and Thermo Q Exactive mass spectrometer, respectively (Thermo Scientific, Waltham, MA, USA). To ensure data quality, pooled quality control (QC) samples were prepared by mixing equal aliquots from all experimental samples and injected regularly throughout the run to monitor instrument stability and signal drift. Blank samples were also included to detect background noise or contamination. We executed dynamic exclusion to eliminate redundant data in the metabolome dataset. After standardizing the quantitative outcomes, we acquired metabolite identification and relative quantitative data. Utilizing the "XCMS" (v4.0) package in R, we performed the identification, filtration, and alignment of the produced MS/MS peaks. The obtained metabolites were annotated using the KEGG, HMDB, and LIPIDMaps databases. After standardization, partial least squares-discriminant analysis (PLS-DA) was conducted with the "ropls" package to determine the differentially expressed metabolites (DEMs) in the comparisons. DEMs were identified based on the threshold of variable importance in the projection (VIP values $\geq$ 1) and a *p*-value of $\leq$0.05. Redundancy Analysis (RDA) was conducted using the vegan (v2.5-7) package in R to assess the relationship between microbial communities and metabolite profiles, with significance tested via permutation.

3. Results

3.1. Soil Properties in Different Mixed Forests

Soil WC exhibited significant differences among the various mixed forests ($p < 0.05$) (Figure 1). Compared to groups A, B and E, mixed forests C and D had notably higher soil WC. Soil pH and OC content also showed significant variations among the different mixed forests ($p < 0.05$). Soil pH was highest in group E, followed by C, B, D, and A. OC content was highest in group D, followed by E, C, B, and A. Soil TN, and AN had the highest content in group D. Soil TK presented the highest in groups D and E. TP was highest in groups A and E, while AP was highest in group E. Group E had the highest AK content, while group A had the lowest.

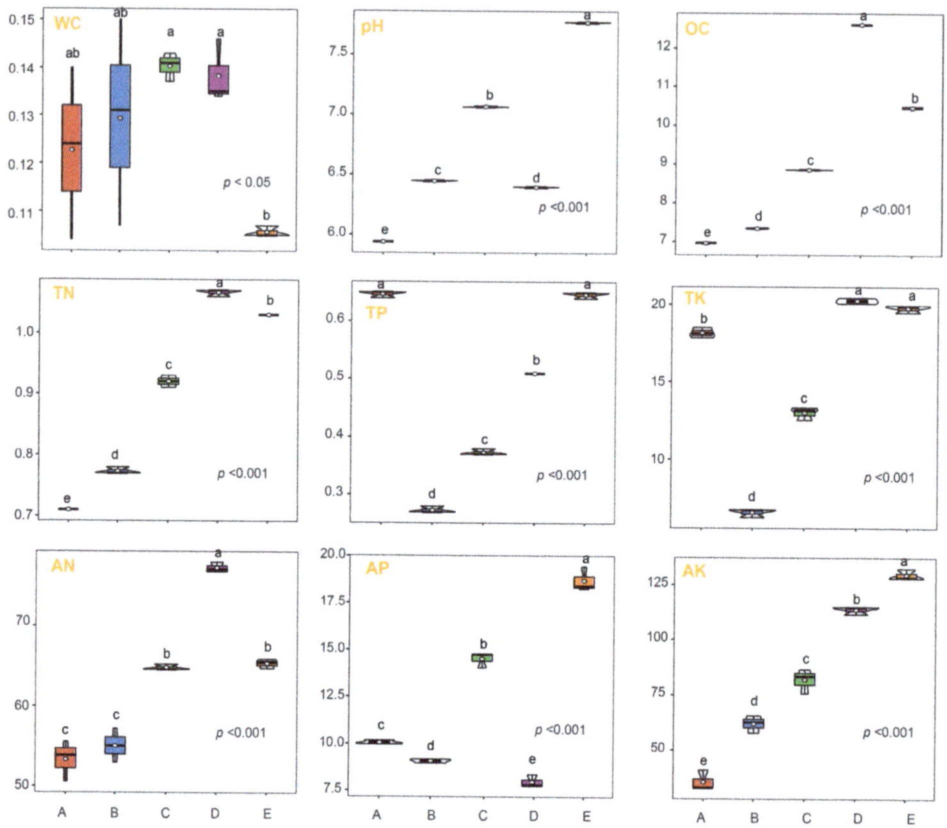

Figure 1. Box plot demonstrating soil property variations across distinct mixed forests. Key soil parameters measured include water content (WC), pH, organic carbon (OC), total nitrogen (TN), total phosphorus (TP), total potassium (TK), available nitrogen (AN), available phosphorus (AP), and available potassium (AK). The depicted forests range from *Eucalyptus* pure forests (group A, *Eucalyptus–Cupressus* 1:0), *Eucalyptus* and *Cupressus* mixed forests in various proportions (groups B, C, D), to *Cupressus* pure forests (group E, *Eucalyptus–Cupressus* 0:1). Different lowercase letters above the boxes indicate significant differences ($p < 0.05$) between groups based on Kruskal–Wallis test followed by Dunn's post hoc comparisons.

3.2. Microbial Community Composition in Different Mixed Forests

Initially, we evaluated the soil microbiome composition and diversity of four forest types: *Eucalypt–Cypress* mixed forest, *Eucalypt–Ficus* mixed forest, *Eucalypt–Ginkgo* mixed forest, and pure *Eucalypt* forest. The results showed that the *Eucalypt–Cypress* mixed forest had higher bacterial and fungal composition and diversity indexes compared to the other three groups (Table S2 and Figure S1). Consequently, we focused our analysis on the *Eucalypt–Cypress* mixed forest with different tree ratios.

The rarefaction curves for bacterial and fungal communities in different soil samples reached a plateau (Figure S2), indicating that the actual bacterial and fungal community compositions in the soil samples were reliably reflected by the sequencing data. The five types of mixed *Eucalyptus* and *Cupressus* forests revealed their affiliations with 33 phyla, 105 classes, 242 orders, 409 families, 777 genera, and 2013 species. Venn diagrams illustrated the unique and shared OTUs among the sample groups, showing significant differences in the number of shared OTUs among the samples (Figure S3). At the phylum level, the dominant bacterial phyla in different forest soils were Actinobacteria, Proteobacteria, Acidobacteria, and Chloroflexi, accounting for approximately 75% of the total abundance (Figure 2A). The relative abundance of Chloroflexi in group C was lower than in other groups, while the relative abundance of Acidobacteria was higher. At the genus level, the dominant bacterial genera in different forest soils included *Acidothermus*, *Rokubacteriales*, and *Gaiella* (Figure 2B). *Acidothermus* had a higher relative abundance in soil samples from groups A and B, while some groups (e.g., *subgroup_6*, *RB41*, and *67-14*) had higher relative abundances in groups C, D, and E.

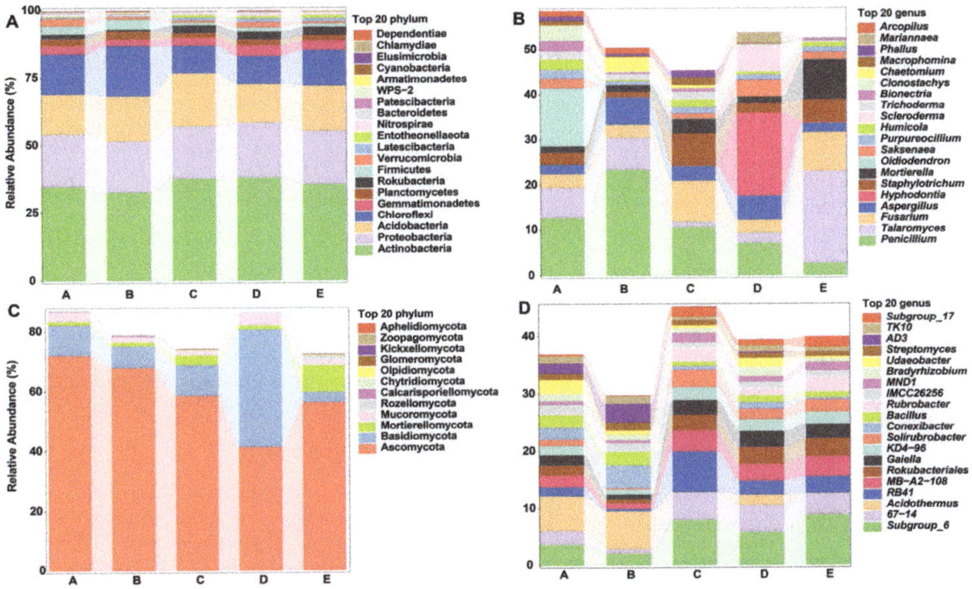

Figure 2. Microbial community taxonomic distribution depicted through bar plots at phylum (**A**,**C**) and genus (**B**,**D**) levels for bacterial and fungal community compositions. The microbial communities are ranked based on their relative abundance across all replicates, showing a decreasing order of prevalence.

For fungi, all annotated fungal OTUs belonged to 15 phyla, 43 classes, 107 orders, 255 families, 475 genera, and 736 species. Group A had the highest number of fungal species, with 420 species, followed by group C with 348 species. Venn diagrams showed that group C had more shared and unique OTUs than the other groups, whereas groups B and E had fewer shared and unique OTUs (Figure S3). At the phylum level, the dominant fungal phyla included Ascomycota, Basidiomycota, Mortierellomycota, and Mucoromycota, constituting approximately 80% of the total abundance (Figure 2C). The Ascomycota fungal phylum in group D had a lower relative abundance compared to other groups, while the Basidiomycota fungal phylum showed a significantly higher relative abundance. At the genus level, dominant fungal genera in different forest soils were *Penicillium*, *Talaromyces*, *Fusarium*, *Aspergillus*, *Hyphodontia*, *Staphylotrichum*, and *Mortieralla*, representing about 30% of the total abundance (Figure 2D).

3.3. Soil Microbial Diversity in Different Mixed Forests

Significant differences were observed in the α-diversity indices among the soil samples from different groups. For bacterial communities, the richness indexes (Chao1, Shannon, and Pielou) were the highest in group D (Figure 3A–C). Adonis analysis demonstrated that there was a highly significant difference ($p < 0.001$) in bacterial community composition among the five sample groups. Interestingly, for fungal communities, the richness index (Chao1) ranked as A > C > B > D > E, the Shannon diversity index ranked as C > E > B > D > A, and the Pielou evenness index ranked as C > E > B > D > A (Figure 3D–F). Adonis analysis also revealed a highly significant difference ($p < 0.001$) in fungal community composition among the five sample groups.

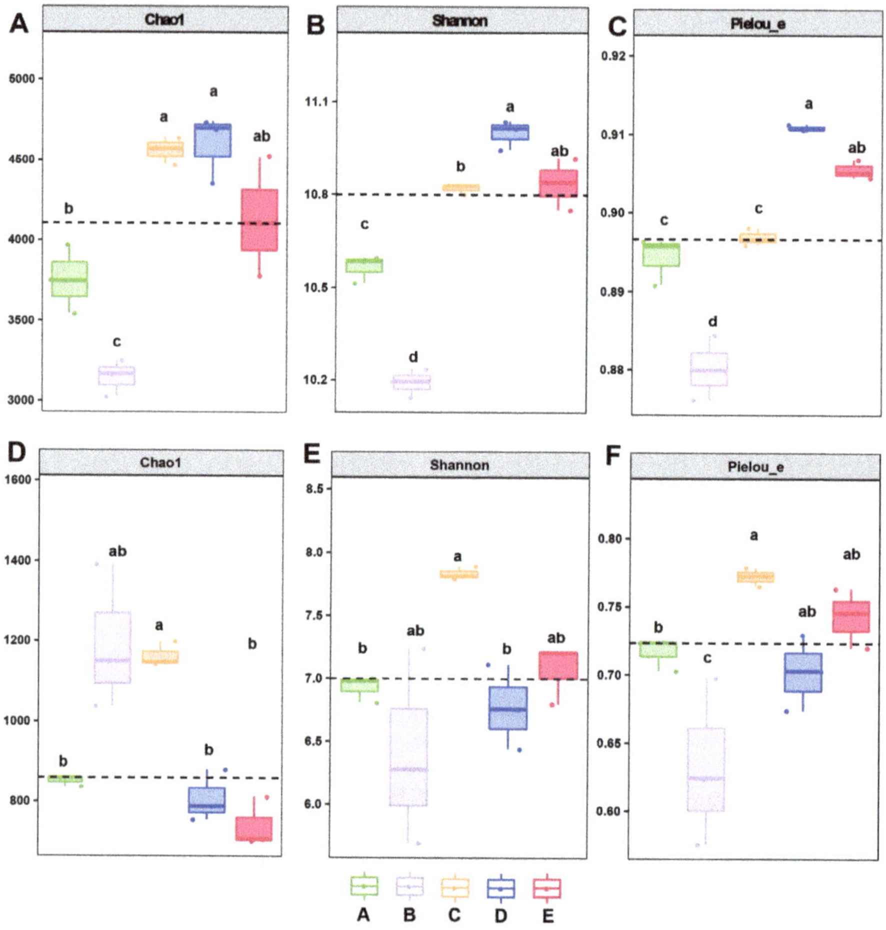

Figure 3. Alpha diversity indices of community groups, showcasing general bacterial (**A–C**) and fungal (**D–F**) alpha-diversity patterns through Chao1, Shannon, and Pielou indices across distinct mixed forests. Different lowercase letters above the boxes indicate significant differences ($p < 0.05$) between groups based on Kruskal–Wallis test followed by Dunn's post hoc comparisons. The dashed line represents the overall mean value of each diversity index across all groups, serving as a reference baseline for comparison.

3.4. Co-Occurrence Network Analyses of Bacteria and Fungi

In this study, network analysis was employed to investigate the co-occurrence patterns within the soil bacterial and fungal communities. For bacteria, the modularity index was 0.56, and the network density was 0.059 (Figure 4A). Notably, the top four taxa, namely *Acidothermus*, *Bacillus*, *Rubrobacter*, and *Entotheonellaceae*, were identified as hub biomarkers due to their high Betweenness scores. Additionally, based on node abundance, a dominant species sub-network was constructed by extracting the top 50 nodes in terms of average abundance (Table S3). In this sub-network, the major dominant nodes were Actinobacteria and Proteobacteria (Figure 4B). For fungi, the network consisted of 44 key nodes, with a modularity index of 0.61 and a network density of 0.023 (Figure 4C). Among these, the top four taxa, *Penicillium*, *Saitozyma*, *Humicola*, and *Fusarium*, were identified as hub biomarkers due to their high Betweenness scores (Figure 4D). In the dominant species sub-network, the primary dominant nodes were Ascomycota and Basidiomycota.

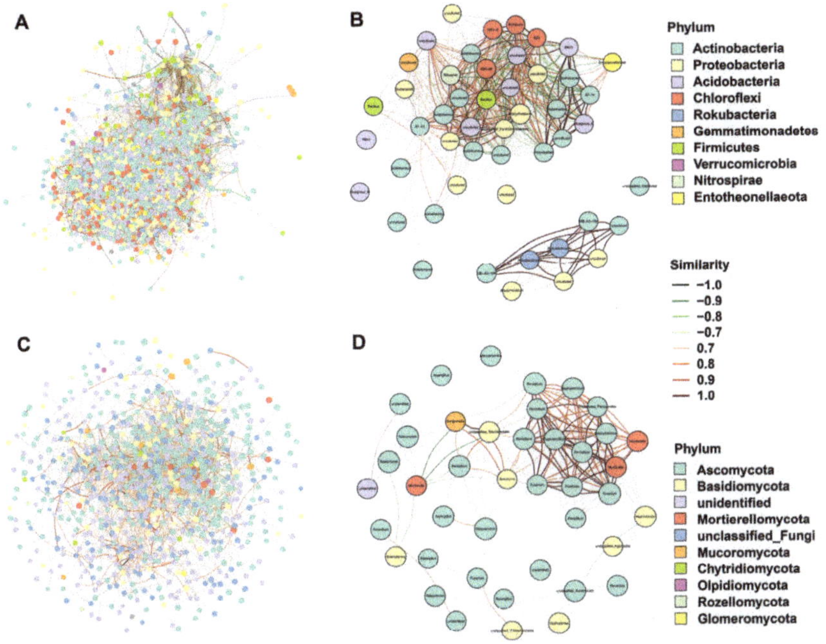

Figure 4. Network diagrams illustrating the modular association of bacteria (**A**,**B**) and fungi (**C**,**D**), along with a subnetwork of dominant species. Nodes represent OTUs in the groups, and node size is proportional to their abundance (log2(CPM/n). The top 10 modules are color-coded, with node size reflecting their abundance. Edges between nodes indicate correlations, with red lines denoting positive correlations and green lines denoting negative correlations.

3.5. Microbial Biomarkers and Functional Analysis in Different Mixed Forests

LDA was employed to uncover distinctive variations in bacterial and fungal taxa. The results unveiled that the highest number of bacterial and fungal taxa were observed in the D group (LDA scores > 4), while the E group exhibited a relatively lower number of taxa. Specifically, the D group featured three bacterial taxa, namely *Acidothermus*, *Conexibacter*, and *AD3* (Figure 5A), alongside four fungal taxa including *Penicillium*, *Aspergillus*, *Chaetomium*, and *Oidiodendron* (Figure 5B). On the other hand, the C group contained four distinctive bacterial taxa (*RB41*, *67_14*, *MB_A2_108*, and *Solirubrobacter*), along with one fungal taxon, *Fusarium*. Subsequently, functional annotation was carried out to assess potential functional disparities among the different sites (Figure 5C–F). The results highlighted alterations in primary functional annotations, particularly biotin biosynthesis and tRNA processing, which exhibited notably high expression levels in the D group. Regarding fungal functions, octane oxidation showed improvement in the C and D groups, while sulfate reduction I displayed elevated activity in the A, B, and C groups but decreased in the E group.

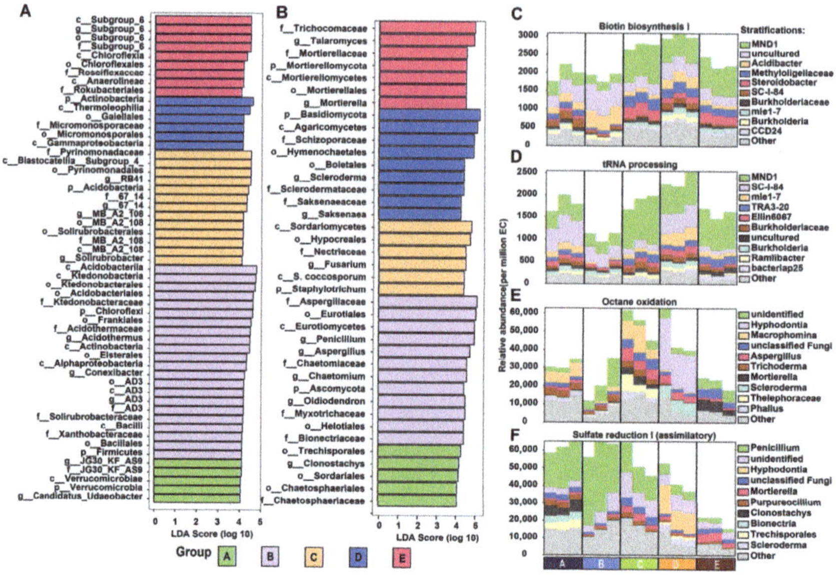

Figure 5. Differential composition and functionality of bacterial and fungal communities within varied forest groups. (**A**,**B**) LDA highlighting distinct variances in bacterial and fungal taxa across five mixed forests. (**C**–**F**) Display of bacterial and fungal community functionalities via the MetaCyc pipeline for microbial function annotation.

3.6. Identification of Metabolites and Functional Annotation of DEMs

This study identified a total of 5726 metabolites in the positive ion mode and 6045 metabolites in the negative ion mode, all passing the quality control using relative standard deviation values (Figure S4). A total of 262 DEMs were detected and identified among the groups. Among these, the largest number of differential metabolites was found in the comparison A vs. D, with 31 up-regulated and 25 down-regulated metabolites (Figure S5). The next highest number of differential metabolites was observed in groups A and E, with 32 up-regulated and 18 down-regulated metabolites.

The focused investigation on the D group, characterized by its exceptional nutrient dynamics and microbial diversity, provides valuable insights into the intricate metabolic pathways that underpin its ecological significance. Notably, in the A vs. D comparison, several pathways exhibited significant alterations, including "phenylpropanoid biosynthesis," "tyrosine metabolism," "arachidonic acid metabolism," and "isoquinoline alkaloid biosynthesis" (Figure 6; Figure S6). Within the context of tyrosine metabolism, the study revealed changes in three up-regulated metabolites, specifically hydroquinone, gentisic acid, and 3,4-Dihydroxyphenylacetaldehyde, as well as two down-regulated metabolites, 4-hydroxycinnamic acid and 4-hydroxyphenylacetaldehyde, impacting the tyrosine metabolic process. Furthermore, isoquinoline alkaloid biosynthesis was influenced by a total of four DEMs, featuring two up-regulated and two down-regulated metabolites. In addition, the D group exerted a significant impact on phenylpropanoid biosynthesis by altering the expression levels of 4-hydroxycinnamic acid, (E)-3-(4-Hydroxyphenyl)-2-propenal, and methyleugenol. Arachidonic acid metabolism was also affected, with changes in four metabolites, including 12-Keto-tetrahydro-leukotriene B4, 15-Deoxy-d-12,14-PGJ2, 5-KETE, and 20-HETE. Interestingly, these four pathways were identified in the D vs. E group comparison, and an additional amino acid pathway, biosynthesis of amino acids, was characterized (Figure 6; Figure S6). This pathway demonstrated changes in metabolites such as 2-aminobenzoic acid, ketoleucine, phosphohydroxypyruvic acid, and N-a-Acetylcitrulline.

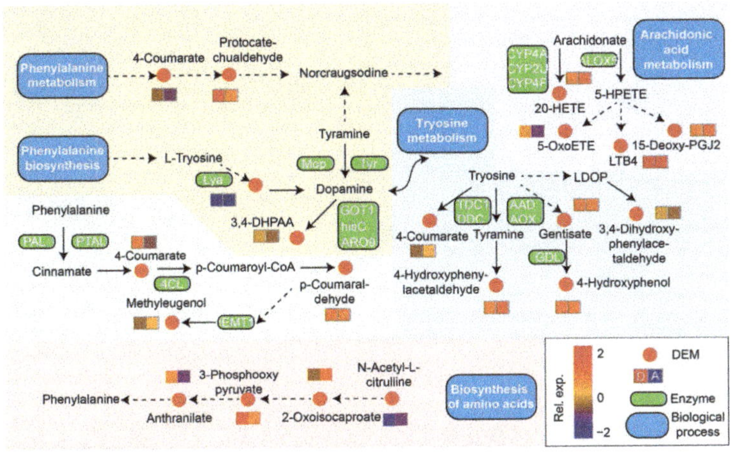

Figure 6. KEGG pathway analysis of differentially expressed metabolites using summarized KO genes and KEGG pathway modules. Red dots represent DEMs, while the heatmaps illustrate the relative abundance of DEMs related to identified metabolic pathways. Solid arrows denote direct interactions between two DEMs, while dashed lines signify unshown KO modules in the pathways.

3.7. Interactions Among Soil Properties, Microbial Communities, and Soil Metabolic Multifunctionality

A correlation analysis was conducted to assess the relationships between α-diversity indices of soil bacterial communities, significant differential species at various taxonomic levels, and soil physicochemical factors (Table 1). The Chao1 richness index of bacterial communities did not exhibit significant correlations with any of the measured soil physicochemical factors ($p > 0.05$). The Shannon diversity index of bacterial communities showed significant positive correlations with AN, AK, and TN content ($p < 0.05$). The Pielou evenness index of bacterial communities displayed significant positive correlations with OC, AN, AK, and TN content ($p < 0.05$). Among the differential bacterial phyla, the relative abundance of Gemmatimonadetes exhibited significant positive correlations with soil OC ($p < 0.05$), while Firmicutes showed significant negative correlations with soil TN content ($p < 0.05$). Additionally, Rokubacteria displayed significant positive correlations with soil AK and TN content ($p < 0.05$). Similarly, the Chao1 richness index, Shannon diversity index, Pielou evenness index of fungal communities did not display significant correlations with any of the measured soil physicochemical factors ($p > 0.05$). Among the differential fungal phyla, the relative abundance of Ascomycota exhibited highly significant negative correlations with soil OC content ($r = -0.967$, $p < 0.01$) and highly significant negative correlations with soil AN content ($r = -0.985$, $p < 0.01$). It also showed significant negative correlations with soil TN content ($p < 0.05$). Calcarisporiellomycota exhibited significant negative correlations with soil TN content ($p < 0.05$). Mortierellomycota displayed a significant positive correlation with soil pH ($p < 0.05$), while Olpidiomycota exhibited a highly significant positive correlation with soil pH ($r = 0.963$, $p < 0.01$). To explore potential microbe-metabolite relationships, we performed RDA, which revealed clear associations between specific microbial genera and key metabolites (Figure S7). For instance, *Alicyclobacillus* and *Rokubacteriales* aligned strongly with metabolites such as kynurenine and X5-KETE in the bacterial dataset, while fungal genera like *Staphylotrichum* and *Ruhlandiella* correlated with X9-S-HPOD and X-E-3-4-Hydroxyphenyl-2-propenal.

Table 1. Pearson correlation coefficients between bacterial (16S) and fungal (ITS) community α-diversity indexes, dominant phyla, and soil physicochemical properties.

		Indexes	WC	pH	OC	AN	AP	AK	TN	TP	TK
16S	α-diversity	Chao1	0.428	0.508	0.769	0.849	0.216	0.775	0.856	−0.285	0.142
		Shannon	0.23	0.583	0.865	0.886 *	0.243	0.904 *	0.938 *	−0.178	0.223
		Pielou	0.063	0.549	0.917 *	0.885 *	0.18	0.945 *	0.948 *	−0.058	0.308
	Phylum	Firmicutes	−0.104	−0.765	−0.741	−0.777	−0.551	−0.872	−0.899 *	0.039	−0.293
		Gemmatimonadetes	0.173	0.278	0.902 *	0.868	−0.124	0.815	0.843	−0.112	0.267
		Acidobacteria	0.057	0.402	−0.273	−0.162	0.648	−0.086	−0.052	0.022	−0.028
		Verrucomicrobia	0.398	−0.47	0.005	−0.006	−0.748	−0.125	−0.148	−0.62	−0.532
		Rokubacteria	0.046	0.762	0.811	0.812	0.477	0.946 *	0.944 *	−0.064	0.272
		Chloroflexi	−0.476	−0.48	−0.734	−0.825	−0.192	−0.737	−0.822	0.335	−0.093
		Actinobacteria	0.527	0.398	0.757	0.856	0.112	0.707	0.815	−0.316	0.132
ITS	α-diversity	Chao1	0.348	−0.444	−0.604	−0.457	−0.098	−0.739	−0.606	−0.006	−0.124
		Shannon	0.35	0.612	0.098	0.217	0.549	0.34	0.346	−0.479	−0.307
		Pielou	0.17	0.788	0.315	0.368	0.598	0.619	0.564	−0.45	−0.252
	Phylum	Ascomycota	−0.192	−0.229	−0.967 **	−0.985 **	0.006	−0.755	−0.892 *	−0.264	−0.68
		Basidiomycota	0.568	−0.349	0.714	0.769	−0.629	0.329	0.502	−0.11	0.298
		Calcarisporiellomycota	−0.151	−0.643	−0.834	−0.877	−0.436	−0.861	−0.935 *	−0.075	−0.45
		Chytridiomycota	0.106	0.352	−0.28	−0.159	0.602	−0.118	−0.07	0.006	−0.034
		Mortierellomycota	−0.76	0.916 *	0.274	0.149	0.918	0.646	0.489	0.458	0.368
		Olpidiomycota	−0.463	0.963 **	0.338	0.293	0.959	0.683	0.592	0.288	0.316
		Rozellomycota	−0.852	0.826	0.313	0.149	0.8	0.647	0.475	0.535	0.424
		Zoopagomycota	0.331	−0.466	−0.361	−0.221	−0.14	−0.598	−0.417	0.208	0.172

Note: */** indicates significant values ($p < 0.05$ or 0.01). Abbreviations: WC, water content; pH, soil pH; OC, organic carbon; AN, available nitrogen; AP, available phosphorus; AK, available potassium; TN, total nitrogen; TP, total phosphorus; TK, total potassium.

Subsequently, we conducted SEM analysis to demonstrate that mixed modes, OC, nutrients, and bacterial and fungal α-diversity were the primary influencing factors on soil metabolic multifunctionality, exerting greater impacts than soil pH, WC, and microbial abundance (Figure 7). Bacterial and fungal diversities had a direct and positive influence on soil metabolic multifunctionality, with standardized path coefficients of 0.89 and 0.75 ($p < 0.01$), respectively. Additionally, soil nutrients indirectly influenced soil microbial diversity, with standardized path coefficients of 0.71 for bacteria and 0.72 for fungi.

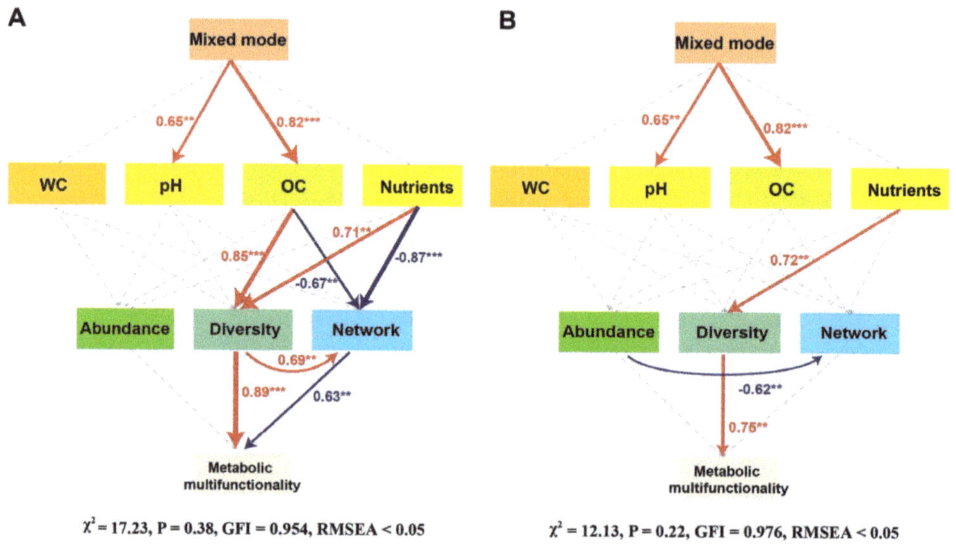

Figure 7. Structural equation model estimating the effects of soil properties, bacterial (**A**) and fungal (**B**) communities on soil metabolic multifunctionality in mixed forest patterns. Arrow width represents the strength of significant standardized path coefficients ($p < 0.05$), with red indicating positive correlation, and blue indicating negative correlation. Pathways with non-significant coefficients are represented by gray lines. *** $p < 0.001$; ** $p < 0.01$.

4. Discussion

4.1. Effects of Tree Species Composition on Soil Physicochemical Properties

Our study's findings elucidate the significant impact of tree species composition on soil properties within mixed forest ecosystems, aligning with our objective to understand how different afforestation patterns influence soil ecology. Numerous studies indicate that establishing mixed forests can improve soil physicochemical properties to some extent and slow down the degradation of forest land in artificial forests [21,22]. During the development of artificial forests, the accumulation of understory litter increases, and its decomposition produces a large amount of acidic substances [23]. Notably, the higher soil water content observed in groups C and D, which have a higher proportion of *Cupressus*, suggests that the water retention capabilities of *Cupressus* are superior to those of *Eucalyptus*. This could be attributed to the physiological or morphological traits of *Cupressus*, such as root depth and leaf structure, which may enhance water retention and reduce evaporation rates [24]. Additionally, the variation in soil pH across the plots, with the highest values in the *Cupressus*-dominated group E, highlights how tree-specific litter and root exudates can modify soil chemical properties, potentially influencing microbial communities and plant nutrient uptake [25]. For instance, alkaline soils, as seen in group E, are known to impact the solubility of phosphates and micronutrients, which can affect plant growth and microbial activity [26]. However, this interpretation remains speculative in the absence of direct chemical characterization of litter or analysis of root exudate composition. Future studies incorporating detailed biochemical profiling of litter and root exudates would help confirm the specific compounds responsible for the observed pH variations. The significant differences in organic carbon (OC) and nitrogen (TN and AN) content, especially higher in group D, indicate a more robust nutrient cycling under mixed forests with a predominant presence of *Cupressus*. This can be associated with the slow decomposition rates of *Cupressus* litter compared to *Eucalyptus*, contributing to a more sustained release of nutrients. Such findings are critical for forest management, suggesting that strategic tree species selection can optimize soil health and enhance ecological resilience. For example, in regions prone to drought or poor soil fertility, integrating *Cupressus* into mixed forests could be advantageous for improving soil moisture and nutrient status, thereby supporting more stable forest ecosystems [27]. Moreover, the marked differences in potassium and phosphorus dynamics across the groups further underline the complex interactions between tree species and soil mineral content, guiding targeted interventions in forest composition to achieve desired ecological outcomes.

4.2. Microbial Community Structure and Ecological Implications

Soil microorganisms play a dominant role in almost all soil ecological processes, such as nutrient mineralization and decomposition, significantly influencing the functionality of forest ecosystems and the sustainability of the soil under the forest [28]. Studies indicate that the mixed planting of tree species with different functional traits can also reduce specific pathogenic microorganisms, positively affecting the health of the ecosystem [29]. Chloroflexi can degrade chitin, cellulose, or hemicellulose in the soil, and its higher relative abundance is associated with the presence of more undecomposed organic matter in the soil. The increase in the abundance of Acidobacteria may indicate changes in the ecological functions of the soil due to environmental stress, serving as an indicator of soil degradation under intensive agriculture scenarios. For soil fungi, the unique fungal composition observed in Group D signifies a distinctive ecological niche shaped by the lower abundance of Ascomycota and a significantly higher prevalence of Basidiomycota compared to other restoration groups. Notably, this fungal profile in Group D is characterized by dominant genera, including *Penicillium*, *Talaromyces*, *Fusarium*, *Aspergillus*,

Hyphodontia, *Staphylotrichum*, and *Mortieralla*. *Penicillium* and *Aspergillus* are well-known for their versatile metabolic capabilities, contributing to nutrient cycling and organic matter decomposition in soil ecosystems [30,31]. *Talaromyces*, with its diverse enzymatic repertoire, participates in the breakdown of complex organic compounds, influencing nutrient availability [32]. *Fusarium*, a common soil inhabitant, plays crucial roles in plant-pathogen interactions and nutrient mobilization [33]. The nuanced exploration of these dominant genera provides a glimpse into their ecological functions, such as nutrient cycling, organic matter decomposition, and potential interactions with plants and other microbes.

4.3. Functional Microbial Adaptations and Ecosystem Processes

Notably, the D group emerged as a hotspot of microbial diversity, boasting a richness of both bacterial and fungal taxa, exemplified by the identification of specific taxa such as *Acidothermus*, *Conexibacter*, and *AD3* among bacteria, and *Penicillium*, *Aspergillus*, *Chaetomium*, and *Oidiodendron* among fungi. In contrast, the E group exhibited a comparatively lower abundance of taxa. The functional annotation analysis provided critical insights into the microbial functions within the forest ecosystems, notably showcasing significant shifts in biotin biosynthesis and tRNA processing pathways, with particular emphasis on the D group. Furthermore, the accentuated modifications in tRNA processing pathways within the D group suggest variations in gene expression and the translation of genetic information into functional proteins, possibly influencing the microbial community's functional potential [34]. These findings signify unique functional adaptations within the microbial communities of the D group, underscoring their distinct metabolic and genetic processing capacities, which may have implications for the overall ecosystem functionality and adaptability compared to other forest groups. Overall, these results illuminate the intricate relationship between microbial taxonomy and functionality, providing a foundation for comprehending the underlying processes in soil ecosystems subjected to diverse restoration approaches.

4.4. Metabolomic Shifts and Microbe–Metabolite Interactions

Leveraging KEGG pathway analysis in a comparative context with the A and E groups unraveled substantial alterations in key pathways. For instance, the up-regulation of metabolites in the tyrosine metabolic process, such as hydroquinone, gentisic acid, and 3,4-Dihydroxyphenylacetaldehyde, suggests an enhanced utilization or synthesis of these compounds, possibly driven by specific microbial activities responding to the nutrient-rich environment [35,36]. Similarly, alterations in pathways could be indicative of adaptations in response to the microbial community's functional dynamics, influenced by the unique characteristics of the D group. The isoquinoline alkaloid biosynthesis pathway, affected by four differentially expressed metabolites, may signify intricate interactions between the microbial community and plant secondary metabolites, potentially linked to ecological roles or defense mechanisms [37,38]. The positive correlation between (E)-3-(4-Hydroxyphenyl)-2-propenal and the *Candidatus_Koribacter* suggests a potentially symbiotic relationship. Conversely, the contrasting correlations of 4-Hydroxycinnamic acid with the *Alicyclobacillus* and its associations with *Rokubacteriales*, *KD4-96*, and *Rubrobacter* hint at differential metabolic responses within the bacterial community. These microbial-metabolite associations echo findings from similar studies, emphasizing the pivotal role of microbial community dynamics in shaping soil ecology.

To enhance our understanding of the intricate dynamics within forest ecosystems, several avenues for future research merit attention. The current study, while providing valuable insights into microbial communities and metabolic pathways in *Eucalyptus–Cupressus* mixed forests, is limited by its single time-point design and restricted geographic scope, which may not fully capture seasonal or site-specific variability. Longitudinal studies tracking temporal changes across seasons and years could reveal how soil microbial and metabolic profiles shift over time, offering a more comprehensive ecological perspective. Additionally, while our correlation analyses between microbial taxa and metabolites revealed meaningful associations, they remain correlational rather than causal. Future research should incorporate mechanistic approaches, such as metagenomics, metatranscriptomics, or stable isotope probing, to unravel the specific functional roles of microbial taxa and validate their contributions to soil biochemical processes in mixed forest systems. From a management perspective, our findings suggest that incorporating *Cupressus* into *Eucalyptus* plantations, particularly at a 1:2 ratio, can improve soil moisture, organic carbon, and nitrogen levels while enhancing microbial diversity and functional potential. This strategy may help mitigate common issues associated with monoculture plantations, such as nutrient depletion and poor soil structure, offering a nature-based solution for building more resilient and ecologically sustainable forest ecosystems in subtropical regions.

5. Conclusions

This study demonstrates that mixing *Eucalyptus* with *Cupressus*, particularly at a 1:2 ratio, can significantly improve soil conditions and promote a more diverse and functionally enriched soil microbiome. The observed increases in organic carbon, nitrogen, and water content, along with shifts in dominant microbial taxa and enriched metabolic pathways, indicate that species composition plays a central role in shaping soil ecological function. While our findings are limited to a single site and time point, they provide clear evidence that incorporating *Cupressus* into *Eucalyptus* plantations may help mitigate issues commonly associated with monoculture plantations, such as soil nutrient depletion, reduced microbial diversity, and poor moisture retention. These results support the application of mixed-species afforestation as a practical strategy to enhance soil health and sustainability in subtropical artificial forest systems. Future work should expand on these findings through multi-season, multi-site studies and deeper functional validation using metagenomics and metabolomics.

Supplementary Materials: The following supporting information can be downloaded at: https://www.mdpi.com/article/10.3390/biology14121667/s1.

Author Contributions: Y.-W.Z., H.-P.D. conceived and designed the study. Y.-Y.L. helped with the experiment design. Y.-W.Z., Y.-Y.L., Y.-X.J., W.-Q.L., Y.P., S.-Q.Y., S.-M.Z. and S.-Q.L. collected the samples. Y.-W.Z., Y.-Y.L., W.-Q.L., analyzed the data and prepared the figures and table. H.-P.D., S.-M.Z. and S.-Q.L. helped with the improvement of the manuscript. All authors have read and agreed to the published version of the manuscript.

Funding: This study was funded by the General Program of the Chongqing Natural Science Foundation (CSTB2024NSCQ-MSX1297).

Institutional Review Board Statement: Not applicable.

Informed Consent Statement: Not applicable.

Data Availability Statement: The sequencing data were submitted to the Sequence Read Archive (SRA) database with BioProject accession number PRJNA1043821 and MTBLS9022. Other data used to support the findings are available from the corresponding author upon request.

Acknowledgments: We would like to thank Jiahui Zhang (Southwest University) for technical support.

Conflicts of Interest: The authors declare that they have no conflicts of interest.

References

1. Shen, C.; Li, L.; Ouyang, L.; Su, M.; Guo, K. *E. urophylla* × *E. grandis* high-quality genome and comparative genomics provide insights on evolution and diversification of eucalyptus. *BMC Genom.* **2023**, *24*, 223. [CrossRef]
2. Kaur, A.; Monga, R. Eucalyptus trees plantation: A review on suitability and their beneficial role. *Int. J. Bio-Resour. Stress Manag.* **2021**, *12*, 16–25. [CrossRef]
3. Molla, G.; Addisie, M.B.; Ayele, G.T. Expansion of eucalyptus plantation on fertile cultivated lands in the North-Western highlands of Ethiopia. *Remote Sens.* **2023**, *15*, 661. [CrossRef]
4. Tomé, M.; Almeida, M.H.; Barreiro, S.; Branco, M.R.; Deus, E.; Pinto, G.; Silva, J.S.; Soares, P.; Rodríguez-Soalleiro, R. Opportunities and challenges of *Eucalyptus plantations* in Europe: The Iberian Peninsula experience. *Eur. J. For. Res.* **2021**, *140*, 489–510. [CrossRef]
5. Medeiros, P.L.D.; Pimenta, A.S.; Miranda, N.D.O.; Melo, R.R.D.; Amorim, J.D.S.; Azevedo, T.K.B.D. The myth that eucalyptus trees deplete soil water—A review. *Forests* **2025**, *16*, 423. [CrossRef]
6. MRahman, A.; Das, A.K.; Al Riyadh, Z.; Suhag, M.; Rahman, M.M. Eucalyptus in Agriculture: Friend or Foe? Analyzing its impact on crop yields, soil dynamics, and farmers' perceptions in Bangladesh. *Agrofor. Syst.* **2024**, *98*, 3109–3128. [CrossRef]
7. Nelson, K.M.; Bisbing, S.; Grossenbacher, D.L.; Ritter, M.; Yost, J.M. Testing an invasion mechanism for *Eucalyptus globulus*: Is there evidence of allelopathy. *Am. J. Bot.* **2021**, *108*, 607–615. [CrossRef]
8. Duff, T.; Porteous, M.G.; Goodger, J.Q. Wildfires, Flammable Oils, and Eucalyptus Trees: The Persistence and Volatility of Terpenes in Excised Leaves. SSRN 4120325. 2022. Available online: https://papers.ssrn.com/sol3/papers.cfm?abstract_id=4120325 (accessed on 6 April 2025).
9. Guerrero, F.; Carmona, C.; Hernández, C.; Toledo, M.; Arriagada, A.; Espinoza, L.; Bergmann, J.; Taborga, L.; Yañez, K.; Carrasco, Y. Drivers of flammability of *Eucalyptus globulus* Labill leaves: Terpenes, essential oils, and moisture content. *Forests* **2022**, *13*, 908. [CrossRef]
10. Williams, R.A. Mitigating biodiversity concerns in *Eucalyptus plantations* located in South China. *J. Biosci. Med.* **2015**, *3*, 1. [CrossRef]
11. Xie, Y.; Arnold, R.J.; Wu, Z.; Chen, S.; Du, A.; Luo, J. Advances in eucalypt research in China. *Front. Agric. Sci. Eng.* **2017**, *4*, 380–390. [CrossRef]
12. Kindt, R.; Dawson, I.K.; Lillesø, J.-P.B.; Muchugi, A.; Pedercini, F.; Roshetko, J.; van Noordwijk, M.; Graudal, L.; Jamnadass, R. The one hundred tree species prioritized for planting in the tropics and subtropics as indicated by database mining. *World Agrofor.* **2021**, *312*, 1–21.
13. Nair, P.R.; Kumar, B.M.; Nair, V.D. Multipurpose trees (MPTs) and other agroforestry species. In *An Introduction to Agroforestry: Four Decades of Scientific Developments*; Springer: Berlin, Germany, 2022; pp. 281–351.
14. Jiang, X.; Yang, J.; Yang, Y.; Yang, J.; Dong, Q.; Zeng, H.; Zhang, K.; Xu, N.; Yuan, J.; Liu, M. Response of Thinning to C: N: P Stoichiometric Characteristics and Seasonal Dynamics of Leaf-Litter-Soil System in *Cupressus funebris* Endl. Artificial Forests in Southwest, China. *Forests* **2024**, *15*, 1435. [CrossRef]
15. Abiyu, A.; Lemenih, M.; Gratzer, G.; Aerts, R.; Teketay, D.; Glatzel, G. Status of native woody species diversity and soil characteristics in an exclosure and in plantations of *Eucalyptus globulus* and *Cupressus lusitanica* in Northern Ethiopia. *Mt. Res. Dev.* **2011**, *31*, 144–152. [CrossRef]
16. Alem, S.; Pavlis, J.; Urban, J.; Kucera, J. Pure and mixed plantations of *Eucalyptus camaldulensis* and *Cupressus lusitanica*: Their growth interactions and effect on diversity and density of undergrowth woody plants in relation to light. *Open J. For.* **2015**, *5*, 375. [CrossRef]
17. Hemkemeyer, M.; Schwalb, S.A.; Heinze, S.; Joergensen, R.G.; Wichern, F. Functions of elements in soil microorganisms. *Microbiol. Res.* **2021**, *252*, 126832. [CrossRef] [PubMed]
18. Mishra, S.; Lin, Z.; Pang, S.; Zhang, W.; Bhatt, P.; Chen, S. Recent advanced technologies for the characterization of xenobiotic-degrading microorganisms and microbial communities. *Front. Bioeng. Biotechnol.* **2021**, *9*, 632059. [CrossRef]
19. Munir, N.; Hanif, M.; Abideen, Z.; Sohail, M.; El-Keblawy, A.; Radicetti, E.; Mancinelli, R.; Haider, G. Mechanisms and strategies of plant microbiome interactions to mitigate abiotic stresses. *Agronomy* **2022**, *12*, 2069. [CrossRef]
20. Pereira, A.P.D.A.; Santana, M.C.; Zagatto, M.R.; Brandani, C.B.; Wang, J.-T.; Verma, J.P.; Singh, B.K.; Cardoso, E.J. Nitrogen-fixing trees in mixed forest systems regulate the ecology of fungal community and phosphorus cycling. *Sci. Total Environ.* **2021**, *758*, 143711. [CrossRef]
21. He, Y.; Han, X.; Wang, X.; Wang, L.; Liang, T. Long-term ecological effects of two artificial forests on soil properties and quality in the eastern Qinghai-Tibet Plateau. *Sci. Total Environ.* **2021**, *796*, 148986. [CrossRef]
22. Qian, J.; Ji, C.; Yang, J.; Zhao, H.; Wang, Y.; Fu, L.; Liu, Q. The advantage of afforestation using native tree species to enhance soil quality in degraded forest ecosystems. *Sci. Rep.* **2024**, *14*, 20022. [CrossRef]
23. Zhang, J.; Zhu, S.; Liu, Y.; Yao, B.; Yu, M.; Ma, J.; Yang, X.; Xue, J.; Xiang, Y.; Li, Y. Impact of mixed plantations on soil physicochemical properties: Variations and controlling factors in China. *For. Ecol. Manag.* **2024**, *568*, 122107. [CrossRef]
24. Li, D.; Fan, S.; He, A.; Yin, F. Forest resources and environment in China. *J. For. Res.* **2004**, *9*, 307–312. [CrossRef]

25. Yao, L.; Jiao, J.; Wu, C.; Jiang, B.; Fan, L. Effects of thinning on the structure of soil microbial communities in a subtropical secondary evergreen broad-leaved forest. *Front. Plant Sci.* **2024**, *15*, 1465237. [CrossRef] [PubMed]
26. Neina, D. The role of soil pH in plant nutrition and soil remediation. *Appl. Environ. Soil Sci.* **2019**, *2019*, 1–9. [CrossRef]
27. Yu, A.; Tang, Z.; Yin, H.; Wu, S.; Xiang, Y.; Yang, J.; Chen, G.; Hou, G.; Fan, C.; Zhao, K. Thinning and replanting enhance soil environmental stability in *Cupressus funebris* plantations: Insights from soil nematodes. *Plant Soil* **2025**, 1–18. [CrossRef]
28. Zhan, C. Microbial decomposition and soil health: Mechanisms and ecological implications. *Mol. Soil Biol.* **2024**, *15*. [CrossRef]
29. Friesen, M.L.; Porter, S.S.; Stark, S.C.; von Wettberg, E.J.; Sachs, J.L.; Martinez-Romero, E. Microbially mediated plant functional traits. *Annu. Rev. Ecol. Evol. Syst.* **2011**, *42*, 23–46. [CrossRef]
30. Pfliegler, W.P.; Pócsi, I.; Győri, Z.; Pusztahelyi, T. The Aspergilli and their mycotoxins: Metabolic interactions with plants and the soil biota. *Front. Microbiol.* **2020**, *10*, 2921. [CrossRef]
31. Vassileva, M.; Mendes, G.D.O.; Deriu, M.A.; Benedetto, G.D.; Flor-Peregrin, E.; Mocali, S.; Martos, V.; Vassilev, N. Fungi, P-solubilization, and plant nutrition. *Microorganisms* **2022**, *10*, 1716. [CrossRef]
32. Yilmaz, N.; López-Quintero, C.A.; Vasco-Palacios, A.M.; Frisvad, J.C.; Theelen, B.; Boekhout, T.; Samson, R.A.; Houbraken, J. Four novel Talaromyces species isolated from leaf litter from Colombian Amazon rain forests. *Mycol. Prog.* **2016**, *15*, 1041–1056. [CrossRef]
33. Binyamin, R.; Nadeem, S.M.; Akhtar, S.; Khan, M.Y.; Anjum, R. Beneficial and pathogenic plant-microbe interactions: A review. *Soil Environ.* **2019**, *38*, 127–150. [CrossRef]
34. Rice, K.C.; Turner, M.E.; Carney, O.N.V.; Gu, T.; Ahn, S.-J. Modification of the Streptococcus mutans transcriptome by LrgAB and environmental stressors. *Microb. Genom.* **2017**, *3*, e000104. [CrossRef]
35. Nowicka, B.; Kruk, J. Occurrence, biosynthesis and function of isoprenoid quinones. *Biochim. Biophys. Acta BBA Bioenerg.* **2010**, *1797*, 1587–1605. [CrossRef] [PubMed]
36. Raza, A. Metabolomics: A systems biology approach for enhancing heat stress tolerance in plants. *Plant Cell Rep.* **2020**, *41*, 741–763. [CrossRef]
37. PDivekar, A.; Narayana, S.; Divekar, B.A.; Kumar, R.; Gadratagi, B.G.; Ray, A.; Singh, A.K.; Rani, V.; Singh, V.; Singh, A.K. Plant secondary metabolites as defense tools against herbivores for sustainable crop protection. *Int. J. Mol. Sci.* **2022**, *23*, 2690. [CrossRef]
38. Jan, R.; Asaf, S.; Numan, M.; Lubna; Kim, K.-M. Plant secondary metabolite biosynthesis and transcriptional regulation in response to biotic and abiotic stress conditions. *Agronomy* **2021**, *11*, 968. [CrossRef]

biology

MDPI

Review

Biogeochemical Cycles in Plant–Soil Systems: Significance for Agriculture, Interconnections, and Anthropogenic Disruptions

Wajid Zaman [1,†], Asma Ayaz [2,†] and Daniel Puppe [3,*]

[1] Department of Life Sciences, Yeungnam University, Gyeongsan 38541, Republic of Korea; wajidzaman@yu.ac.kr
[2] Faculty of Sports Science, Ningbo University, Ningbo 315211, China; asma@nbu.edu.cn
[3] Leibniz Centre for Agricultural Landscape Research (ZALF), 15374 Müncheberg, Germany
* Correspondence: daniel.puppe@zalf.de
† These authors contributed equally to this work.

Simple Summary

Biogeochemical cycling is essential for maintaining the balance of nutrients/elements in ecosystems. Carbon, nitrogen, phosphorus, sulfur, and silicon cycles facilitate nutrient/element transfer and storage ensuring that plants and soil organisms receive the components necessary for life. This review provides a comprehensive overview of interconnected biogeochemical cycles in terrestrial ecosystems with a focus on agricultural plant–soil systems. The review aims to explore underlying mechanisms and interactions and to derive implications for ecosystem dynamics and services. Moreover, the negative impacts of human activities on biogeochemical cycles are addressed, and mitigation strategies and sustainable management practices are presented. Key findings reveal that while each cycle operates through distinct processes, their coupling is essential for maintaining ecosystem balance and productivity. The disruptions caused by human activities like industrial agriculture or deforestation pose significant challenges to the stability of these cycles. At the same time, advancements in technology, particularly artificial intelligence, remote sensing, and soil health monitoring, offer transformative opportunities to study and manage these cycles with greater precision and efficiency. These innovations can help to identify hotspots of nutrient/element deficiencies or disruptions, predict ecosystem responses to environmental changes, and guide future research and policy development regarding sustainable management practices.

Abstract

Biogeochemical cycles are fundamental to the functioning of plant–soil systems, driving the availability and transfer of essential nutrients (like carbon (C), nitrogen (N), phosphorus (P), and sulfur (S)) as well as beneficial elements (like silicon (Si)). These interconnected cycles regulate ecosystem productivity, biodiversity, and resilience, forming the basis of critical ecosystem services. This review explores the mechanisms and dynamics of biogeochemical C, N, P, S, and Si cycles, emphasizing their roles in nutrient/element cycling, plant growth, and soil health, especially in agricultural plant–soil systems. The coupling between these cycles, facilitated mainly by microbial communities, highlights the complexity of nutrient/element interactions and corresponding implications for ecosystem functioning and stability. Human activities including industrial agriculture, deforestation, and pollution disrupt the underlying natural processes leading to nutrient/element imbalances, soil degradation, and susceptibility to climate impacts. Technological advancements such as artificial intelligence, remote sensing, and real-time soil monitoring offer innovative solutions for studying and managing biogeochemical cycles. These tools enable precise

Biology **2025**, *14*, 433

https://doi.org/10.3390/biology14040433

nutrient/element management, identification of ecosystem vulnerabilities, and the development of sustainable practices. Despite significant progress, research gaps remain, particularly in understanding the interlinkages between biogeochemical cycles and their responses to global change. This review underscores the need for integrated approaches that combine interdisciplinary research, technological innovation, and sustainable land-use strategies to mitigate human-induced disruptions and enhance ecosystem resilience. By addressing these challenges, biogeochemical processes and corresponding critical ecosystem services can be safeguarded, ensuring the sustainability of plant–soil systems in the face of environmental change.

Keywords: ecosystem services; global change; nutrient dynamics; carbon sequestration; microbial communities; ecosystem resilience; sustainable agriculture

1. Introduction

Biogeochemical cycles are fundamental processes that regulate the flow of essential nutrients (like carbon (C), nitrogen (N), phosphorus (P), and sulfur (S)) as well as beneficial elements (like silicon (Si)) through the biosphere, lithosphere, hydrosphere, and atmosphere. These cycles are crucial for sustaining life on Earth by ensuring the availability of nutrients/elements required for growth, reproduction, and ecosystem functioning [1]. In plant–soil systems, biogeochemical cycling plays a pivotal role by mediating nutrient/element exchange between plants and soil, influencing the composition and productivity of ecosystems. This interplay of biotic and abiotic components underscores the intricate relationships that define ecological stability and resilience [2]. The concept of biogeochemical cycles emerged as scientists sought to understand how nutrients/elements move through natural systems, tracing elemental pathways across different environmental compartments and uncovering their interactions with biological processes [3].

The historical context of research into biogeochemical cycling reveals a growing recognition of its complexity and ecological importance [4]. Early studies primarily focused on individual cycles, such as the C and N cycles, examining their contributions to soil fertility and atmospheric regulation [5]. Over time, advances in molecular biology, isotopic techniques, and remote sensing expanded our ability to investigate these cycles on multiple scales, from microbial processes in soil to global fluxes of C and N. The increasing availability of data has enabled the identification of interconnected feedback loops and the recognition of human-induced disruptions, such as deforestation and industrial agriculture, which have altered natural nutrient/element flows [6–8].

Biogeochemical cycling is indispensable for maintaining ecosystem health, as it supports key functions like primary productivity, nutrient/element recycling, and soil formation [4,9,10]. The nutrient/element exchange facilitated by these cycles sustains plant growth, which, in turn, influences the structure and function of terrestrial and aquatic ecosystems. This relationship is exemplified by the interdependence between plants, soil, and microbial communities [11]. Plants absorb nutrients/elements from the soil to fuel their metabolic activities, while microbial communities decompose organic matter, releasing nutrients/elements back into the soil in forms accessible to plants. This dynamic feedback system ensures the continuity of life-sustaining processes, highlighting the importance of conserving biogeochemical integrity in the face of environmental changes [12].

Understanding the significance of biogeochemical cycles extends beyond their ecological implications, as they are also central to addressing global challenges such as climate change, food security, and biodiversity loss [13]. For instance, the C cycle's role in regulating atmospheric carbon dioxide (CO_2) levels is critical for mitigating climate change [14]. Similarly, the N and P cycles influence agricultural productivity and water quality, underscoring the need for sustainable nutrient management practices. The Si cycle is closely linked to the C cycle on a global scale and the positive effects of Si on plant performance, crop production, and ecosystem functioning are well-documented in the literature [15–17]. Biogeochemical cycles also contribute to the resilience of ecosystems by supporting species diversity and enabling ecosystems to recover from disturbances [18,19]. Therefore, any imbalance in biogeochemical cycling, whether due to natural events or human activities, poses significant risks to ecosystem stability and human well-being [20,21].

This review provides a comprehensive overview of interconnected biogeochemical cycles in terrestrial ecosystems with a focus on agricultural plant–soil systems. The review aims to explore underlying mechanisms and interactions and to derive implications for ecosystem dynamics and services. By delving into the details of selected nutrient/element, i.e., C, N, P, S, and Si, cycles, this review elucidates their roles in maintaining ecosystem health and how they are connected. Additionally, this review seeks to address the negative impacts of human activities on biogeochemical cycles, offering insights into mitigation strategies and sustainable management practices. Through a synthesis of existing knowledge and identification of research gaps, this review finally aims to foster a deeper understanding of biogeochemical C, N, P, S, and Si cycling, guiding future research and policy development in ecosystem conservation and management.

2. A Short Overview of Carbon, Nitrogen, Phosphorus, Sulfur, and Silicon Cycles in Plant–Soil Systems

Biogeochemical cycles are essential for maintaining the balance of nutrients/elements in ecosystems, especially within plant–soil systems where these cycles drive the availability of critical elements [22,23]. C, N, P, S, and Si cycles comprise key processes that facilitate nutrient/element transfer and storage, ensuring that plants and soil organisms receive the necessary components for growth, energy production, and structural development [2,24]. Each of these cycles operates through unique mechanisms and pathways, interacting with other cycles in complex ways that underscore the intricate relationships between biotic and abiotic factors [25,26]. The following sections provide a short examination of each cycle discussing selected underlying mechanisms, its significance in plant nutrition, and specific challenges. For reasons of clarity and consistency, every section is divided into three subsections. In each first subsection, a short nutrient/element profile is given, which briefly summarizes information on the element's natural plant availability, its significance in plant nutrition, molecular transformation processes, element limitations and management, and consequences of land-use for the biogeochemical cycle of this element. In the second and third subsections, selected aspects from the first subsection are then presented in more detail.

2.1. Carbon Cycle

2.1.1. Carbon Profile

(i) Natural plant availability: While the vast majority of inorganic C in terrestrial plants is sequestered from atmospheric CO_2 by photosynthesis, C can also be absorbed (in the form of CO_2, H_2CO_3 (carbonic acid), HCO_3^- (bicarbonate), or CO_3^{2-} (carbonate), depending on soil solution pH) and fixed by roots to some extent [27,28]. (ii) Significance in plant nutrition: Essential. Component of all organic compounds, which represent the basis of all known life. (iii) Transformation processes: Photosynthesis, decomposition, and respiration (see Section 2.1.2). (iv) Limitations and management: Not limited. C sequestration in soil represents a promising pathway of climate change mitigation (see

Section 2.1.2). (v) Consequences of land-use for biogeochemical C cycling: increased CO_2 emissions and decreased C sequestration (see Section 2.1.3).

2.1.2. Carbon Respiration and Sequestration in Soils

The C cycle is fundamental to plant–soil systems, mediating the storage and transfer of C between the atmosphere, soil, and vegetation [29]. While soil respiration by plant roots, microorganisms (bacteria, fungi), and soil animals is a key source of global CO_2 release, CO_2 is also sequestered in soils [30]. On a global scale, the uptake of C from the atmosphere by land plant photosynthesis and the C release rates by soil respiration are about the same size [31]. However, heterotrophic soil respiration has been found to substantially increase, driven by global warming, which might have severe consequences for the global C budget on a short time scale (i.e., up to hundreds of years; on a long time scale, the C budget is assumed to be in equilibrium) [32]. C sequestration in soil occurs through two primary biological mechanisms: photosynthesis and decomposition. During photosynthesis, plants absorb CO_2 from the atmosphere, converting it into organic compounds that are stored in plant tissues [33]. When plants and organic materials decompose, microorganisms break down these compounds, integrating C into soil organic matter (SOM) [34]. This SOM, along with humus, serves as a major reservoir for C, enhancing soil fertility and water retention [35]. Agricultural practices to enhance biological C sequestration in soils comprise, e.g., conservation agriculture, cover cropping, or the recycling of biomass (see Section 4.2 for more details). On a multimillion-year time scale, atmospheric CO_2 concentrations are largely controlled by the removal of CO_2 from the atmosphere through chemical soil weathering [36]. Enhancing this chemical weathering, e.g., by liming, might be another promising approach to increase the uptake of CO_2 from the atmosphere on a global scale [37].

2.1.3. Impact of Land-Use on Carbon Cycling

Land-use significantly disrupts C dynamics in the soil. Deforestation reduces C sequestration by removing plant biomass that stores C, while agricultural practices such as tilling accelerate the decomposition of organic matter, leading to increased C release [38]. Urbanization exacerbates these effects by reducing vegetative cover and compacting soils, further limiting their capacity to store C [39]. All these human activities highly impact the storage of C within plant biomass with consequences for C sequestration in soils and finally the release of C to the atmosphere (see Section 4.2 for a detailed overview). Key processes of C movement in plant–soil systems, i.e., C sequestration, storage, and release, and the impact of land-use are summarized in Figure 1.

2.2. Nitrogen Cycle

2.2.1. Nitrogen Profile

(i) Natural plant availability: While the vast majority of plant available N in soil originates from the conversion of atmospheric dinitrogen (N_2) into ammonia (NH_3) by bacteria (biological N fixation), some atmospheric N_2 is abiologically fixed by lightning (see Section 2.2.2). (ii) Significance in plant nutrition: Essential. As it is a component of diverse metabolites and structural compounds like proteins, nucleic acids, chlorophylls, phytohormones, and secondary metabolites, N plays a vital role in plant physiology. (iii) Transformation processes: N fixation and nitrification/denitrification (see Section 2.2.2). (iv) Limitations and management: Not limited in natural ecosystems, where the N cycle is in equilibrium. However, in agricultural plant–soil systems, the harvest-related substantial annual N losses have to be compensated via N fertilization (see Section 2.2.3). (v) Consequences of land-use for biogeochemical N cycling: emission of the greenhouse gas nitrous oxide (N_2O), deterioration of soil health, and eutrophication (see Section 2.2.3).

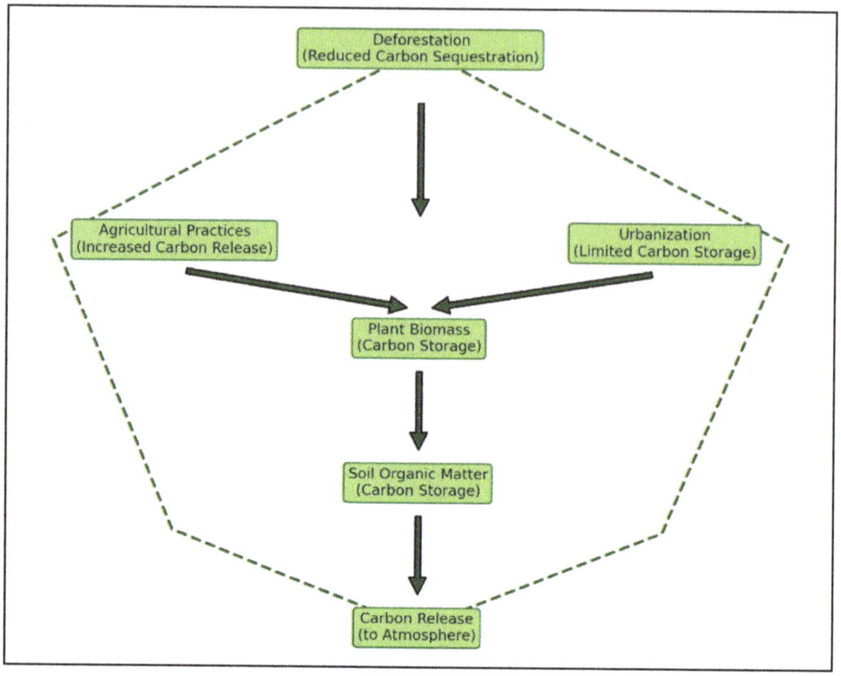

Figure 1. Impact of land-use on C dynamics in plant–soil systems.

2.2.2. Nitrogen Fixation, Nitrification, and Denitrification

The N cycle plays a critical role in plant–soil systems by converting inert atmospheric N into biologically available forms [40]. N fixation is primarily achieved by symbiotic bacteria such as *Rhizobium* (associated with leguminous plants) and free-living bacteria such as *Azotobacter*. These microorganisms transform atmospheric N_2 into NH_3, which plants can absorb [41]. Abiotic processes, including lightning and industrial fixation, also contribute to N availability, albeit to a lesser extent [42]. Lightning provides the energy to produce NO_x (nitrogen oxides) in a reaction of N_2 and oxygen in the atmosphere [43]. When NO_x cools down afterwards it further reacts with oxygen forming nitrogen dioxide (NO_2), which in turn is converted to nitric acid (HNO_3) during chemical reactions with ozone and water [44]. In soil, this acid finally makes NO_3^- (nitrate), a form readily absorbed by plants [45]. Following N fixation, nitrification and denitrification are key processes that regulate the transformation and movement of N in soil [46]. Nitrification involves the microbial oxidation of ammonia to nitrite (NO_2^-) and subsequently to plant available NO_3^-. Conversely, denitrification converts nitrate into gaseous N forms (N_2 or N_2O), releasing it back into the atmosphere [47].

2.2.3. Nitrogen Supply in Agricultural Plant–Soil Systems

On the one hand, intensive N fertilization has led to improved food supply, because significantly more crops can be grown on a limited area of land as N is the main yield-determining nutrient. On the other hand, the oversupply with reactive N (i.e., all N compounds except for N_2) in many agricultural plant–soil systems has resulted in ecological threats like the emission of the greenhouse gas N_2O, deterioration of soil health, or eutrophication (see Section 5.1.) [48,49]. Annual harvest-related N removal from agricultural plant–soil systems has to be replaced to maintain productivity, because there is no potentially available N in the rocks from which most soils developed. To ensure maximal crop yields and minimal N losses in any agricultural plant–soil system, the N fertilization rate should be tailored to the N demand of a specific crop [50]. In this context, synthetic fertilizers are most often the means of choice [51], especially in intensive agricultural systems, because they are readily available, easy to transport and to apply, and relatively cost-effective [52]. During industrial N fertilizer production, N_2 is fixed in the Haber–Bosch process, which uses hydrogen (H_2) from natural gas (CH_4) to react with N_2 under high temperature and pressure to form NH_3. Apart from synthetic N fertilizers, there are numerous other N sources used in agricultural plant–soil systems, for example, manure or compost. Table 1 provides a comparative overview of N sources, highlighting their relative contributions to soil fertility and crop productivity.

Table 1. Comparative overview of N sources and their contribution to soil fertility.

Nitrogen Source	Soil Fertility Contribution	Crop Productivity Contribution	Advantages	Limitations	Reference
Synthetic Fertilizers	High immediate nutrient availability	High yield response	Quick release, tailored compositions	Risk of leaching, environmental harm	[53]
Manure	Slow nutrient release	Moderate to high, depending on quality	Organic matter improvement	Variable nutrient content	[54]
Compost	Slow and steady nutrient release	Moderate	Enhances soil structure	Requires time for production	[55]
Cover Crops	Long-term improvement	Indirect, through soil health	Erosion control, organic matter boost	Requires land-use during growth period	[56]
Leguminous Plants	Biological nitrogen fixation	High in compatible systems	Self-sustaining nitrogen source	Limited to suitable crops	[57]
Biofertilizers	Variable, depends on microbial activity	Variable	Environmentally friendly	Requires optimal conditions	[58]
Organic-Inorganic Mix	Balanced nutrient availability	High yield response	Improves nutrient-use efficiency	Complex management	[59]
High-Quality Organic Resources	Moderate to high	High, particularly in low-fertility soils	Reduces dependency on synthetic inputs	Requires high-quality material	[60]

2.3. *Phosphorus Cycle*

2.3.1. Phosphorus Profile

(i) Natural plant availability: Only inorganic orthophosphates in soils (i.e., $H_2PO_4^-$ and HPO_4^{2-}) are plant available. (ii) Significance in plant nutrition: Essential. P plays a major role in the structural framework of deoxyribonucleic acid (DNA) and ribonucleic acid (RNA), is used for the transport of cellular energy with adenosine triphosphate (ATP), and represents a main structural component of cell membranes in the form of phospholipids. (iii) Transformation processes: Weathering and mineralization (see Section 2.3.2). (iv) Limitations and management: Global P resources are limited, which is why a sustainable use of P fertilizers is urgently needed in future cropping systems (see Section 2.3.3). (v) Consequences of land-use for biogeochemical P cycling: deterioration of soil health and eutrophication (see Section 2.3.3).

2.3.2. Availability of Phosphorus in Soils

Unlike N and C cycles, the P cycle lacks a significant gaseous phase, relying primarily on the weathering of phosphate-bearing rocks to release P into the soil [61]. Once in the soil,

P undergoes mineralization, where organic P is converted into inorganic forms accessible to plants. However, P availability is often limited by its strong affinity for soil particles, which immobilizes it in forms unavailable for uptake [62]. The negatively charged phosphate ions are, for example, adsorbed on positively charged soil constituents such as iron (Fe) and aluminum (Al) oxides, carboxyl groups (organic matter), or silanol groups (clay minerals). Moreover, phosphates are immobilized during mineral precipitation in combination with metals such as calcium (Ca), Fe, and Al [63,64]. Soil amendments such as biochar, rock phosphate, and phosphate-solubilizing microorganisms enhance P accessibility. These interventions are particularly important in agricultural systems where P deficiencies can limit crop yields [30]. Figure 2 illustrates the dynamics of P cycling in plant–soil systems in relation to abiotic and biotic factors.

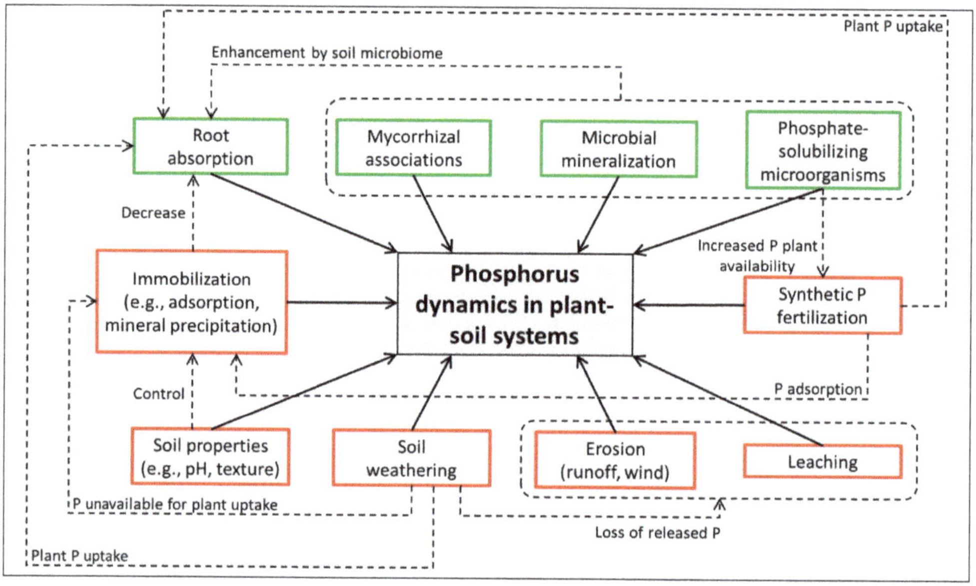

Figure 2. Overview of factors that influence (solid arrows) P dynamics in plant–soil systems. Biotic and abiotic factors are shown in green and red boxes, respectively. Dashed arrows show interactions between biotic and abiotic factors.

2.3.3. Global P Resources and Strategies to Reduce the Need for Synthetic P Fertilizers in Agricultural Plant–Soil Systems

Synthetic P fertilizers are mainly manufactured from mined rock phosphate, which is a geographically restricted and finite resource. Currently, China, Morocco, and the United States are the main phosphate producers worldwide, accounting for about 70% of the total global production [65]. The largest phosphate resources on Earth can be found in the Western Phosphate Fields (United States) and the Mediterranean Sedimentary Phosphate Province (e.g., Morocco, Spain, Algeria, Tunisia) with deposits that are estimated to last for more than 1000 years [66]. Despite these huge resources, factors like growing P demand, geopolitical constraints, increasing prices, and ecological awareness against the background of global change call for a sustainable management of global P resources [67]. In this context, agriculture plays a pivotal role regarding the demand for synthetic P fertilizers as most global phosphates are consumed during the production of these fertilizers. The overuse of synthetic P fertilizers results in a deterioration of soil health and P runoff from agricultural lands to water bodies, where it causes eutrophication (see Section 5.1.). To avoid P fertilizer overuse, farmers should assess the P status (i.e., P plant availability) of their soils as well as specific crop P demands before determining P fertilizer rates. In this context, precision agriculture [68,69] and the use of bio-inoculants (bacteria and fungi) [70] represent promising strategies to increase the P use efficiency of crops.

2.4. *Sulfur Cycle*

2.4.1. Sulfur Profile

(i) Natural plant availability: Sulfate (SO_4^{2-}) as the most common inorganic form of S in soils is readily available to plants. Moreover, atmospheric hydrogen sulfide (H_2S) and sulfur dioxide (SO_2) can also be used by plants as a source of S, if root S uptake is limited [71]. (ii) Significance in plant nutrition: Essential. As S is a component of several specific biomolecules (e.g., amino acids and cofactors), S plays a crucial role in biochemical functioning of all living organisms. (iii) Transformation processes: Weathering, mineralization, and inorganic transformations (see Section 2.4.2). (iv) Limitations and management: As SO_4^{2-} contents in soils greatly vary and intensified agriculture leads to substantial S losses, the need for S fertilization is strongly increasing (see Section 2.4.3). (v) Consequences of land-use for biogeochemical S cycling: acidification of terrestrial and aquatic ecosystems and air pollution (see Section 2.4.3).

2.4.2. Sulfur Transformation Processes

S can be abundantly found in rocks and soil minerals. Rock and mineral weathering, geothermal vents, and volcanic eruptions contribute to the release of S into the environment. Human activities like fossil fuel combustion and agriculture largely influence the biogeochemical S cycle [72]. The content of SO_4^{2-} in soils greatly varies in relation to factors like climate, vegetation, soil processes (e.g., leaching, adsorption–desorption, precipitation, and oxidation-reduction reactions), and soil microbial activity. S cycling in plant–soil systems involves the transformation of S compounds through microbial and chemical processes [73]. In aerobic conditions, microorganisms oxidize S compounds to SO_4^{2-}, the most plant-available form of S [74]. In anaerobic environments, sulfate-reducing bacteria convert sulfate into H_2S, completing the S cycle [75].

2.4.3. Sulfur Management in Agricultural Plant–Soil Systems

S is vital for plant nutrition, as it forms the backbone of essential amino acids like cysteine and methionine. These amino acids are critical for protein synthesis and enzymatic functions. S inputs to agricultural plant–soil systems in the 19th and 20th centuries were mainly in the form of atmospheric wet ("acid rain") and dry depositions, which were driven by industrial SO_2 emissions and had negative environmental impacts, i.e., acidification of terrestrial and aquatic ecosystems and air pollution [76]. Due to stricter emission regulations in many countries worldwide, industrial S emissions are decreasing [72]. However, this emission decrease in turn has increased S deficiencies in soils, and thus the need for S fertilization in intensified cropping systems with environmental consequences that might be comparable to the ones described above [77]. This need for S fertilization is exacerbated by S losses from agricultural soils driven by, e.g., topsoil erosion and leaching [78,79]. To address S deficiencies in agricultural systems, S-rich fertilizers and soil amendments are commonly applied [80]. Table 2 summarizes various S-related soil amendments, their sources, and their impacts on plant health.

Table 2. Overview of S containing soil amendments and corresponding sources, effects on plant–soil systems, and advantages as well as limitations of application.

Soil Amendment	Source	Impact on Plant–Soil System	Advantages	Limitations	Reference
Elemental Sulfur	Naturally mined sulfur deposits	Lowers soil pH, improves nutrient availability	Long-term sulfur supply, pH adjustment	Requires microbial oxidation for effect	[55]
Gypsum ($CaSO_4 \cdot 2H_2O$)	By-product of industrial processes or mined	Supplies calcium and sulfur; improves soil structure	Reduces aluminum toxicity in acidic soils	Limited to soils needing calcium	[54]
Ammonium Sulfate	By-product of fertilizer manufacturing	Rapid sulfur and nitrogen source	Quick nutrient release	Potential to acidify soils	[53]
Sulfur-Coated Urea	Industrially coated nitrogen fertilizer	Slow-release sulfur and nitrogen source	Provides consistent nutrient availability	Expensive to produce	[59]
Organic Matter	Plant residues, manure, compost	Gradual sulfur release through decomposition	Improves soil organic matter and soil fertility	Variable sulfur content	[58]
Biosolids	Treated sewage sludge	Supplies sulfur and organic matter	Recycling waste material	May contain "heavy metals" or contaminants	[81]
Potassium Sulfate	By-product of potash mining	Provides potassium and sulfur	Improves potassium levels	Limited to crops needing potassium	[82]
Sulfuric Acid	Industrial sulfur by-product	Lowers soil pH quickly in alkaline soils	Rapid correction of high soil pH	Risk of over-acidification	[57]

2.5. Silicon Cycle

2.5.1. Silicon Profile

(i) Natural plant availability: Only dissolved silica (i.e., silicic acid, H_4SiO_4) in soils is plant available. (ii) Significance in plant nutrition: Beneficial. As silica accumulation in higher plants has been found to enhance their resistance against abiotic and biotic stress, Si is considered as quasi-essential for plants, especially for plant species from the grass family (Poaceae) [83]. (iii) Transformation processes: Weathering and biosilicification (see Section 2.5.2). (iv) Limitations and management: Substantial Si losses in agricultural plant–soil systems due to annual harvest-related Si exports, especially driven by cereal crop harvesting (see Section 2.5.3). (v) Consequences of land-use for biogeochemical Si cycling: anthropogenic desilication and changes in land–ocean Si fluxes (see Section 2.5.3).

2.5.2. Biosilicification and Its Role in Silicon Cycling

Si has been found to be a beneficial element for plants, enhancing their resistance to stress, and thus positively affecting plant performance and ecosystem functioning, especially under stress conditions [9,84,85]. The original source of bioavailable Si (i.e., dissolved silica in the form of H_4SiO_4) on a geological time scale is mineral weathering. In soils, H_4SiO_4 follows various pathways including abiotic and biotic ones. While abiotic pathways comprise (i) leaching influenced by rainfall and irrigation and (ii) immobilization through soil processes like adsorption, precipitation, and complexation, the biotic pathway is represented by the uptake of H_4SiO_4 by living organisms [86]. The process by which inorganic H_4SiO_4 is utilized by plants, protists, and animals to create biogenic silica (BSi), i.e., amorphous hydrated silica ($SiO_2 \cdot nH_2O$), is known as biosilicification [87]. As BSi is much more soluble compared to silicate minerals, biosilicification and BSi dissolution regulate H_4SiO_4 concentrations in soils over biological time scales [88,89].

In fact, various organisms can accumulate BSi. However, while particular attention has been paid to phytogenic (plants) silica pools [90–92], less is known about protozoic (testate amoebae), protophytic (diatoms), zoogenic (sponges), fungal (fungi), and bacterial (bacteria) Si pools in soils (Table 3) [93]. While protozoic Si pools in forest soils are relatively small compared to phytogenic ones, biosilicification rates of testate amoebae can equal or even exceed the annual Si uptake rates of trees in terrestrial ecosystems [94,95]. As testate amoebae belong to the earliest colonizers of new areas [96], they might also play a crucial role in establishing biological Si cycling in initial ecosystems [95,97]. Protophytic and zoogenic Si pools in soils were quantified in few studies, but there are no quantitative data for corresponding biosilicification rates by diatoms and sponges in terrestrial ecosystems, respectively [98,99]. Regarding fungal and bacterial biosilicification, there is no quantitative data (Si pools, annual Si uptake rates) in the literature to the best of our knowledge.

Table 3. Information on biosilicification (Si pools and annual Si uptake rates) by testate amoebae, diatoms, and sponges in soils of terrestrial ecosystems. Note different units for Si pool sizes.

Year	Ecosystem	Organism	Si Pool Size	Biosilicification Rate	Reference
2007	Various forests	Testate amoebae	Up to 0.8 kg Si ha^{-1}	Up to 106 kg Si ha^{-1} yr^{-1}	[100]
2013	Beech forest 'Beerenbusch'	"	1.9 kg Si ha^{-1}	17 kg Si ha^{-1} yr^{-1}	[101]
2014	Initial ecosystem states (different artificial catchments)	"	Up to 0.7 kg Si ha^{-1}	Up to 16 kg Si ha^{-1} yr^{-1}	[97]
2015	Various forests	"	Up to 4.7 kg Si ha^{-1}	Up to 80 kg Si ha^{-1} yr^{-1}	[94]
2016	Initial ecosystem states (artificial catchment 'Chicken Creek')	"	Up to 0.06 kg Si ha^{-1}	--	[98]
"	"	Diatoms	Up to 0.3 kg Si ha^{-1}	--	"
"	"	Sponges	Up to 0.2 kg Si ka^{-1}	--	"
"	Various habitats in a nature reserve (artificial catchment 'Mere Sands Wood')	Testate amoebae	Up to 82 ng Si g^{-1} dm	--	[102]
"	"	Diatoms	Up to 58 ng Si g^{-1} dm	--	"
2017	Initial ecosystem states (artificial catchment 'Chicken Creek')	Testate amoebae	Up to 0.4 kg Si ha^{-1}	--	[99]
"	"	Diatoms	Up to 1.6 kg Si ha^{-1}	--	"
"	"	Sponges	Up to 0.5 kg Si ha^{-1}	--	"
2020	Floodplain (Watarase retarding basin)	Testate amoebae	Up to 2.9 µg Si g^{-1} dm	--	[103]

Table 3. *Cont.*

Year	Ecosystem	Organism	Si Pool Size	Biosilicification Rate	Reference
"	"	Diatoms	Up to 12.8 μg Si g^{-1} dm	--	"
"	Peatland and cropland sites (Dajiuhu National Wetland Park)	Testate amoebae	Up to 5.3 μg Si g^{-1} dm	--	[104]
2021	Natural and cultivated *Sphagnum* sites	"	Up to 0.1 μg Si per 150 testate amoeba shells	--	[105]
2022	Various peatlands	"	Up to 97 ng Si per 150 testate amoeba shells	--	[106]

" = ditto; -- = no data available; dm = dry mass.

2.5.3. The Threat of Anthropogenic Desilication and How to Prevent It

Human activities like deforestation and damming have profound impacts on Si cycling in terrestrial ecosystems with consequences for Si fluxes from land to oceans, and thus for the global biogeochemical Si cycle (see Section 5.1). Moreover, annual Si exports by crop harvesting have been recognized to lead to substantial Si losses in agricultural plant–soil systems known as anthropogenic desilication [107–109]. Certain agricultural practices may enhance Si availability (H_4SiO_4) in soils like the use of prescribed burning [110], the application of Si-rich fertilizers and/or amorphous silica [17,111], the application of biochar [112], the recycling of crop residues [113,114], and liming [115]. However, further research is needed to determine how much of the released H_4SiO_4 is (i) taken up by plants, (ii) immobilized through adsorption and complexation processes in the soil, or (iii) lost through leaching.

All these practices can prevent anthropogenic desilication and ensure a sufficient Si supply for plants leading to an improved crop stress resilience [85,111]. Additionally, practices like biochar application or crop residue recycling can promote C sequestration in agricultural soils by enhancing weathering (e.g., using silicate rock powders), stabilizing soil organic C (e.g., through crop residue recycling or biochar application), and increasing phytolith-occluded C storage (e.g., by enhancing cereal crop production and corresponding residue recycling) [116,117].

3. Interactions Between Biogeochemical Cycles

Biogeochemical cycles rarely operate in isolation. Instead, they are interconnected, with processes in one cycle influencing and being influenced by others. This interconnectedness enhances the complexity of nutrient/element dynamics within plant–soil systems, as each cycle interacts with others to maintain ecosystem stability and productivity [118]. The coupling between cycles such as C, N, and P, along with the pivotal role played by microbial communities, creates a dynamic feedback network. These interactions determine nutrient/element availability, ecosystem resilience, and plant productivity [119]. A deeper understanding of these interactions is critical for managing ecosystem sustainably, particularly in the context of anthropogenic disruptions like land-use change and climate variability [120].

3.1. Underlying Mechanisms

The coupling of biogeochemical cycles refers to the feedback mechanisms that connect processes within different cycles [121]. For instance, the C, N, and P cycles are intricately linked through plant growth, microbial activity, and soil processes [122]. Photosynthesis, a key process in the C cycle, requires N for chlorophyll synthesis and P for energy transfer via ATP molecules [123]. N fixation, a major process in the N cycle, is often influenced by the availability of organic C, which serves as an energy source for N-fixing bacteria [124]. Similarly, P availability impacts microbial activity and root exudates, further influencing C dynamics in the soil [125]. Since S is also an essential element for organisms (e.g., as a constituent of many proteins and cofactors), the S cycle is closely linked to the other biogeochemical cycles by biological productivity as well. The global Si and C cycles are closely linked by weathering processes and diatom growth in the oceans [126,127]. While soil weathering principally removes CO_2 from the atmosphere, tectonic activities can expose ancient organic C that was sequestered from the atmosphere over millions of years and causes CO_2 release when this organic C is oxidized [128]. Finally, volcanic eruptions worldwide emit gases rich in C and S, i.e., mainly CO_2 and SO_2, but also N species such as NO_x into the atmosphere [129,130], while volcanic ash is rich in Si (volcanic glass) and also contains certain amounts of P [131,132]. Gaseous S and N species in the atmosphere (i) are transformed into aerosol acid particles (SO_4^{2-} and NO_x) that eventually fall to the ground in the form of dry depositions or (ii) react with water, oxygen, and other chemicals to form sulfuric (H_2SO_4) and nitric (HNO_3) acids, which are then deposited via precipitation (wet depositions) [130,133].

However, at this point, it should be noted that the processes stated above naturally represent only some examples of biogeochemical cycle connections, because the elements C, N, P, S, and Si are quite abundant in the Earth's system, and corresponding interactions between the Earth's spheres (i.e., the litho-, hydro-, bio-, and atmosphere) are just too complex to be depicted in full detail. While all these natural processes are globally in equilibrium over a long time scale, human activities substantially disturb these cycles over a short time scale (i.e., decades to hundreds of years) with far reaching consequences for element cycling and availability in plant–soil systems, and thus for ecosystem resilience and plant productivity. Figure 3 provides an overview of natural key processes (e.g., photosynthesis, N fixation, and rock weathering) and human disturbances, which are driving nutrient/element exchange between the Earth's spheres.

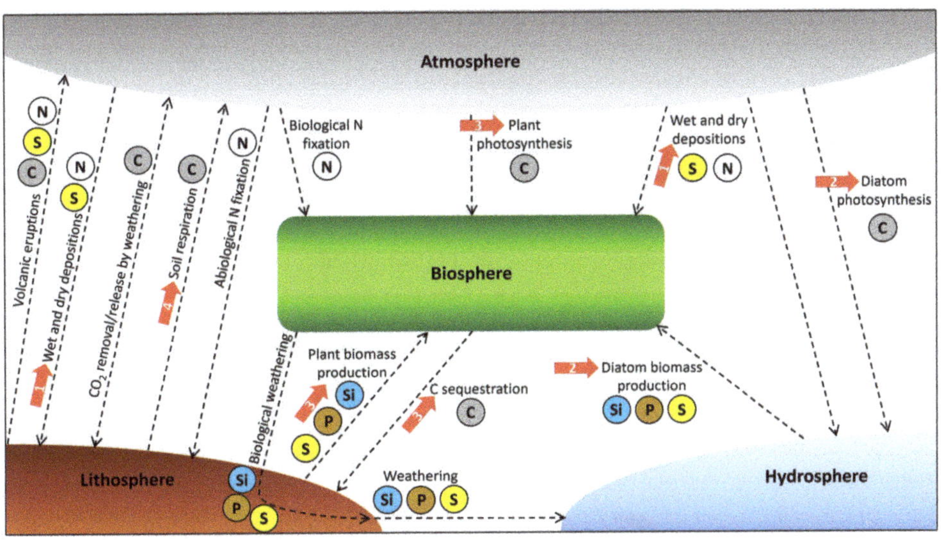

Figure 3. Schematic overview of natural key processes that link biogeochemical C, N, P, S, and Si cycles on a global scale. Dashed arrows indicate directional interactions between biogeochemical cycles

such as photosynthesis, N fixation, weathering, or C sequestration, which connect the different Earth's spheres (i.e., the litho-, hydro-, bio-, and atmosphere). Biogeochemical cycles that are mainly involved in these interactions are indicated by colored element symbols. Principle element flux directions are indicated by arrow directions. Red solid arrows show disturbances resulting from human activities. These disturbances are represented by numbers in the red arrows: 1 = fossil fuel combustion, 2 = eutrophication, 3 = land-use (agriculture, deforestation), and 4 = global warming. Note that the Earth's spheres naturally are inextricably linked with each other and that the separated visualization in this figure is for illustration purposes only.

The feedback mechanisms of biogeochemical cycles have profound implications for plant productivity and ecosystem functioning. When nutrient/element availability is balanced across these cycles, plants can optimize growth and yield. However, imbalances, such as N or P limitations, can disrupt C sequestration and reduce ecosystem productivity [134]. For example, excess N from fertilizers can lead to P depletion, altering microbial communities and reducing soil health (see also Section 4.3) [135].

3.2. *Role of Microbial Communities*

Microbial communities are central to the interactions between biogeochemical cycles, acting as catalysts for nutrient/element transformations and exchange. These communities engage in symbiotic relationships with plants, facilitating nutrient/element acquisition and cycling [136]. For example, N-fixing bacteria like *Rhizobium* form symbiotic associations with leguminous plants, converting atmospheric N into ammonia, which the plants can use [137]. Similarly, mycorrhizal fungi establish mutualistic relationships with plant roots, enhancing P uptake in exchange for C from the plant [138]. Modern molecular biological techniques are well-suited to understand the underlying processes of the cross-talk between plants and soil microorganisms, and thus to regulate these interactions for the benefit of plant nutrition [139–142].

The enzymatic processes carried out by microbes are critical for driving nutrient/element availability and cycling [143]. Microbial enzymes decompose organic matter, releasing N, P, and C compounds into the soil [144]. Specific microbial species also mediate processes such as nitrification, denitrification, and P solubilization, bridging the gaps between different biogeochemical cycles. Furthermore, certain bacteria (e.g., *Proteus mirabilis*) and fungi can enhance the dissolution of silica via acidic metabolites (bio-weathering) [87]. Diatoms might also play a role in enhancing bio-weathering [145], while there is no information on this aspect regarding testate amoebae.

These activities not only sustain plant growth but also regulate nutrient/element losses and emissions, such as N leaching or nitrous oxide release [146]. To highlight their contributions, Table 4 presents a summary of key microbial species involved in nutrient cycling, their functional roles, and the processes they drive. For instance, *Azotobacter* and *Bradyrhizobium* are pivotal in N fixation, while phosphate-solubilizing bacteria like *Pseudomonas* and *Bacillus* enhance P availability. By bridging cycles, microbes play a crucial role in maintaining nutrient balance and ensuring the resilience of plant–soil systems [147]. The intricate interactions between biogeochemical cycles, facilitated by microbial communities, underscore the complexity of nutrient dynamics in ecosystems. These interactions not only regulate the flow of essential nutrients but also enhance ecosystem productivity and resilience [148].

Table 4. Key microbial species and their functional roles in nutrient cycling.

Microbial Species (Scientific Name)	Functional Role	Nutrient Cycling Process	Key Outputs/Impacts	Reference
Azotobacter vinelandii	Nitrogen fixation	Atmospheric N_2 → Ammonia	Enhances soil nitrogen availability for plants.	[149]
Bradyrhizobium japonicum	Symbiotic nitrogen fixation	Forms nodules on legumes	Supplies nitrogen directly to host plants.	[150]
Pseudomonas fluorescens	Phosphate solubilization	Converts insoluble phosphorus	Increases bioavailability of phosphorus for plant uptake.	[151]
Bacillus subtilis	Phosphate solubilization	Organic phosphorus mineralization	Supports plant growth by enhancing soil phosphorus levels.	[152]
Nitrosomonas europaea	Nitrification	Ammonia → Nitrites	Facilitates conversion of nitrogen into usable forms, influencing nutrient cycling.	[153]
Nitrobacter winogradskyi	Nitrification	Nitrites → Nitrates	Ensures availability of nitrate for plant uptake but increases leaching risks.	[154]
Paraburkholderia phytofirmans	Plant growth promotion	Enhances phosphorus and nitrogen	Improves nutrient acquisition, fostering plant growth.	[155]
Rhizobium leguminosarum	Symbiotic nitrogen fixation	Forms nodules on legumes	Converts atmospheric nitrogen for host plants, improving soil fertility.	[156]
Frankia spp.	Nitrogen fixation in actinorhizal plants	Atmospheric N_2 → Ammonia	Supports nitrogen levels in non-leguminous plants.	[157]
Desulfovibrio desulfuricans	Sulfate reduction	Sulfate (SO_4^{2-}) → Hydrogen sulfide (H_2S)	Contributes to sulfur cycling in anaerobic environments, impacting soil and water chemistry.	[158]
Thiobacillus thioparus	Sulfur oxidation	Elemental sulfur → Sulfate (SO_4^{2-})	Increases soil sulfate levels, promoting plant sulfur uptake.	[159]

4. Implications for Ecosystem Services

Biogeochemical cycles are not just fundamental processes for nutrient/element dynamics but also pivotal drivers of ecosystem services, which encompass the benefits that ecosystems provide to humanity. These cycles directly influence soil fertility, C sequestration, climate mitigation, biodiversity, and ecosystem resilience [160]. By mediating the availability of essential nutrients and beneficial elements, they support primary productivity, regulate greenhouse gas fluxes, and maintain ecosystem stability under varying environmental conditions [161]. Understanding and managing the implications of these cycles is crucial for promoting sustainable agricultural practices, enhancing climate resilience, and preserving biodiversity.

4.1. Soil Fertility and Crop Production

Soil fertility, a cornerstone of agricultural productivity, is intrinsically tied to the efficiency of biogeochemical cycles. These cycles regulate the availability of macronutrients like N, P, and potassium (K), which are critical for crop growth and yield [162]. Sustainable practices such as crop rotation, cover cropping, and the use of organic fertilizers enhance nutrient cycling, ensuring a steady supply of these nutrients [163]. For example, N fixation by legumes in rotation systems replenishes soil N, reducing the need for synthetic fertilizers and improving soil health [164]. Continuous crop straw recycling has been found to replenish plant available Si in agricultural soils and to a reduce the need for N fertilization rates by about 69% in the long term [113]. Additionally, regular long-term incorporation of crop straw is a promising strategy to alleviate soil erosion [165] and to enhance C sequestration in agricultural soils [116,166]. As dissolved Si and P compete for equivalent adsorption sites in soil, Si supply can also help to release previously plant unavailable phosphates in agricultural soils resulting in a reduced need for P fertilization [111]. Due to the manifold reported positive effects of Si for agricultural soils, Si supply has been suggested as crucial for soil health, especially in plant–soil systems that are prone to drought and soil degradation [167].

The impact of biogeochemical processes on yield is profound. Proper nutrient/element cycling ensures optimal root development, photosynthesis, and energy transfer within plants, translating into higher crop productivity [168]. Conversely, disruptions in these cycles, often caused by intensive farming or soil degradation, can lead to nutrient/element imbalances, reduced yields, and long-term soil infertility [169]. Figure 4 presents pathways of nutrient/element transfer from soil to crops, highlighting the role of sustainable practices in enhancing these processes.

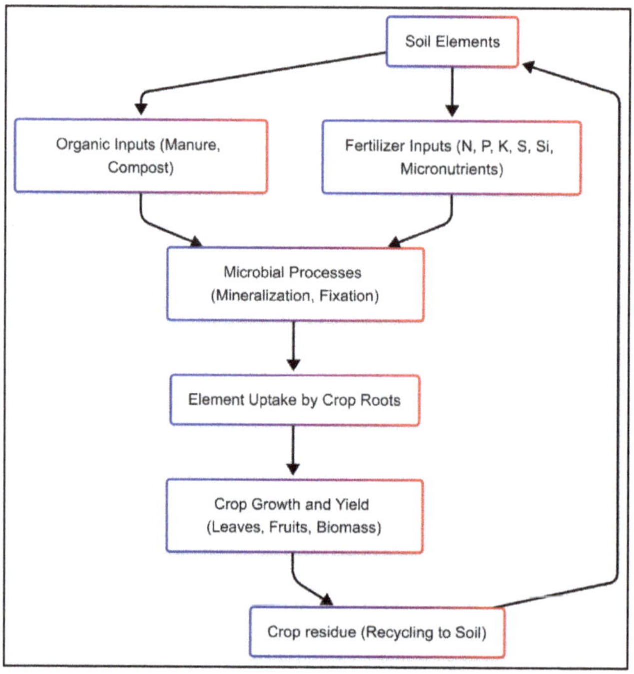

Figure 4. Nutrient/element transfer pathway from soil to crops.

4.2. Carbon Sequestration and Climate Mitigation

Soil serves as a major C sink, playing a critical role in mitigating climate change by sequestering atmospheric CO_2 [170]. Biogeochemical processes such as photosynthesis and the decomposition of organic matter drive soil C storage, with SOM acting as a stable reservoir of C [171]. Enhanced C sequestration not only reduces greenhouse gas concentrations but also improves soil structure, water retention, and fertility, creating a positive feedback loop for agricultural productivity [172].

Strategies for maximizing soil C storage include conservation tillage, agroforestry, and the application of biochar. These practices reduce soil disturbance, increase organic C inputs, and promote microbial activity, which stabilizes C in the soil [173]. Moreover, crop straw recycling has been found a promising strategy to replenish plant available Si (H_4SiO_4) and to enhance C sequestration in agricultural soils in the long term [113,114]. In this context, a certain amount of C is stored in phytoliths (i.e., silica bodies formed in terrestrial plants), which are transferred to soil with plant litter. As phytoliths are relatively stable and can remain in soils up to centuries or even millennia, C sequestration via soil phytoliths has been recently discussed as a promising long-term C sink [174,175]. Table 5 provides a comparative analysis of soil management practices, outlining their potential for C sequestration and their co-benefits for soil health and ecosystem services. By adopting such strategies, farmers and land managers can contribute to both climate mitigation and sustainable development goals.

Table 5. Overview of soil management practices and corresponding C sequestration potentials.

Soil Management Practice	Description	Carbon Sequestration Potential	Impact on Soil Health	Reference
Conservation Tillage	Reduced tillage to minimize soil disturbance.	Moderate to high	Improves soil structure, reduces erosion, and enhances organic matter retention.	[149]
Cover Cropping	Planting cover crops during off-season periods.	High	Increases organic carbon inputs and reduces nutrient leaching.	[176]
Compost Addition	Application of compost to soils.	High	Enhances microbial activity, nutrient availability, and organic carbon.	[177]
Agroforestry	Integration of trees with agricultural crops.	Very high	Promotes biodiversity, reduces soil erosion, and increases carbon storage.	[178]
Biochar Amendment	Adding pyrolyzed biomass to soil.	High	Increases soil carbon stability, improves water retention, and supports microbial growth.	[179]
Crop straw recycling	Application of chopped straw to soil.	Moderate to high (long-term effects)	Replenishes plant available Si, reduces the need for N fertilizers, and increases organic carbon inputs in the long term.	[113]
Crop Rotation	Alternating crops to improve soil nutrient balance.	Moderate	Reduces pest buildup, enhances nitrogen use efficiency, and improves soil structure.	[154]
Integrated Livestock Management	Combining livestock and crop systems.	Moderate to high	Enhances nutrient recycling and boosts organic matter input through manure.	[155]

Table 5. *Cont.*

Soil Management Practice	Description	Carbon Sequestration Potential	Impact on Soil Health	Reference
No-Tillage	Avoiding plowing entirely to maintain soil integrity.	High	Reduces erosion, improves water infiltration, and increases organic matter retention.	[156]
Perennial Grass Systems	Using perennial grasses for soil coverage.	Very high	Reduces erosion, improves soil structure, and enhances long-term carbon storage.	[157]

4.3. Biodiversity and Ecosystem Resilience

Biodiversity and ecosystem resilience are deeply influenced by the efficiency of nutrient/element cycling within ecosystems [9,180]. Healthy biogeochemical cycles promote plant diversity by ensuring the availability of diverse nutrients/elements needed for different species. This diversity, in turn, supports complex food webs and enhances ecosystem functionality [181]. For example, P cycling facilitates the growth of nutrient-demanding species, while N cycling supports legumes and other N-fixing plants, contributing to species coexistence and diversity [182]. Si uptake by plants might play an important role in influencing plant community assembly and ecosystem structure, thus affecting plant biodiversity patterns and ecosystem functioning as well as resilience [183].

In nutrient-poor soils, resilience mechanisms such as microbial symbioses and adaptive root systems play a crucial role in maintaining ecosystem stability. In this context, changes in the root architecture, the formation of cluster roots, or symbiotic associations with mycorrhizae or N-fixing bacteria represent key processes [184]. These mechanisms allow plants to access limited nutrients, ensuring ecosystem productivity even under stressful conditions [185]. Regarding crop production, the phenotypic characterization of root adaptations allows plant breeders to develop improved cultivars, which can ensure yield stability and nutritional security under global change [186]. Moreover, the application of Si to soils has been found to be a promising strategy to enhance the resilience of soil microbial communities by, e.g., changing soil pH, improving nutrient and water availability, altering root exudation patterns and plant physiology, and stimulating the abundance, diversity, and functional potential of key microbial groups [187]. Si uptake by legumes (Fabaceae) might also promote the symbiotic interactions between N-fixing bacteria (rhizobia) inside the legume root nodules and their host plant [188]. Furthermore, enhancing the biological N fixation by introducing selected, adapted diazotrophic bacteria into agricultural soils might be a promising approach to significantly reduce the high demand for synthetic N fertilizers in future cropping systems [189]. Figure 5 provides an overview of how nutrient/element cycling enhances ecosystem resilience, showing pathways through which biogeochemical cycles support biodiversity and mitigate stress. This resilience is vital for ecosystem recovery from disturbances, such as droughts or human interventions, underscoring the need to conserve these natural processes [190]. The implications of biogeochemical cycles for ecosystem services are far-reaching, influencing food security, climate stability, and biodiversity conservation [191].

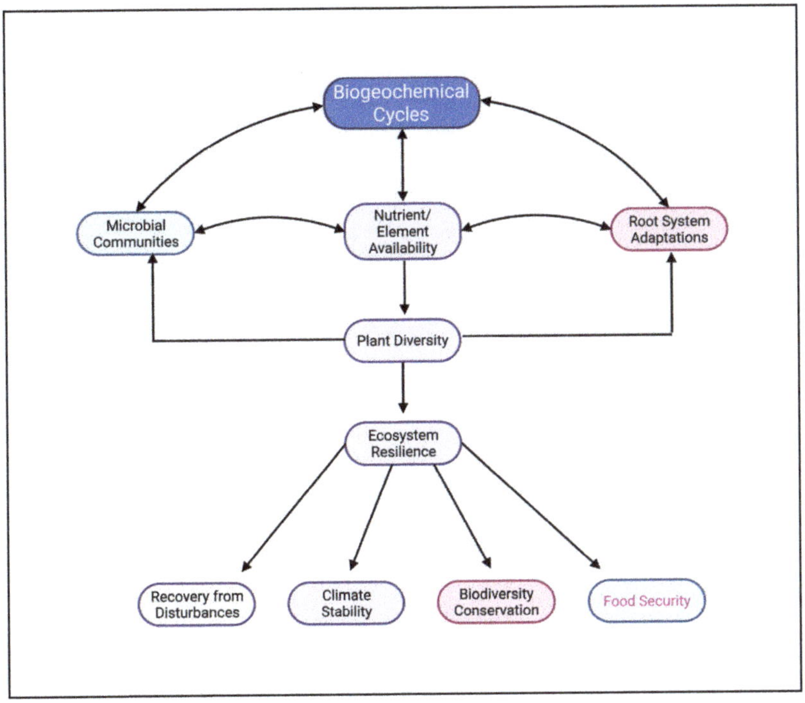

Figure 5. Interactions between biogeochemical cycles, plant biodiversity, and ecosystem resilience in plant–soil systems.

5. Challenges and Future Perspectives

The intricate functioning of biogeochemical cycles is increasingly under threat from human activities, posing significant challenges to ecosystem stability and the delivery of critical ecosystem services [192]. Addressing these challenges requires a combination of mitigation strategies, technological innovations, and forward-thinking research.

5.1. Anthropogenic Impacts on Biogeochemical Cycles

Industrial agriculture and urban development have profoundly disrupted natural biogeochemical cycles, altering nutrient/element flows and ecosystem dynamics [193]. Intensified agriculture often characterized by an overuse of synthetic fertilizers and monocropping has led to imbalances in N, P, and Si cycles with far-reaching consequences [18,109]. In general, a continuous overuse of synthetic fertilizers can decline the contents of SOM, change soil pH, reduce soil fertility as well as microbial activity, and thus result in an overall decrease in agricultural soil quality/health [194,195]. More specifically, excessive N application, for instance, contributes to soil acidification, nitrate leaching, and the release of N_2O, a potent greenhouse gas [196]. Similarly, P runoff from agricultural lands causes eutrophication in freshwater systems, leading to harmful algal blooms and biodiversity loss. Land-use change (e.g., deforestation) has been found to substantially affect Si pools in terrestrial ecosystems with consequences for Si fluxes from land to ocean [8,197,198]. Further human activities like damming also influence the global biogeochemical Si cycle and thus the growth of various organisms (e.g., diatoms) in aquatic ecosystems, as they need dissolved Si to form their siliceous shells [199,200]. Beyond agriculture, urbanization and industrial activities further exacerbate these issues by increasing C emissions and thus global warming, which in turn has strong effects on global biogeochemical cycles [200–202].

Pollution is another major driver of biogeochemical cycle disruption. Chemical pollutants, such as "heavy metals" (metal(loid)s) and pesticides, affect microbial communities and soil health, impairing the efficiency of nutrient cycling processes [203]. Eutrophication, caused by nutrient overloading, particularly N and P, depletes oxygen levels in aquatic systems, creating dead zones and disrupting aquatic ecosystems [204]. Table 6 provides an overview of major anthropogenic threats to biogeochemical cycles, alongside potential mitigation strategies, such as precision agriculture, sustainable land management, and pollution control measures.

Table 6. Overview of major anthropogenic threats and mitigation strategies.

Threat	Impact on Biogeochemical Cycles	Potential Mitigation Strategies	Reference
Excessive Fertilizer Use	Disrupts nitrogen and phosphorus cycles; causes eutrophication.	Precision agriculture, optimized fertilizer application, and crop-specific nutrient management.	[149]
Deforestation	Reduces carbon sequestration and alters nitrogen and silicon cycling.	Reforestation, afforestation, and agroforestry practices.	[198,205]
Desilication	Loss of Si from agricultural plant–soil systems.	Crop straw recycling, application of amorphous silica.	[111,113]
Industrial Pollution	Releases "heavy metals" and toxic compounds, affecting microbial activity and soil health.	Pollution control measures, phytoremediation, and stricter industrial regulations.	[206,207]
Urbanization	Alters land-use, leading to loss of soil organic matter and nutrient imbalances.	Urban green spaces, soil restoration projects, and sustainable urban planning.	[208]
Overgrazing by Livestock	Depletes soil nutrients and increases erosion, disrupting nutrient cycling.	Rotational grazing, controlled stocking rates, and land rehabilitation.	[209]
Waste Mismanagement	Accumulation of organic waste disrupts carbon and nitrogen cycles.	Composting, recycling, and waste-to-energy technologies.	[154]
Mining Activity	Causes soil degradation and disrupts phosphorus and sulfur cycles.	Land reclamation, sustainable mining practices, and ecosystem restoration.	[155]
Climate Change	Accelerates nutrient leaching and alters carbon, nitrogen, and water cycles.	Carbon capture technologies, renewable energy, and climate-smart agriculture.	[156]
Aquatic Pollution	Disturbs nutrient cycling in water bodies, leading to hypoxia.	Wetland restoration, buffer strips, and controlled effluent discharge.	[157]

5.2. Technological Advancements

The advent of advanced technologies offers promising solutions for studying and managing biogeochemical cycles more effectively. Artificial intelligence (AI) and remote sensing technologies have revolutionized our ability to monitor and model nutrient dynamics across scales [210]. AI algorithms can process vast datasets from satellite imagery, ground sensors, and climate models, providing insights into patterns of nutrient flows, hotspots of disruption, and potential areas for intervention. Remote sensing, coupled with geospatial analysis, enables the mapping of nutrient deficiencies and soil health metrics at regional and global scales, facilitating targeted and efficient management strategies [211].

Moreover, remote sensing can detect biomass heterogeneities at landscape scales, offering a promising tool for quantifying lifelike Si or C aboveground plant stocks, which are directly related to soil properties and nutrient/element availability [212,213].

Advancements in soil health monitoring techniques further complement these technological strides. Tools such as automated soil sensors and molecular diagnostics allow for real-time analysis of soil properties, including nutrient content, microbial activity, and organic matter levels. These innovations not only enhance the precision of nutrient management but also help predict the impacts of land-use changes and climate variability on biogeochemical cycles [214,215].

5.3. Research Gaps and Future Directions

Despite significant progress in understanding biogeochemical cycles, critical knowledge gaps remain, particularly regarding their interlinkages and responses to global change. For instance, while individual biogeochemical cycles are well-studied, the feedback loops and synergies between these cycles require further investigation. Understanding how disruptions in one cycle affect others, particularly under scenarios of climate change and land-use transformation, is essential for developing holistic management approaches. To mitigate anthropogenic disturbances of biogeochemical cycles, natural resources should be the focus of future research. In this context, closing element cycles in agricultural plant–soil systems to the highest extent possible (e.g., by crop residue recycling), using the potentials of (adapted) soil microorganisms and improved plant cultivars, and applying soil conservation methods might be the means of choice.

Future research agendas must prioritize the integration of multiple disciplines to address underlying complexities. This includes coupling biogeochemical research with ecosystem modeling, socio-economic studies, and policy analysis to develop comprehensive frameworks for sustainable management. Additionally, long-term field experiments and multi-scale studies are needed to capture the variability of biogeochemical processes and their responses to both natural and anthropogenic drivers. There is also a pressing need to explore innovative solutions, such as microbial engineering and biogeochemical restoration, to enhance nutrient/element cycling and mitigate the impacts of human activities. Addressing the challenges facing biogeochemical cycles requires a multi-pronged approach that combines the mitigation of anthropogenic impacts, leveraging technological advancements, and closing critical research gaps.

6. Conclusions

The intricate and interdependent nature of biogeochemical cycles forms the backbone of nutrient/element dynamics in plant–soil systems, shaping ecosystem health, productivity, and resilience. This review highlights the critical roles played by the C, N, P, S, and Si cycles, emphasizing their unique mechanisms and interactions. These cycles govern the availability and movement of essential nutrients/beneficial elements, sustaining plant growth, microbial activity, and soil health. Key findings reveal that while each cycle operates through distinct processes, their coupling is essential for maintaining ecosystem balance and productivity. The influence of these cycles extends beyond nutrient/element dynamics, impacting climate regulation, biodiversity conservation, and agricultural sustainability.

The disruptions caused by human activities, including industrial agriculture, deforestation, and pollution, pose significant challenges to the stability of these cycles. Excessive nutrient inputs, eutrophication, and C imbalances have led to ecosystem degradation and reduced functionality, emphasizing the urgent need for mitigation strategies. At the same time, advancements in technology, particularly AI, remote sensing, and soil health monitoring, offer transformative opportunities to study and manage these cycles with greater precision and efficiency. These innovations can help to identify hotspots of nutrient/element deficiencies or disruptions, predict ecosystem responses to environmental changes, and guide sustainable management practices.

The implications of biogeochemical cycles for plant–soil systems underscore the necessity of integrated approaches in both research and practice. The interconnectedness of these cycles demands a holistic perspective that considers their feedback mechanisms, synergies, and cascading effects. Future studies must prioritize interdisciplinary research that bridges biology, geology, chemistry, and environmental science to develop comprehensive models of nutrient/element dynamics. These models should include land-use policies, conservation strategies, and agricultural practices to address current challenges and build resilience against future disruptions.

Plant–soil system sustainability hinges on our ability to understand and manage biogeochemical cycles effectively. By synthesizing knowledge across disciplines, leveraging technological advancements, and implementing integrated approaches, the functionality of these cycles and corresponding ecosystem services can be safeguarded. Such efforts are vital for mitigating the impacts of climate change, enhancing food security, and preserving biodiversity, ensuring a sustainable future for both natural ecosystems and human societies.

Author Contributions: Conceptualization, W.Z. and D.P.; writing—original draft preparation, writing—review and editing, visualization, W.Z., A.A. and D.P.; supervision, D.P. All authors have read and agreed to the published version of the manuscript.

Funding: The authors declare that no funds, grants, or other support were received during the preparation of this manuscript.

Data Availability Statement: No data were used for the research described in the article.

Conflicts of Interest: The authors have no relevant financial or non-financial interests to disclose.

References

1. Luo, L.; Meng, H.; Gu, J. Microbial extracellular enzymes in biogeochemical cycling of ecosystems. *J. Environ. Manag.* **2017**, *197*, 539–549. [CrossRef] [PubMed]
2. Janes-Bassett, V.; Davies, J.; Rowe, E.; Tipping, E. Simulating long-term carbon nitrogen and phosphorus biogeochemical cycling in agricultural environments. *Sci. Total Environ.* **2020**, *714*, 136599. [CrossRef] [PubMed]
3. Auguères, A.-S.; Loreau, M. Can Organisms Regulate Global Biogeochemical Cycles? *Ecosystems* **2015**, *18*, 813–825. [CrossRef]
4. Beare, M.; Coleman, D.; Crossley, D.; Hendrix, P.; Odum, E. A hierarchical approach to evaluating the significance of soil biodiversity to biogeochemical cycling. *Plant Soil* **1995**, *170*, 5–22. [CrossRef]
5. Lu, X.; Vitousek, P.; Mao, Q.; Gilliam, F.; Luo, Y.; Turner, B.; Zhou, G.; Mo, J. Nitrogen deposition accelerates soil carbon sequestration in tropical forests. *Proc. Natl. Acad. Sci. USA* **2021**, *118*, e2020790118. [CrossRef]
6. Xu, Z. *Unravelling the Biogeochemical Cycles of Carbon and Nutrients in Forest Ecosystems: Innovative Approaches with Advanced Stable Isotope and NMR Techniques*; Springer: Berlin/Heidelberg, Germany, 2009. [CrossRef]
7. Di, H.; Condron, L.; Frossard, E. Isotope techniques to study phosphorus cycling in agricultural and forest soils: A review. *Biol. Fertil. Soils* **1997**, *24*, 1–12. [CrossRef]
8. Struyf, E.; Smis, A.; Van Damme, S.; Garnier, J.; Govers, G.; Van Wesemael, B.; Conley, D.J.; Batelaan, O.; Frot, E.; Clymans, W.; et al. Historical land use change has lowered terrestrial silica mobilization. *Nat. Commun.* **2010**, *1*, 129. [CrossRef]
9. Katz, O.; Puppe, D.; Kaczorek, D.; Prakash, N.B.; Schaller, J. Silicon in the Soil-Plant Continuum: Intricate Feedback Mechanisms within Ecosystems. *Plants* **2021**, *10*, 652. [CrossRef]
10. Schaller, J.; Puppe, D.; Kaczorek, D.; Ellerbrock, R.; Sommer, M. Silicon Cycling in Soils Revisited. *Plants* **2021**, *10*, 295. [CrossRef]
11. Paredes, S.; Lebeis, S. Giving back to the community: Microbial mechanisms of plant–soil interactions. *Funct. Ecol.* **2016**, *30*, 1043–1052. [CrossRef]
12. Arcand, M.; Helgason, B.; Lemke, R. Microbial crop residue decomposition dynamics in organic and conventionally managed soils. *Appl. Soil Ecol.* **2016**, *107*, 347–359. [CrossRef]
13. Shibata, H.; Branquinho, C.; McDowell, W.; Mitchell, M.; Monteith, D.; Tang, J.; Arvola, L.; Cruz, C.; Cusack, D.; Halada, L.; et al. Consequence of altered nitrogen cycles in the coupled human and ecological system under changing climate: The need for long-term and site-based research. *AMBIO* **2015**, *44*, 178–193. [CrossRef] [PubMed]

14. Isson, T.; Planavsky, N.; Coogan, L.; Stewart, E.; Ague, J.; Bolton, E.; Zhang, S.; McKenzie, N.; Kump, L. Evolution of the Global Carbon Cycle and Climate Regulation on Earth. *Glob. Biogeochem. Cycles* **2020**, *34*, e2018GB006061. [CrossRef]
15. Puppe, D.; Kaczorek, D.; Schaller, J. Biological impacts on silicon availability and cycling in agricultural plant-soil systems. In *Silicon and Nano-Silicon in Environmental Stress Management and Crop Quality Improvement*; Elsevier: Amsterdam, The Netherlands, 2022; pp. 309–324.
16. Street-Perrott, F.A.; Barker, P.A. Biogenic silica: A neglected component of the coupled global continental biogeochemical cycles of carbon and silicon. *Earth Surf. Process. Landf.* **2008**, *33*, 1436–1457. [CrossRef]
17. Haynes, R.J. Significance and Role of Si in Crop Production. *Adv. Agron.* **2017**, *146*, 83–166. [CrossRef]
18. Guignard, M.; Leitch, A.; Acquisti, C.; Eizaguirre, C.; Elser, J.; Hessen, D.; Jeyasingh, P.; Neiman, M.; Richardson, A.; Soltis, P.; et al. Impacts of Nitrogen and Phosphorus: From Genomes to Natural Ecosystems and Agriculture. *Front. Ecol. Evol.* **2017**, *5*, 1–9. [CrossRef]
19. Smith, P.; Cotrufo, M.; Rumpel, C.; Paustian, K.; Kuikman, P.; Elliott, J.; McDowell, R.; Griffiths, R.; Asakawa, S.; Bustamante, M.; et al. Biogeochemical cycles and biodiversity as key drivers of ecosystem services provided by soils. *Soil* **2015**, *1*, 665–685. [CrossRef]
20. Olalekan, R.M.; Ilesanmi, A.; Alima, O.; Omini, D.; Raimi, A.-A.G. Exploring How Human Activities Disturb the Balance of Biogeochemical Cycles: Evidence from the Carbon, Nitrogen and Hydrologic Cycles. *SSRN Electron. J.* **2021**, *2*, 24–44. [CrossRef]
21. Tong, S.; Bambrick, H.; Beggs, P.J.; Chen, L.; Hu, Y.; Ma, W.; Steffen, W.; Tan, J. Current and future threats to human health in the Anthropocene. *Environ. Int.* **2022**, *158*, 106892. [CrossRef]
22. Fontaine, S.; Abbadie, L.; Aubert, M.; Barot, S.; Bloor, J.; Derrien, D.; Duchene, O.; Gross, N.; Henneron, L.; Roux, X.L.; et al. Plant–soil synchrony in nutrient cycles: Learning from ecosystems to design sustainable agrosystems. *Glob. Chang. Biol.* **2023**, *30*, e17034. [CrossRef]
23. Schlesinger, W.H.; Cole, J.J.; Finzi, A.C.; Holland, E.A. Introduction to coupled biogeochemical cycles. *Front. Ecol. Environ.* **2011**, *9*, 5–8. [CrossRef]
24. Conley, D.J. Terrestrial ecosystems and the global biogeochemical silica cycle. *Glob. Biogeochem. Cycles* **2002**, *16*, 68-1–68-8. [CrossRef]
25. Dodds, W.; Whiles, M. *Nitrogen, Sulfur, Phosphorus, and Other Nutrients*; Elsevier: Amsterdam, The Netherlands, 2010. [CrossRef]
26. Van Cappellen, P. Biomineralization and global biogeochemical cycles. *Rev. Mineral. Geochem.* **2003**, *54*, 357–381. [CrossRef]
27. Stemmet, M.; De Bruyn, J.; Zeeman, P. The uptake of carbon dioxide by plant roots. *Plant Soil* **1962**, *17*, 357–364. [CrossRef]
28. Shimono, H.; Kondo, M.; Evans, J.R. Internal transport of CO_2 from the root-zone to plant shoot is pH dependent. *Physiol. Plant.* **2019**, *165*, 451–463. [CrossRef]
29. Bardgett, R.; De Deyn, G.; Ostle, N. Plant–soil interactions and the carbon cycle. *J. Ecol.* **2009**, *97*, 838–839. [CrossRef]
30. Kudeyarov, V. Soil respiration and carbon sequestration: A review. *Eurasian Soil Sci.* **2023**, *56*, 1191–1200. [CrossRef]
31. Friedlingstein, P.; O'sullivan, M.; Jones, M.W.; Andrew, R.M.; Hauck, J.; Landschützer, P.; Le Quéré, C.; Li, H.; Luijkx, I.T.; Olsen, A. Global carbon budget 2024. *Earth Syst. Sci. Data* **2025**, *17*, 965–1039. [CrossRef]
32. Nissan, A.; Alcolombri, U.; Peleg, N.; Galili, N.; Jimenez-Martinez, J.; Molnar, P.; Holzner, M. Global warming accelerates soil heterotrophic respiration. *Nat. Commun.* **2023**, *14*, 3452. [CrossRef]
33. Lal, R.; Negassa, W.; Lorenz, K. Carbon sequestration in soil. *Curr. Opin. Environ. Sustain.* **2015**, *15*, 79–86. [CrossRef]
34. Kögel-Knabner, I. The macromolecular organic composition of plant and microbial residues as inputs to soil organic matter. *Soil Biol. Biochem.* **2002**, *34*, 139–162. [CrossRef]
35. Kästner, M.; Miltner, A.; Thiele-Bruhn, S.; Liang, C. Microbial Necromass in Soils—Linking Microbes to Soil Processes and Carbon Turnover. *Front. Environ. Sci.* **2021**, *9*, 756378. [CrossRef]
36. Berner, R.A. Weathering, plants, and the long-term carbon cycle. *Geochim. Cosmochim. Acta* **1992**, *56*, 3225–3231. [CrossRef]
37. Knapp, W.J.; Tipper, E.T. The efficacy of enhancing carbonate weathering for carbon dioxide sequestration. *Front. Clim.* **2022**, *4*, 928215. [CrossRef]
38. Padbhushan, R.; Kumar, U.; Sharma, S.; Rana, D.; Kumar, R.; Kohli, A.; Kumari, P.; Parmar, B.; Kaviraj, M.; Sinha, A.; et al. Impact of Land-Use Changes on Soil Properties and Carbon Pools in India: A Meta-analysis. *Front. Environ. Sci.* **2022**, *9*, 794866. [CrossRef]
39. Edmondson, J.; Davies, Z.; McHugh, N.; Gaston, K.; Leake, J. Organic carbon hidden in urban ecosystems. *Sci. Rep.* **2012**, *2*, srep00963. [CrossRef] [PubMed]
40. Reid, A.; Webster, G.; Krouse, H. Nitrogen movement and transformation in soils. *Plant Soil* **1969**, *31*, 224–237. [CrossRef]
41. Masson-Boivin, C.; Sachs, J. Symbiotic nitrogen fixation by rhizobia-the roots of a success story. *Curr. Opin. Plant Biol.* **2018**, *44*, 7–15. [CrossRef]
42. Yansheng, C.; Fengliang, Z.; Zhongyi, Z.; Tongbin, Z.; Hua-Yun, X. Biotic and abiotic nitrogen immobilization in soil incorporated with crop residue. *Soil Tillage Res.* **2020**, *202*, 104664. [CrossRef]
43. Hill, R.; Rinker, R.; Wilson, H.D. Atmospheric nitrogen fixation by lightning. *J. Atmos. Sci.* **1980**, *37*, 179–192. [CrossRef]

44. Singh, H.B. Reactive nitrogen in the troposphere. *Environ. Sci. Technol.* **1987**, *21*, 320–327. [CrossRef] [PubMed]
45. Beeckman, F.; Motte, H.; Beeckman, T. Nitrification in agricultural soils: Impact, actors and mitigation. *Curr. Opin. Biotechnol.* **2018**, *50*, 166–173. [CrossRef] [PubMed]
46. Lehnert, N.; Dong, H.; Harland, J.; Hunt, A.; White, C. Reversing nitrogen fixation. *Nat. Rev. Chem.* **2018**, *2*, 278–289. [CrossRef]
47. Verstraete, W.; Focht, D. *Biochemical Ecology of Nitrification and Denitrification*; Springer: Berlin/Heidelberg, Germany, 1977. [CrossRef]
48. Withers, P.J.; Neal, C.; Jarvie, H.P.; Doody, D.G. Agriculture and eutrophication: Where do we go from here? *Sustainability* **2014**, *6*, 5853–5875. [CrossRef]
49. Schulte-Uebbing, L.F.; Beusen, A.H.; Bouwman, A.F.; De Vries, W. From planetary to regional boundaries for agricultural nitrogen pollution. *Nature* **2022**, *610*, 507–512. [CrossRef]
50. Meisinger, J.; Schepers, J.; Raun, W. Crop nitrogen requirement and fertilization. *Nitrogen Agric. Syst.* **2008**, *49*, 563–612.
51. Smil, V. Nitrogen in crop production: An account of global flows. *Glob. Biogeochem. Cycles* **1999**, *13*, 647–662. [CrossRef]
52. Robertson, G.P.; Vitousek, P.M. Nitrogen in agriculture: Balancing the cost of an essential resource. *Annu. Rev. Environ. Resour.* **2009**, *34*, 97–125. [CrossRef]
53. Kolbe, H. Comparative analysis of soil fertility, productivity and sustainability of organic farming in Central Europe—Part 1: Effect of medium manifestations on conversion, fertilizer types and cropping systems. *Agronomy* **2022**, *12*, 2001. [CrossRef]
54. Shafi, M.; Shah, A.; Bakht, J.; Shah, M.; Mohammad, W. Integrated effect of inorganic and organic nitrogen sources on soil fertility and productivity of maize. *J. Plant Nutr.* **2012**, *35*, 524–537. [CrossRef]
55. Herencia, J.F.; Ruiz-Porras, J.C.; Melero, S.; Garcia-Galavis, P.A.; Morillo, E.; Maqueda, C. Comparison between organic and mineral fertilization for soil fertility levels, crop macronutrient concentrations, and yield. *Agron. J.* **2007**, *99*, 973–983. [CrossRef]
56. Mugendi, D.N.; Waswa, B.S.; Mucheru-Muna, M.W.; Kimetu, J.M.; Palm, C. Comparative analysis of the current and potential role of legumes in integrated soil fertility management in East Africa. In *Fighting Poverty in Sub-Saharan Africa: The Multiple Roles of Legumes in Integrated Soil Fertility Management*; Springer: Dordrecht, The Netherlands, 2011; pp. 151–173.
57. Mapfumo, P. Comparative analysis of the current and potential role of legumes in integrated soil fertility management in southern Africa. In *Fighting Poverty in Sub-Saharan Africa: The Multiple Roles of Legumes in Integrated Soil Fertility Management*; Springer: Dordrecht, The Netherlands, 2011; pp. 175–200.
58. Hammad, H.M.; Khaliq, A.; Abbas, F.; Farhad, W.; Fahad, S.; Aslam, M.; Shah, G.M.; Nasim, W.; Mubeen, M.; Bakhat, H.F. Comparative effects of organic and inorganic fertilizers on soil organic carbon and wheat productivity under arid region. *Commun. Soil Sci. Plant Anal.* **2020**, *51*, 1406–1422. [CrossRef]
59. Azam, F. Comparative effects of organic and inorganic nitrogen sources applied to a flooded soil on rice yield and availability of N. *Plant Soil* **1990**, *125*, 255–262. [CrossRef]
60. Mtambanengwe, F.; Mapfumo, P.; Vanlauwe, B. Comparative short-term effects of different quality organic resources on maize productivity under two different environments in Zimbabwe. In *Advances in Integrated Soil Fertility Management in Sub-Saharan Africa: Challenges and Opportunities*; Springer: Dordrecht, The Netherlands, 2007; pp. 575–588. [CrossRef]
61. Föllmi, K. The phosphorus cycle, phosphogenesis and marine phosphate-rich deposits. *Earth-Sci. Rev.* **1996**, *40*, 55–124. [CrossRef]
62. Amadou, I.; Faucon, M.; Houben, D. Role of Soil Minerals on Organic Phosphorus Availability and Phosphorus Uptake by Plants. *Geoderma* **2022**, *428*, 116125. [CrossRef]
63. Hinsinger, P. Bioavailability of soil inorganic P in the rhizosphere as affected by root-induced chemical changes: A review. *Plant Soil* **2001**, *237*, 173–195. [CrossRef]
64. Shen, J.; Yuan, L.; Zhang, J.; Li, H.; Bai, Z.; Chen, X.; Zhang, W.; Zhang, F. Phosphorus dynamics: From soil to plant. *Plant Physiol.* **2011**, *156*, 997–1005. [CrossRef]
65. U.S.G.S. *Mineral Commodity Summaries*; U.S. Geological Survey: Reston, VA, USA, 2023; 210p.
66. Scholz, R.W.; Wellmer, F.-W.; Mew, M.; Steiner, G. The dynamics of increasing mineral resources and improving resource efficiency: Prospects for mid-and long-term security of phosphorus supply. *Resour. Conserv. Recycl.* **2025**, *213*, 107993. [CrossRef]
67. Chowdhury, R.B.; Moore, G.A.; Weatherley, A.J.; Arora, M. Key sustainability challenges for the global phosphorus resource, their implications for global food security, and options for mitigation. *J. Clean. Prod.* **2017**, *140*, 945–963. [CrossRef]
68. Hedley, C. The role of precision agriculture for improved nutrient management on farms. *J. Sci. Food Agric.* **2015**, *95*, 12–19. [CrossRef]
69. Pätzold, S.; Leenen, M.; Frizen, P.; Heggemann, T.; Wagner, P.; Rodionov, A. Predicting plant available phosphorus using infrared spectroscopy with consideration for future mobile sensing applications in precision farming. *Precis. Agric.* **2020**, *21*, 737–761. [CrossRef]
70. Owen, D.; Williams, A.P.; Griffith, G.W.; Withers, P.J. Use of commercial bio-inoculants to increase agricultural production through improved phosphrous acquisition. *Appl. Soil Ecol.* **2015**, *86*, 41–54. [CrossRef]
71. Aghajanzadeh, T.; Hawkesford, M.J.; De Kok, L.J. Atmospheric H_2S and SO_2 as sulfur sources for Brassica juncea and Brassica rapa: Regulation of sulfur uptake and assimilation. *Environ. Exp. Bot.* **2016**, *124*, 1–10. [CrossRef]

72. Zhong, Q.; Shen, H.; Yun, X.; Chen, Y.; Ren, Y.a.; Xu, H.; Shen, G.; Du, W.; Meng, J.; Li, W. Global sulfur dioxide emissions and the driving forces. *Environ. Sci. Technol.* **2020**, *54*, 6508–6517. [CrossRef] [PubMed]
73. Kertesz, M.; Mirleau, P. The role of soil microbes in plant sulphur nutrition. *J. Exp. Bot.* **2004**, *55*, 1939–1945. [CrossRef]
74. Xia, Y.; Lü, C.; Hou, N.; Xin, Y.; Liu, J.; Liu, H.; Xun, L. Sulfide production and oxidation by heterotrophic bacteria under aerobic conditions. *ISME J.* **2017**, *11*, 2754–2766. [CrossRef] [PubMed]
75. Li, W.; Dong, X.; Lu, R.; Zhou, Y.; Zheng, P.-F.; Feng, D.; Wang, Y. Microbial ecology of sulfur cycling near the sulfate-methane transition of deep-sea cold seep sediments. *Environ. Microbiol.* **2021**, *23*, 6844–6858. [CrossRef]
76. Galloway, J. Acid deposition: Perspectives in time and space. *Water Air Soil Pollut.* **1995**, *85*, 15–24. [CrossRef]
77. Hinckley, E.-L.S.; Crawford, J.T.; Fakhraei, H.; Driscoll, C.T. A shift in sulfur-cycle manipulation from atmospheric emissions to agricultural additions. *Nat. Geosci.* **2020**, *13*, 597–604. [CrossRef]
78. Sharma, R.K.; Cox, M.S.; Oglesby, C.; Dhillon, J.S. Revisiting the role of sulfur in crop production: A narrative review. *J. Agric. Food Res.* **2024**, *15*, 101013. [CrossRef]
79. Kovar, J.L.; Grant, C.A. Nutrient cycling in soils: Sulfur. In *Soil Management: Building a Stable Base for Agriculture*; Wiley: Hoboken, NJ, USA, 2011; pp. 103–115.
80. Gilbert, F. The place of sulfur in plant nutrition. *Bot. Rev.* **1951**, *17*, 671–691. [CrossRef]
81. Elgarahy, A.M.; Eloffy, M.G.; Priya, A.K.; Yogeshwaran, V.; Yang, Z.; Elwakeel, K.Z.; Lopez-Maldonado, E.A. Biosolids management and utilizations: A review. *J. Clean. Prod.* **2024**, *451*, 141974. [CrossRef]
82. Gahane, D.; Mandavgane, S.A. Biogenic potassium: Sources, method of recovery, and sustainability assessment. *Rev. Chem. Eng.* **2024**, *40*, 707–722. [CrossRef]
83. Epstein, E. Silicon. *Annu. Rev. Plant Biol.* **1999**, *50*, 641–664. [CrossRef] [PubMed]
84. Ma, J.F. Role of silicon in enhancing the resistance of plants to biotic and abiotic stresses. *Soil Sci. Plant Nutr.* **2004**, *50*, 11–18. [CrossRef]
85. Epstein, E. Silicon: Its manifold roles in plants. *Ann. Appl. Biol.* **2009**, *155*, 155–160. [CrossRef]
86. Sommer, M.; Kaczorek, D.; Kuzyakov, Y.; Breuer, J. Silicon pools and fluxes in soils and landscapes—A review. *J. Plant Nutr. Soil Sci.* **2006**, *169*, 310–329. [CrossRef]
87. Ehrlich, H.; Demadis, K.D.; Pokrovsky, O.S.; Koutsoukos, P.G. Modern views on desilicification: Biosilica and abiotic silica dissolution in natural and artificial environments. *Chem. Rev.* **2010**, *110*, 4656–4689. [CrossRef]
88. Dürr, H.H.; Meybeck, M.; Hartmann, J.; Laruelle, G.G.; Roubeix, V. Global spatial distribution of natural riverine silica inputs to the coastal zone. *Biogeosciences* **2011**, *8*, 597–620. [CrossRef]
89. Struyf, E.; Conley, D.J. Emerging understanding of the ecosystem silica filter. *Biogeochemistry* **2012**, *107*, 9–18. [CrossRef]
90. Bartoli, F. The biogeochemical cycle of silicon in two temperate forest ecosystems. In *Ecological Bulletins*; Oikos Editorial Office: Lund, Sweden, 1983; pp. 469–476.
91. Gérard, F.; Mayer, K.U.; Hodson, M.J.; Ranger, J. Modelling the biogeochemical cycle of silicon in soils: Application to a temperate forest ecosystem. *Geochim. Cosmochim. Acta* **2008**, *72*, 741–758. [CrossRef]
92. Yang, X.; Song, Z.; Yu, C.; Ding, F. Quantification of different silicon fractions in broadleaf and conifer forests of northern China and consequent implications for biogeochemical Si cycling. *Geoderma* **2020**, *361*, 114036. [CrossRef]
93. Puppe, D. Review on protozoic silica and its role in silicon cycling. *Geoderma* **2020**, *365*, 114224. [CrossRef]
94. Puppe, D.; Ehrmann, O.; Kaczorek, D.; Wanner, M.; Sommer, M. The protozoic Si pool in temperate forest ecosystems—Quantification, abiotic controls and interactions with earthworms. *Geoderma* **2015**, *243–244*, 196–204. [CrossRef]
95. Puppe, D.; Wanner, M.; Sommer, M. Data on euglyphid testate amoeba densities, corresponding protozoic silicon pools, and selected soil parameters of initial and forested biogeosystems. *Data Brief* **2018**, *21*, 1697–1703. [CrossRef]
96. Wanner, M.; Elmer, M.; Sommer, M.; Funk, R.; Puppe, D. Testate amoebae colonizing a newly exposed land surface are of airborne origin. *Ecol. Indic.* **2015**, *48*, 55–62. [CrossRef]
97. Puppe, D.; Kaczorek, D.; Wanner, M.; Sommer, M. Dynamics and drivers of the protozoic Si pool along a 10-year chronosequence of initial ecosystem states. *Ecol. Eng.* **2014**, *70*, 477–482. [CrossRef]
98. Puppe, D.; Höhn, A.; Kaczorek, D.; Wanner, M.; Sommer, M. As time goes by—Spatiotemporal changes of biogenic Si pools in initial soils of an artificial catchment in NE Germany. *Appl. Soil Ecol.* **2016**, *105*, 9–16. [CrossRef]
99. Puppe, D.; Höhn, A.; Kaczorek, D.; Wanner, M.; Wehrhan, M.; Sommer, M. How big is the influence of biogenic silicon pools on short-term changes in water-soluble silicon in soils? Implications from a study of a 10-year-old soil–plant system. *Biogeosciences* **2017**, *14*, 5239–5252. [CrossRef]
100. Aoki, Y.; Hoshino, M.; Matsubara, T. Silica and testate amoebae in a soil under pine–oak forest. *Geoderma* **2007**, *142*, 29–35. [CrossRef]
101. Sommer, M.; Jochheim, H.; Höhn, A.; Breuer, J.; Zagorski, Z.; Busse, J.; Barkusky, D.; Meier, K.; Puppe, D.; Wanner, M.; et al. Si cycling in a forest biogeosystem—The importance of transient state biogenic Si pools. *Biogeosciences* **2013**, *10*, 4991–5007. [CrossRef]

102. Creevy, A.L.; Fisher, J.; Puppe, D.; Wilkinson, D.M. Protist diversity on a nature reserve in NW England—With particular reference to their role in soil biogenic silicon pools. *Pedobiologia* **2016**, *59*, 51–59. [CrossRef]
103. Wanner, M.; Birkhofer, K.; Fischer, T.; Shimizu, M.; Shimano, S.; Puppe, D. Soil Testate Amoebae and Diatoms as Bioindicators of an Old Heavy Metal Contaminated Floodplain in Japan. *Microb. Ecol.* **2020**, *79*, 123–133. [CrossRef]
104. Qin, Y.; Puppe, D.; Payne, R.; Li, L.; Li, J.; Zhang, Z.; Xie, S. Land-use change effects on protozoic silicon pools in the Dajiuhu National Wetland Park, China. *Geoderma* **2020**, *368*, 114305. [CrossRef]
105. Qin, Y.; Puppe, D.; Zhang, L.; Sun, R.; Li, P.; Xie, S. How Does Sphagnum Growing Affect Testate Amoeba Communities and Corresponding Protozoic Si Pools? Results from Field Analyses in SW China. *Microb. Ecol.* **2021**, *82*, 459–469. [CrossRef]
106. Qin, Y.; Puppe, D.; Li, H.; Li, H.; Mazei, Y.; Tsyganov, A.N.; Man, B.; Huang, X.; Gu, Y.; Xie, S. Peatland degradation in Asia threatens the biodiversity of testate amoebae (Protozoa) with consequences for protozoic silicon cycling. *Geoderma* **2022**, *420*, 115870. [CrossRef]
107. Guntzer, F.; Keller, C.; Poulton, P.R.; McGrath, S.P.; Meunier, J.-D. Long-term removal of wheat straw decreases soil amorphous silica at Broadbalk, Rothamsted. *Plant Soil* **2012**, *352*, 173–184. [CrossRef]
108. Meunier, J.; Guntzer, F.; Kirman, S.; Keller, C. Terrestrial plant-Si and environmental changes. *Mineral. Mag.* **2008**, *72*, 263–267. [CrossRef]
109. Vandevenne, F.; Struyf, E.; Clymans, W.; Meire, P. Agricultural silica harvest: Have humans created a new loop in the global silica cycle? *Front. Ecol. Environ.* **2012**, *10*, 243–248. [CrossRef]
110. Schaller, J.; Puppe, D. Heat improves silicon availability in mineral soils. *Geoderma* **2021**, *386*, 114909. [CrossRef]
111. Schaller, J.; Webber, H.; Ewert, F.; Stein, M.; Puppe, D. The transformation of agriculture towards a silicon improved sustainable and resilient crop production. *npj Sustain. Agric.* **2024**, *2*, 27. [CrossRef]
112. Li, Z.; Delvaux, B. Phytolith-rich biochar: A potential Si fertilizer in desilicated soils. *GCB Bioenergy* **2019**, *11*, 1264–1282. [CrossRef]
113. Puppe, D.; Kaczorek, D.; Schaller, J.; Barkusky, D.; Sommer, M. Crop straw recycling prevents anthropogenic desilication of agricultural soil–plant systems in the temperate zone–Results from a long-term field experiment in NE Germany. *Geoderma* **2021**, *403*, 115187. [CrossRef]
114. Yang, X.; Song, Z.; Qin, Z.; Wu, L.; Yin, L.; Van Zwieten, L.; Song, A.; Ran, X.; Yu, C.; Wang, H. Phytolith-rich straw application and groundwater table management over 36 years affect the soil-plant silicon cycle of a paddy field. *Plant Soil* **2020**, *454*, 343–358. [CrossRef]
115. Haynes, R.J. What effect does liming have on silicon availability in agricultural soils? *Geoderma* **2019**, *337*, 375–383. [CrossRef]
116. Berhane, M.; Xu, M.; Liang, Z.; Shi, J.; Wei, G.; Tian, X. Effects of long-term straw return on soil organic carbon storage and sequestration rate in North China upland crops: A meta-analysis. *Glob. Chang. Biol.* **2020**, *26*, 2686–2701. [CrossRef] [PubMed]
117. Song, Z.; Müller, K.; Wang, H. Biogeochemical silicon cycle and carbon sequestration in agricultural ecosystems. *Earth-Sci. Rev.* **2014**, *139*, 268–278. [CrossRef]
118. Ostle, N.; Smith, P.; Fisher, R.; Woodward, I.; Fisher, J.; Smith, J.; Galbraith, D.; Levy, P.; Meir, P.; McNamara, N.; et al. Integrating plant–soil interactions into global carbon cycle models. *J. Ecol.* **2009**, *97*, 851–863. [CrossRef]
119. Burgin, A.; Yang, W.; Hamilton, S.; Silver, W. Beyond carbon and nitrogen: How the microbial energy economy couples elemental cycles in diverse ecosystems. *Front. Ecol. Environ.* **2011**, *9*, 44–52. [CrossRef]
120. Widdig, M.; Heintz-Buschart, A.; Schleuss, P.; Guhr, A.; Borer, E.; Seabloom, E.; Spohn, M. Effects of nitrogen and phosphorus addition on microbial community composition and element cycling in a grassland soil. *Soil Biol. Biochem.* **2020**, *151*, 108041. [CrossRef]
121. Finzi, A.; Austin, A.; Cleland, E.; Frey, S.; Houlton, B.; Wallenstein, M. Responses and feedbacks of coupled biogeochemical cycles to climate change: Examples from terrestrial ecosystems. *Front. Ecol. Environ.* **2011**, *9*, 61–67. [CrossRef]
122. Waring, B.; Weintraub, S.; Sinsabaugh, R. Ecoenzymatic stoichiometry of microbial nutrient acquisition in tropical soils. *Biogeochemistry* **2013**, *117*, 101–113. [CrossRef]
123. Walker, A.; Walker, A.; Beckerman, A.; Gu, L.; Kattge, J.; Cernusak, L.; Domingues, T.; Scales, J.; Wohlfahrt, G.; Wullschleger, S.; et al. The relationship of leaf photosynthetic traits—Vcmax and Jmax—To leaf nitrogen, leaf phosphorus, and specific leaf area: A meta-analysis and modeling study. *Ecol. Evol.* **2014**, *4*, 3218–3235. [CrossRef]
124. Tang, B.; Rocci, K.; Lehmann, A.; Rillig, M. Nitrogen increases soil organic carbon accrual and alters its functionality. *Glob. Chang. Biol.* **2023**, *29*, 1971–1983. [CrossRef]
125. Mehnaz, K.; Corneo, P.; Keitel, C.; Dijkstra, F. Carbon and phosphorus addition effects on microbial carbon use efficiency, soil organic matter priming, gross nitrogen mineralization and nitrous oxide emission from soil. *Soil Biol. Biochem.* **2019**, *134*, 175–186. [CrossRef]
126. Treguer, P.J.; De La Rocha, C.L. The world ocean silica cycle. *Ann. Rev. Mar. Sci.* **2013**, *5*, 477–501. [CrossRef] [PubMed]
127. Derry, L.A.; Kurtz, A.C.; Ziegler, K.; Chadwick, O.A. Biological control of terrestrial silica cycling and export fluxes to watersheds. *Nature* **2005**, *433*, 728–731. [CrossRef]

128. Zondervan, J.R.; Hilton, R.G.; Dellinger, M.; Clubb, F.J.; Roylands, T.; Ogrič, M. Rock organic carbon oxidation CO_2 release offsets silicate weathering sink. *Nature* **2023**, *623*, 329–333. [CrossRef] [PubMed]
129. Textor, C.; Graf, H.-F.; Timmreck, C.; Robock, A. Emissions from volcanoes. In *Emissions of Atmospheric Trace Compounds*; Springer: Berlin/Heidelberg, Germany, 2004; pp. 269–303.
130. Mather, T.; Allen, A.; Davison, B.; Pyle, D.; Oppenheimer, C.; McGonigle, A. Nitric acid from volcanoes. *Earth Planet. Sci. Lett.* **2004**, *218*, 17–30. [CrossRef]
131. Newman, E. Phosphorus inputs to terrestrial ecosystems. *J. Ecol.* **1995**, *83*, 713–726. [CrossRef]
132. Nanzyo, M. Unique properties of volcanic ash soils. *Glob. Environ. Res.-Engl. Ed.* **2002**, *6*, 99–112.
133. Zhu, Y.; Toon, O.B.; Jensen, E.J.; Bardeen, C.G.; Mills, M.J.; Tolbert, M.A.; Yu, P.; Woods, S. Persisting volcanic ash particles impact stratospheric SO_2 lifetime and aerosol optical properties. *Nat. Commun.* **2020**, *11*, 4526. [CrossRef] [PubMed]
134. Zhu, Q.; Riley, W.; Tang, J.; Bouskill, N. Plant responses to elevated CO_2 under competing hypotheses of nitrogen and phosphorus limitations. *Ecol. Appl.* **2024**, *34*, e2967. [CrossRef] [PubMed]
135. Sun, F.; Tariq, A.; Chen, H.; He, Q.; Guan, Y.; Pan, K.; Chen, S.; Li, J.; Zhao, C.; Wang, H.; et al. Effect of nitrogen and phosphorus application on agricultural soil food webs. *Arch. Agron. Soil Sci.* **2017**, *63*, 1176–1186. [CrossRef]
136. Netherway, T. From Forests to Microbiomes: The Mediation of Plant-Soil Systems by Root-Symbiotic Fungi. Ph.D. Thesis, Swedish University of Agricultural Sciences, Uppsala, Sweden, 2023.
137. Franche, C.; Lindström, K.; Elmerich, C. Nitrogen-fixing bacteria associated with leguminous and non-leguminous plants. *Plant Soil* **2009**, *321*, 35–59. [CrossRef]
138. Genre, A.; Lanfranco, L.; Perotto, S.; Bonfante, P. Unique and common traits in mycorrhizal symbioses. *Nat. Rev. Microbiol.* **2020**, *18*, 649–660. [CrossRef] [PubMed]
139. Ohkama-Ohtsu, N.; Wasaki, J. Recent progress in plant nutrition research: Cross-talk between nutrients, plant physiology and soil microorganisms. *Plant Cell Physiol.* **2010**, *51*, 1255–1264. [CrossRef]
140. Liu, Y.; Shi, A.; Chen, Y.; Xu, Z.; Liu, Y.; Yao, Y.; Wang, Y.; Jia, B. Beneficial microorganisms: Regulating growth and defense for plant welfare. *Plant Biotechnol. J.* **2025**, *23*, 986–998. [CrossRef]
141. Joshi, D.; Kaushik, A.; Kumar, R.; Arya, A.; Santoyo, G.; Singh, V.K.; Kashyap, N.; Solanki, M.K.; Kumari, M.; Bhardwaj, N. Improving Plant Performance Through Microbiome Manipulation: The Potential Role of Current Bioengineering Approaches. *Bacteria* **2025**, *4*, 12. [CrossRef]
142. Santana, M.M.; Dias, T.; Gonzalez, J.M.; Cruz, C. Transformation of organic and inorganic sulfur–adding perspectives to new players in soil and rhizosphere. *Soil Biol. Biochem.* **2021**, *160*, 108306. [CrossRef]
143. Sinsabaugh, R. Enzymic analysis of microbial pattern and process. *Biol. Fertil. Soils* **2004**, *17*, 69–74. [CrossRef]
144. Ren, Q.; Song, H.-S.; Yuan, Z.; Ni, X.; Li, C. Changes in Soil Enzyme Activities and Microbial Biomass after Revegetation in the Three Gorges Reservoir, China. *Forests* **2018**, *9*, 249. [CrossRef]
145. Brehm, U.; Gorbushina, A.; Mottershead, D. The role of microorganisms and biofilms in the breakdown and dissolution of quartz and glass. *Palaeogeogr. Palaeoclimatol. Palaeoecol.* **2005**, *219*, 117–129. [CrossRef]
146. White, J.; Reddy, K. Nitrification and denitrification rates of Everglades wetland soils along a phosphorus-impacted gradient. *J. Environ. Qual.* **2003**, *32*, 2436–2443. [CrossRef] [PubMed]
147. Mohanty, P.; Singh, P.; Chakraborty, D.; Mishra, S.; Pattnaik, R. Insight Into the Role of PGPR in Sustainable Agriculture and Environment. *Front. Sustain. Food Syst.* **2021**, *5*, 667150. [CrossRef]
148. Rousk, J.; Bengtson, P. Microbial regulation of global biogeochemical cycles. *Front. Microbiol.* **2014**, *5*, 103. [CrossRef]
149. Hobbie, S.E. Effects of plant species on nutrient cycling. *Trends Ecol. Evol.* **1992**, *7*, 336–339. [CrossRef]
150. Favero, V.O.; De Carvalho, R.H.; Motta, V.M.; Leite, A.B.C.; Coelho, M.; Xavier, G.; Rumjanek, N.; Urquiaga, S. Bradyrhizobium as the Only Rhizobial Inhabitant of Mung Bean (*Vigna radiata*) Nodules in Tropical Soils: A Strategy Based on Microbiome for Improving Biological Nitrogen Fixation Using Bio-Products. *Front. Plant Sci.* **2021**, *11*, 602645. [CrossRef]
151. Ghosh, P.; Rathinasabapathi, B. Phosphorus solubilization and plant growth enhancement by arsenic-resistant bacteria. *Chemosphere* **2015**, *134*, 1–6. [CrossRef]
152. Jensen, C.N.G.; Pang, J.K.Y.; Gottardi, M.; Kračun, S.; Svendsen, B.; Nielsen, K.F.; Kovács, Á.; Moelbak, L.; Fimognari, L.; Husted, S.; et al. Bacillus subtilis promotes plant phosphorus (P) acquisition through P solubilization and stimulation of root and root hair growth. *Physiol. Plant.* **2024**, *176*, e14338. [CrossRef]
153. Shrestha, N.; Hadano, S.; Kamachi, T.; Okura, I. Dinitrogen production from ammonia by Nitrosomonas europaea. *Appl. Catal. A Gen.* **2002**, *237*, 33–39. [CrossRef]
154. Hopkins, D.W.; Dungait, J.A.J. Soil Microbiology and Nutrient Cycling. In *Soil Microbiology and Sustainable Crop Production*; Dixon, G.R., Tilston, E.L., Eds.; Springer: Dordrecht, The Netherlands, 2010; pp. 59–80.
155. Zhang, M.; O'Connor, P.J.; Zhang, J.; Ye, X. Linking soil nutrient cycling and microbial community with vegetation cover in riparian zone. *Geoderma* **2021**, *384*, 114801. [CrossRef]

156. Yadav, A.N.; Kour, D.; Kaur, T.; Devi, R.; Yadav, A.; Dikilitas, M.; Abdel-Azeem, A.M.; Ahluwalia, A.S.; Saxena, A.K. Biodiversity, and biotechnological contribution of beneficial soil microbiomes for nutrient cycling, plant growth improvement and nutrient uptake. *Biocatal. Agric. Biotechnol.* **2021**, *33*, 102009. [CrossRef]
157. Haghverdi, K.; Kooch, Y. Effects of diversity of tree species on nutrient cycling and soil-related processes. *CATENA* **2019**, *178*, 335–344. [CrossRef]
158. Singh, S.B.; Carroll-Portillo, A.; Lin, H.C. Desulfovibrio in the Gut: The Enemy within? *Microorganisms* **2023**, *11*, 1772. [CrossRef] [PubMed]
159. Fortin, D.; Davis, B.; Beveridge, T.J. Role of *Thiobacillus* and sulfate-reducing bacteria in iron biocycling in oxic and acidic mine tailings. *FEMS Microbiol. Ecol.* **1996**, *21*, 11–24. [CrossRef]
160. Schulze, E.; Beck, E.; Buchmann, N.; Clemens, S.; Müller-Hohenstein, K.; Scherer-Lorenzen, M. *Global Biogeochemical Cycles*; Springer: Berlin/Heidelberg, Germany, 2019; pp. 827–841. [CrossRef]
161. Cleveland, C.; Townsend, A.; Taylor, P.; Álvarez-Clare, S.; Bustamante, M.; Chuyong, G.; Dobrowski, S.; Grierson, P.; Harms, K.; Houlton, B.; et al. Relationships among net primary productivity, nutrients and climate in tropical rain forest: A pan-tropical analysis. *Ecol. Lett.* **2011**, *14*, 939–947. [CrossRef]
162. Basu, S.; Kumar, G.; Chhabra, S.; Prasad, R. *Role of Soil Microbes in Biogeochemical Cycle for Enhancing Soil Fertility*; Elsevier: Amsterdam, The Netherlands, 2021; pp. 149–157.
163. Blanco, J.; Zavala, M.; Imbert, J.; Castillo, F. Sustainability of forest management practices: Evaluation through a simulation model of nutrient cycling. *For. Ecol. Manag.* **2005**, *213*, 209–228. [CrossRef]
164. Kebede, E. Contribution, Utilization, and Improvement of Legumes-Driven Biological Nitrogen Fixation in Agricultural Systems. *Front. Sustain. Food Syst.* **2021**, *5*, 767998. [CrossRef]
165. Peng, X.; Zhu, Q.H.; Xie, Z.B.; Darboux, F.; Holden, N.M. The impact of manure, straw and biochar amendments on aggregation and erosion in a hillslope Ultisol. *Catena* **2016**, *138*, 30–37. [CrossRef]
166. Song, Z.; McGrouther, K.; Wang, H. Occurrence, turnover and carbon sequestration potential of phytoliths in terrestrial ecosystems. *Earth-Sci. Rev.* **2016**, *158*, 19–30. [CrossRef]
167. Schaller, J.; Kleber, M.; Puppe, D.; Stein, M.; Sommer, M.; Rillig, M.C. The importance of reactive silica for maintaining soil health. *Plant Soil* **2025**, 1–12. [CrossRef]
168. Wu, A.; Hammer, G.; Doherty, A.; Von Caemmerer, S.; Farquhar, G. Quantifying impacts of enhancing photosynthesis on crop yield. *Nat. Plants* **2019**, *5*, 380–388. [CrossRef]
169. Lal, R. Soil degradation as a reason for inadequate human nutrition. *Food Secur.* **2009**, *1*, 45–57. [CrossRef]
170. Bossio, A.; Cook-Patton, S.; Ellis, P.; Fargione, J.; Sanderman, J.; Smith, P.; Wood, S.; Wood, S.; Zomer, R.; Unger, M.; et al. The role of soil carbon in natural climate solutions. *Nat. Sustain.* **2020**, *3*, 391–398. [CrossRef]
171. Crowther, T.; Hoogen, J.; Wan, J.; Mayes, M.; Mayes, M.; Keiser, A.; Keiser, A.; Mo, L.; Averill, C.; Averill, C.; et al. The global soil community and its influence on biogeochemistry. *Science* **2019**, *365*, 772. [CrossRef]
172. Jarecki, M.; Lal, R. Crop Management for Soil Carbon Sequestration. *Crit. Rev. Plant Sci.* **2003**, *22*, 471–502. [CrossRef]
173. Adekiya, A.; Alori, E.; Ogunbode, T.; Sangoyomi, T.; Oriade, O. Enhancing Organic Carbon Content in Tropical Soils: Strategies for Sustainable Agriculture and Climate Change Mitigation. *Open Agric. J.* **2023**, *17*, e18743315282476. [CrossRef]
174. Hodson, M.J. The Relative Importance of Cell Wall and Lumen Phytoliths in Carbon Sequestration in Soil: A Hypothesis. *Front. Earth Sci.* **2019**, *7*, 167. [CrossRef]
175. de Tombeur, F.; Hodson, M.J.; Saunders, M.; Clode, P.L. How important is carbon sequestration in phytoliths within the soil? *Plant Soil* **2024**, *505*, 185–198. [CrossRef]
176. Porwollik, V.; Rolinski, S.; Heinke, J.; Von Bloh, W.; Schaphoff, S.; Müller, C. The role of cover crops for cropland soil carbon, nitrogen leaching, and agricultural yields—A global simulation study with LPJmL (V. 5.0-tillage-cc). *Biogeosciences* **2021**, *19*, 957–977. [CrossRef]
177. Kutos, S.; Stricker, E.; Cooper, A.; Ryals, R.; Creque, J.; Machmuller, M.; Kroegar, M.; Silver, W. Compost amendment to enhance carbon sequestration in rangelands. *J. Soil Water Conserv.* **2023**, *78*, 163–177. [CrossRef]
178. Montagnini, F.; Nair, P. Carbon sequestration: An underexploited environmental benefit of agroforestry systems. *Agrofor. Syst.* **2004**, *61–62*, 281–295. [CrossRef]
179. Zhu, X.; Mao, L.; Chen, B. Driving forces linking microbial community structure and functions to enhanced carbon stability in biochar-amended soil. *Environ. Int.* **2019**, *133 Pt B*, 105211. [CrossRef]
180. Swift, M.; Andrén, O.; Brussaard, L.; Briones, M.; Coûteaux, M.; Ekschmitt, K.; Kjøller, A.; Loiseau, P.; Smith, P. Global change, soil biodiversity, and nitrogen cycling in terrestrial ecosystems: Three case studies. *Glob. Chang. Biol.* **1998**, *4*, 729–743. [CrossRef]
181. Wan, N.F.; Zheng, X.-R.; Fu, L.; Kiær, L.; Zhang, Z.; Chaplin-Kramer, R.; Dainese, M.; Tan, J.; Qiu, S.; Hu, Y.-Q.; et al. Global synthesis of effects of plant species diversity on trophic groups and interactions. *Nat. Plants* **2020**, *6*, 503–510. [CrossRef] [PubMed]
182. Siddique, I.; Vieira, I.; Schmidt, S.; Lamb, D.; De Carvalho, C.J.R.; Figueiredo, R.; Blomberg, S.; Davidson, E. Nitrogen and phosphorus additions negatively affect tree species diversity in tropical forest regrowth trajectories. *Ecology* **2010**, *91*, 2121–2131. [CrossRef]

183. Moura, R.F.; Sternberg, M.; Vorst, C.; Katz, O. Plant silicon content as a proxy for understanding plant community properties and ecosystem structure. *Ecosphere* **2024**, *15*, e4907. [CrossRef]
184. Kleinert, A.; Benedito, V.; Morcillo, R.; Dames, J.; Cornejo-Rivas, P.; Zuniga-Feest, A.; Delgado, M.; Muñoz, G. Morphological and symbiotic root modifications for mineral acquisition from nutrient-poor soils. In *Root Biology*; Springer: Cham, Switzerland, 2018; pp. 85–142.
185. Griffiths, B.; Philippot, L. Insights into the resistance and resilience of the soil microbial community. *FEMS Microbiol. Rev.* **2013**, *37*, 112–129. [CrossRef]
186. Rao, I.M.; Miles, J.W.; Beebe, S.E.; Horst, W.J. Root adaptations to soils with low fertility and aluminium toxicity. *Ann. Bot.* **2016**, *118*, 593–605. [CrossRef]
187. Etesami, H. Enhancing soil microbiome resilience: The mitigating role of silicon against environmental stresses. *Front. Agron.* **2024**, *6*, 1465165. [CrossRef]
188. Putra, R.; Powell, J.R.; Hartley, S.E.; Johnson, S.N. Is it time to include legumes in plant silicon research? *Funct. Ecol.* **2020**, *34*, 1142–1157. [CrossRef]
189. Zayed, O.; Hewedy, O.A.; Abdelmoteleb, A.; Ali, M.; Youssef, M.S.; Roumia, A.F.; Seymour, D.; Yuan, Z.-C. Nitrogen journey in plants: From uptake to metabolism, stress response, and microbe interaction. *Biomolecules* **2023**, *13*, 1443. [CrossRef]
190. Wilcox, K.; Koerner, S.; Hoover, D.; Borkenhagen, A.; Burkepile, D.; Collins, S.; Hoffman, A.; Kirkman, K.; Knapp, A.; Strydom, T.; et al. Rapid recovery of ecosystem function following extreme drought in a South African savanna-grassland. *Ecology* **2020**, *101*, e02983. [CrossRef]
191. Telo da Gama, J. The Role of Soils in Sustainability, Climate Change, and Ecosystem Services: Challenges and Opportunities. *Ecologies* **2023**, *4*, 552–567. [CrossRef]
192. Moiseenko, T. Evolution of biogeochemical cycles under anthropogenic loads: Limits impacts. *Geochem. Int.* **2017**, *55*, 841–860. [CrossRef]
193. Lorenz, K.; Lal, R. Biogeochemical C and N cycles in urban soils. *Environ. Int.* **2009**, *35*, 1–8. [CrossRef]
194. Iqbal, S.; Riaz, U.; Murtaza, G.; Jamil, M.; Ahmed, M.; Hussain, A.; Abbas, Z. Chemical fertilizers, formulation, and their influence on soil health. In *Microbiota and Biofertilizers: A Sustainable Continuum for Plant Soil Health*; Springer: Cham, Switzerland, 2021; pp. 1–15.
195. Howe, J.A.; McDonald, M.D.; Burke, J.; Robertson, I.; Coker, H.; Gentry, T.J.; Lewis, K.L. Influence of fertilizer and manure inputs on soil health: A review. *Soil Secur.* **2024**, *16*, 100155. [CrossRef]
196. Zhao, H.; Li, X.; Jiang, Y. Response of Nitrogen Losses to Excessive Nitrogen Fertilizer Application in Intensive Greenhouse Vegetable Production. *Sustainability* **2019**, *11*, 1513. [CrossRef]
197. Clymans, W.; Struyf, E.; Govers, G.; Vandevenne, F.; Conley, D.J. Anthropogenic impact on amorphous silica pools in temperate soils. *Biogeosciences* **2011**, *8*, 2281–2293. [CrossRef]
198. Conley, D.J.; Likens, G.E.; Buso, D.C.; Saccone, L.; Bailey, S.W.; Johnson, C.E. Deforestation causes increased dissolved silicate losses in the Hubbard Brook Experimental Forest. *Glob. Chang. Biol.* **2008**, *14*, 2548–2554. [CrossRef]
199. Maavara, T.; Dürr, H.H.; Van Cappellen, P. Worldwide retention of nutrient silicon by river damming: From sparse data set to global estimate. *Glob. Biogeochem. Cycles* **2014**, *28*, 842–855. [CrossRef]
200. Laruelle, G.G.; Roubeix, V.; Sferratore, A.; Brodherr, B.; Ciuffa, D.; Conley, D.; Dürr, H.; Garnier, J.; Lancelot, C.; Le Thi Phuong, Q. Anthropogenic perturbations of the silicon cycle at the global scale: Key role of the land-ocean transition. *Glob. Biogeochem. Cycles* **2009**, *23*, GB4031. [CrossRef]
201. Sharpley, A.; McDowell, R.; Kleinman, P. Phosphorus loss from land to water: Integrating agricultural and environmental management. *Plant Soil* **2001**, *237*, 287–307. [CrossRef]
202. Niklaus, P.A. Climate change effects on biogeochemical cycles, nutrients, and water supply. In *Agroecosystems in a Changing Climate*; CRC Press: Boca Raton, FL, USA, 2007; pp. 11–52.
203. Aponte, H.; Meli, P.; Butler, B.; Paolini, J.; Matus, F.; Merino, C.; Cornejo, P.; Kuzyakov, Y. Meta-analysis of heavy metal effects on soil enzyme activities. *Sci. Total Environ.* **2020**, *737*, 139744. [CrossRef]
204. Díaz, R.; Nestlerode, J.; Díaz, M. *A Global Perspective on the Effects of Eutrophication and Hypoxia on Aquatic Biota and Water Quality*; U.S. Environmental Protection Agency: Washington, DC, USA, 2019.
205. Hong, S.; Cong, N.; Ding, J.; Piao, S.; Liu, L.; Peñuelas, J.; Chen, A.; Quine, T.; Zeng, H.; Houlton, B. Effects of Afforestation on Soil Carbon and Nitrogen Accumulation Depend on Initial Soil Nitrogen Status. *Glob. Biogeochem. Cycles* **2022**, *37*, e2022GB007490. [CrossRef]
206. Zagural'skaya, L.; Zaybchenko, S. Impact of industrial pollution on soil microbial activity in boreal forests of the Kostomusksha region. *Eurasian Soil Sci.* **1994**, *26*, 67–74.
207. Puppe, D.; Kaczorek, D.; Stein, M.; Schaller, J. Silicon in Plants: Alleviation of Metal (loid) Toxicity and Consequential Perspectives for Phytoremediation. *Plants* **2023**, *12*, 2407. [CrossRef]

208. Chai, L.; Huang, M.; Fan, H.; Wang, J.; Jiang, D.; Zhang, M.; Huang, Y. Urbanization altered regional soil organic matter quantity and quality: Insight from excitation emission matrix (EEM) and parallel factor analysis (PARAFAC). *Chemosphere* **2019**, *220*, 249–258. [CrossRef]
209. Adimassu, Z.; Tamene, L.; Degefie, D. The influence of grazing and cultivation on runoff, soil erosion, and soil nutrient export in the central highlands of Ethiopia. *Ecol. Process.* **2020**, *9*, 23. [CrossRef]
210. Toming, K.; Liu, H.; Soomets, T.; Uuemaa, E.; Nõges, T.; Kutser, T. Estimation of the Biogeochemical and Physical Properties of Lakes Based on Remote Sensing and Artificial Intelligence Applications. *Remote Sens.* **2024**, *16*, 464. [CrossRef]
211. Reddy, T.; Shiva, S.S.; Reddy, R.; Meghana, S.; Prasanna, K.; Sai, T.; Reddy, S. Design and Developing AI-Driven Agro-sage for Optimal Precision Agriculture. In Proceedings of the 2024 5th International Conference on Smart Electronics and Communication (ICOSEC), Trichy, India, 18–20 September 2024; pp. 1538–1542.
212. Wehrhan, M.; Puppe, D.; Kaczorek, D.; Sommer, M. Spatial patterns of aboveground phytogenic Si stocks in a grass-dominated catchment—Results from UAS-based high-resolution remote sensing. *Biogeosciences* **2021**, *18*, 5163–5183. [CrossRef]
213. Wehrhan, M.; Rauneker, P.; Sommer, M. UAV-Based Estimation of Carbon Exports from Heterogeneous Soil Landscapes--A Case Study from the CarboZALF Experimental Area. *Sensors* **2016**, *16*, 255. [CrossRef] [PubMed]
214. Kashyap, B.; Kumar, R. Sensing Methodologies in Agriculture for Soil Moisture and Nutrient Monitoring. *IEEE Access* **2021**, *9*, 14095–14121. [CrossRef]
215. Sivakumar, V.; Vimal, S.; Baskar, V.; Murugan, S.; Vadivel, M. IoT and GIS Integration for Real-Time Monitoring of Soil Health and Nutrient Status. In Proceedings of the 2023 International Conference on Self Sustainable Artificial Intelligence Systems (ICSSAS), Erode, India, 18–20 October 2023; pp. 1265–1270. [CrossRef]

MDPI AG
Grosspeteranlage 5
4052 Basel
Switzerland
Tel.: +41 61 683 77 34

Biology Editorial Office
E-mail: biology@mdpi.com
www.mdpi.com/journal/biology

www.ingramcontent.com/pod-product-compliance
Lightning Source LLC
Chambersburg PA
CBHW061440210926
51850CB00008B/1561
9783725867943